Leckie
the education publisher
for Scotland

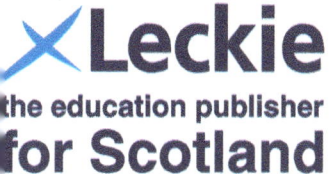

Third Level MATHS

Benchmark edition

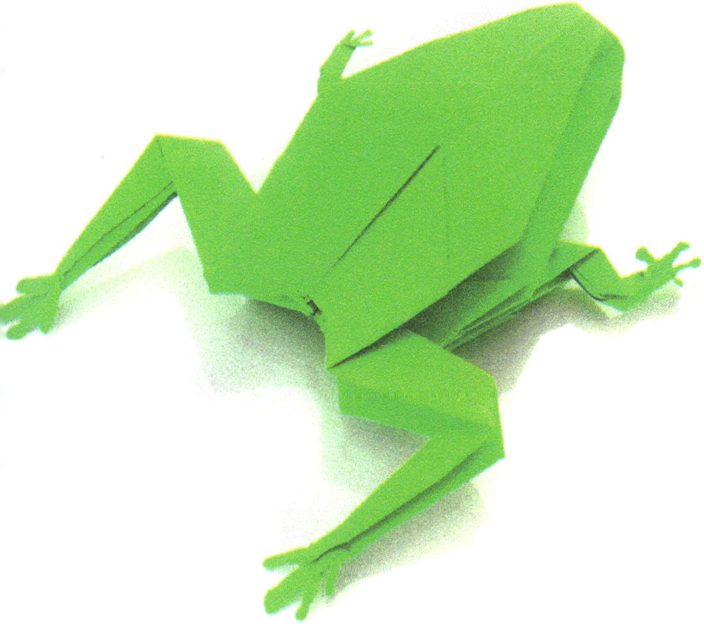

Student Book

John Boath, Robin Christie, Claire Crossman, Craig Lowther and Ian MacAndie with Kevin Evans, Keith Gordon, Trevor Senior and Brian Speed

Text © John Boath, Robin Christie, Claire Crossman, Craig Lowther and Ian MacAndie
with Kevin Evans, Keith Gordon, Trevor Senior and Brian Speed
Design and layout © 2020 Leckie

001/29042020

10 9 8 7 6 5 4

All rights reserved. No part of this publication may be reproduced, stored in a retrieval system, or transmitted in any form or by any means, electronic, mechanical, photocopying, recording or otherwise, without the prior written permission of the Publisher or a licence permitting restricted copying in the United Kingdom issued by the Copyright Licensing Agency Ltd., 90 Tottenham Court Road, London W1T 4LP.

Without limiting the exclusive rights of any author, contributor or the publisher of this publication, any unauthorised use of this publication to train generative artificial intelligence (AI) technologies is expressly prohibited. HarperCollins also exercise their rights under Article 4(3) of the Digital Single Market Directive 2019/790 and expressly reserve this publication from the text and data mining exception.

The right of John Boath, Robin Christie, Claire Crossman, Craig Lowther, Ian MacAndie, Kevin Evans, Keith Gordon, Trevor Senior and Brian Speed to be identified as authors of this Work has been asserted by them in accordance with sections 77 and 78 of the Copyright, Designs and Patents Act 1988.

ISBN 9780008407766

Published by
Leckie
An imprint of HarperCollins Publishers
1 Robroyston Gate, Glasgow, G33 1JN

HarperCollins Publishers
Macken Huse, 39/40 Mayor Street Upper, Dublin 1, DO1 C9W8, Ireland
T: 0844 576 8126 F: 0844 576 8131
leckiescotland@harpercollins.co.uk   leckiescotland.co.uk

Publisher: Sarah Mitchell
Project manager: Gillian Bowman

Special thanks to
Planman Technologies and Jouve (page layout and illustration)
Project One Publishing Solutions (project management, copy editing and proofreading)
Caleb O'Loan (answer generation and checking)

Printed in the UK by Ashford Colour Ltd

A CIP Catalogue record for this book is available from the British Library.

Acknowledgements
Leckie is grateful to the following for their permission to
reproduce their material: Education Scotland © Crown Copyright 2019, for the charts on pages 348–358.

Whilst every effort has been made to trace the copyright holders, in cases where this has been unsuccessful, or if any have inadvertently been overlooked, the Publishers would gladly receive any information enabling them to rectify any error or omission at the first opportunity.

This book contains FSC™ certified paper and other controlled
sources to ensure responsible forest management.

For more information visit: www.harpercollins.co.uk/green

Contents

Contents 3
Introduction 8

Number, money and measure

Estimation and rounding

1. **I can round a number using an appropriate degree of accuracy, having taken into account the context of the problem.** MNU 3-01a 11
   Rounding 11
   Estimates 12
   Using a calculator and rounding 14

Number and number processes

2. **I can use a variety of methods to solve number problems in familiar contexts, clearly communicating my processes and solutions.** MNU 3-03a 16
   Multiplying and dividing by 10, 100 and 1000 16
   Whole numbers 18
   More multiplying and dividing whole numbers 20
   Number problems 21
   Using a calculator 23
   Long multiplication 24
   Long division 24
   More decimals 26
   More number problems 27

3. **I can continue to recall number facts quickly and use them accurately when making calculations.** MNU 3-03b 30
   Quick methods and estimation 30
   Mental checks 32
   Extending knowledge of number facts 33
   Multiplying decimals 35
   Working backwards 36
   Number problems 37

4. **I can use my understanding of numbers less than zero to solve simple problems in context.** MNU 3-04a 40
   The number line and inequalities 40
   Negative numbers in real life 42
   Adding negative numbers 45
   Subtracting negative numbers 47

Multiples, factors and primes

5. **I have investigated strategies for identifying common multiples and common factors, explaining my ideas to others, and can apply my understanding to solve related problems.** MTH 3-05a 50
   Common multiples 50
   Common factors 53

6. **I can apply my understanding of factors to investigate and identify when a number is prime.** MTH 3-05b 57
   Prime factors 57

Contents

Powers and roots

7 Having explored the notation and vocabulary associated with whole number powers and the advantages of writing numbers in this form, I can evaluate powers of whole numbers mentally or using technology. MTH 3-06a 60

Powers and index form 60

Fractions, decimal fractions and percentages

8 I can solve problems by carrying out calculations with a wide range of fractions, decimal fractions and percentages, using my answers to make comparisons and informed choices for real-life situations. MNU 3-07a 64

Adding and subtracting decimals 64
Multiplying decimals 66
Dividing decimals 69
Fractions of quantities 70
Multiplying fractions 71
Expressing one quantity as a percentage of another 73
Percentages of quantities 74
Percentage increase and decrease 76
Real-life problems 78

9 By applying my knowledge of equivalent fractions and common multiples, I can add and subtract commonly used fractions. MTH 3-07b 81

Adding and subtracting fractions 81
Adding and subtracting fractions with a common denominator 82
Adding and subtracting fractions with different denominators 83

10 Having used practical, pictorial and written methods to develop my understanding, I can convert between whole or mixed numbers and fractions. MTH 3-07c 87

Converting mixed numbers to improper fractions 87
Converting improper fractions to mixed numbers 88

11 I can show how quantities that are related can be increased or decreased proportionally and apply this to solve problems in everyday contexts. MNU 3-08a 91

Proportion 91
Ratio 94
Simplifying ratios 96
Using equivalent ratios 97
Dividing quantities in a given ratio 99

Money

12 When considering how to spend my money, I can source, compare and contrast different contracts and services, discuss their advantages and disadvantages, and explain which offer best value to me. MNU 3-09a 102

Services and contracts 102
Borrowing money 106

13 I can budget effectively, making use of technology and other methods, to manage money and plan for future expenses. MNU 3-09b 108

Budgeting 108
Exchange rates 113

Time

14 Using simple time periods, I can work out how long a journey will take, the speed travelled at or distance covered, using my knowledge of the link between time, speed and distance. MNU 3-10a 117

Calculating speed 117
Calculating distance 119
Calculating time 121
Calculating speed, distance and time: mixed questions 122
Distance-time graphs 123

Contents

Measurement

15 I can solve practical problems by applying my knowledge of measure, choosing the appropriate units and degree of accuracy for the task and using a formula to calculate area or volume when required. **MNU 3-11a** 126

> Perimeter and area of rectangles 126
> Area of a triangle 129
> Area of a parallelogram 131
> Area of a trapezium 133
> Task: Design a bedroom 137
> Volume of a cuboid 138
> Metric units for area and volume 141
> Appropriate units and accuracy 143
> Investigation: Units and accuracy 143

16 Having investigated different routes to a solution, I can find the area of compound 2D shapes and the volume of compound 3D objects, applying my knowledge to solve practical problems. **MTH 3-11b** 145

> Perimeter and area of compound 2D shapes 145
> Area of compound 2D shapes involving triangles 148
> Volume of compound 3D objects 149

Mathematics – its impact on the world, past, present and future

17 I have worked with others to research a famous mathematician and the work they are known for, or investigated a mathematical topic, and have prepared and delivered a short presentation. **MTH 3-12a** 152

> Introducing the task 152
> Famous mathematicians 152
> Mathematical topics 153
> Researching your topic 153
> Presenting your findings 154
> Assessment 155

Patterns and relationships

18 Having explored number sequences, I can establish the set of numbers generated by a given rule and determine a rule for a given sequence, expressing it using appropriate notation. **MTH 3-13a** 156

> Sequences and rules 156
> Finding missing terms 158
> The nth term formula 159
> More difficult sequences 161

Expressions and equations

19 I can collect like algebraic terms, simplify expressions and evaluate using substitution. **MTH 3-14a** 165

> Algebraic terms and expressions 165
> Simplifying expressions 168
> Collecting like terms 169
> Substituting into expressions 173
> Substitutng into formulae 174

20 Having discussed ways to express problems or statements using mathematical language, I can construct, and use appropriate methods to solve, a range of simple equations. **MTH 3-15a** 177

> Solving equations 177
> Solving equations using the 'cover-up' method 177
> Solving equations using inverse mapping 179
> Doing the same thing to both sides 180
> Solving equations with letters on both sides 182
> Using algebra and diagrams to solve problems 183

Contents

21 I can create and evaluate a simple formula representing information contained in a diagram, problem or statement. **MTH 3-15b** 188

    Formulae 188
    Evaluating a formula 189
    Finding one of the other variables in a formula 192
    Creating a formula from a table of values 193
    Creating your own formula to represent a statement or problem 195

Shape, position and movement

Properties of 2D shapes and 3D objects

22 Having investigated a range of methods, I can accurately draw 2D shapes using appropriate mathematical instruments and methods. **MTH 3-16a** 199

    Constructing triangles 199
    Constructing other 2D shapes 202
    Constructing nets of 3D objects 205

Angle, symmetry and transformation

23 I can name angles and find their sizes using my knowledge of the properties of a range of 2D shapes and the angle properties associated with intersecting and parallel lines. **MTH 3-17a** 207

    Naming angles 207
    Calculating missing angles 208
    Parallel and perpendicular lines 211
    Alternate and corresponding angles 213
    Angles in a triangle 216
    Angles in a quadrilateral 219
    Interior and exterior angles of polygons 222

24 Having investigated navigation in the world, I can apply my understanding of bearings and scale to interpret maps and plans and create accurate plans, and scale drawings of routes and journeys. **MTH 3-17b** 226

    Scale drawings 226
    Map scales 229
    Bearings 232

25 I can apply my understanding of scale when enlarging or reducing pictures and shapes, using different methods, including technology. **MTH 3-17c** 237

    Scale factor 237
    Enlargements and reductions 242
    Using technology 245

26 I can use my knowledge of the coordinate system to plot and describe the location of a point on a grid. **MTH 3-18a** 248

    Coordinates 248
    Using coordinates 250

27 I can illustrate the lines of symmetry for a range of 2D shapes and apply my understanding to create and complete symmetrical pictures and patterns. **MTH 3-19a** 254

    Line symmetry 254
    Reflections 257
    Reflections in two mirror lines 260

Information handling

Data and analysis

28 I can work collaboratively, making appropriate use of technology, to source information presented in a range of ways, interpret what it conveys and discuss whether I believe the information to be robust, vague or misleading. **MNU 3-20a** 265

    Information from charts 265
    Interpreting graphs and diagrams 273

Contents

29 When analysing information or collecting data of my own, I can use my understanding of how bias may arise and how sample size can affect precision, to ensure that the data allows for fair conclusions to be drawn. MTH 3-20b 277

    Sample size 277
    Bias 279
    Collecting data for statistical surveys 280

30 I can display data in a clear way using a suitable scale, by choosing appropriately from an extended range of tables, charts, diagrams and graphs, making effective use of technology. MTH 3-21a 283

    Creating graphs, charts and diagrams 283
    Grouped frequencies 288
    Conversion graphs 290
    Pie charts 293
    Choosing data displays and communicating findings 296

Ideas of chance and uncertainty

31 I can find the probability of a simple event happening and explain why the consequences of the event, as well as its probability, should be considered when making choices. MNU 3-22a 299

    Probability and the probability scale 299
    Calculating the probability of an event not happening 304
    Listing all the possible outcomes 307
    Making choices and decisions based on chance and uncertainty 308

Answers

Answers 311

Benchmarks

Numeracy and mathematics learning experiences and outcomes for Second, Third and Fourth Level 348

Introduction

About this book

This book provides a resource to practise and assess your understanding of the maths covered at Third Level. There is a separate chapter for each of the Curriculum for Excellence Outcomes and Experiences, and most chapters use the same features to help you progress. You will find a range of worked examples to show you how to tackle problems, and an extensive set of exercises designed to help you develop the whole range of mathematical skills needed at Third Level.

You should not be trying to work through the book from page 1 to page 310. Your teacher will choose a range of topics throughout the school year and teach them in the order they think works best for your class, so you will use different parts of the book at different times of the year.

In this updated edition we have added new examples and exercises to ensure full coverage of the Benchmarks for Numeracy and Mathematics published by Education Scotland in 2017.

In addition to all the original content we have added; questions to help you practice rounding numbers to three decimal places, expressing numbers in terms of powers, calculating time durations and using a given probability to calculate an expected outcome.

By completing the exercises in the book, talking about and sharing your thinking and working when solving problems in mathematics and using the learning intention and prior knowledge sections at the start of each chapter, you will develop a deeper understanding of mathematical skills and knowledge as you progress through the broad general education and onto future examination courses.

Features

Chapter title

The chapter title shows the Curriculum for Excellence Outcome and Experience, and the CfE code for that Outcome and Experience.

1

> I can round a number using an appropriate degree of accuracy, having taken into account the context of the problem.
>
> **MNU 3-01a**

This chapter will show you how to:

Each chapter opens with a list of topics covered in the chapter, and tells you what you should be able to do when you have worked your way through the whole chapter.

> **This chapter will to show you how to:**
> - round a number to an appropriate degree of accuracy
> - check that answers to problems are accurate using estimation

Introduction

You should already know:

After the list of topics covered in the chapter, there is a list of topics you should already know before you start the chapter. These topics will sometimes be a continuation of topics you covered at Second Level, but there are some topics (such as Chapter 7 **Powers and roots**), which don't continue Second Level work. In these cases, the **You should already know** list shows the basic maths skills you should have before you start work. In some cases, the **You should already know** list will show some of the skills you have developed in previous Third Level work.

> **You should already know:**
> - The place value column headings
> - the basic rules of rounding
> - why estimating answers is a useful life skill.

Example

Each new topic is demonstrated with at least one worked **Example**, which shows how to go about tackling the questions in the following **Exercise**. Each **Example** breaks the question down into steps, so you can see what calculations are involved and how to go work out the best way of answering the question.

> **Example 11·5**   Five pens cost £3·25. How much do 8 pens cost?
>
> First, work out the cost of 1 pen: £3·25 ÷ 5 = £0·65.
>
> Hence, 8 pens cost 8 × £0·65 = £5·20.

Exercise

The most important parts of the book are the **Exercises**. The questions in the **Exercises** are carefully graded in difficulty, so you should be developing your skills as you work through an **Exercise**. If you find the questions difficult, look back at the **Example** for ideas on what to do.

Exercise 1A

1 Round each of the following to the nearest:

   i 1000   ii 100   iii 10

   a 4563   b 3247   c 5992   d 35 234   e 456
   f 9982·4 g 43 263 h 5237·34 i 9043   j 5006

Challenge

A lot of the **Exercises** have **Challenge** activities at the end. These are generally more difficult than most of the questions in the **Exercise**, and they are usually more about solving problems without much guidance, so they require you to work out how to do a problem, and then to do it. Don't worry if you find the **Challenge** activities difficult – they're intended to provide you with a *challenge*, and if you can do them, it means you're well on the way to having a very good understanding of the maths involved.

> **Challenge**   The diagram shows one side of an **isosceles** triangle.
>
> a Find all the possible positions on this grid for the other vertex.
>
> b Find the area of each triangle you find.
>
>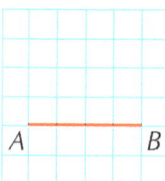

Introduction

End of chapter summary

Each chapter closes with a summary of learning statements showing what you should be able to do when you complete the chapter. The summary identifies **Key questions** ★ for each learning statement. You can use the **End-of-chapter summary** and the Key questions to check you have a good understanding of the topics covered in the chapter.

- By working on this topic I understand how to round numbers appropriately and how to use rounding to estimate and check answers.
- I can round to a required degree of accuracy. ★ Exercise 1A Q2

 ## Key questions

Most Exercises include a **Key question** (and some have more than one **Key question**). These questions are a guide to the depth and strength of your understanding of the topic. If you can answer a **Key question** without having to look over notes or look back at previous Exercise questions, you have a good grasp of the concepts underlying the topic and should feel confident you can use and apply your new skills and understanding.

 ## Problem-solving questions

Maths is an essential skill in solving a whole range of real-life problems. **Problem-solving questions** present real-life contexts for mathematical calculations, and help you learn how to apply your skills outside the classroom.

 ## Using a calculator

Using a calculator properly is a useful skill. **Calculator questions** have been designed to help you develop confidence in using your calculator accurately. In addition to the four basic mathematical operations of addition, subtraction, multiplication and division, you need to know how to use the brackets and powers functions on your calculator.

 Although it is useful to be able to use a calculator properly, it is even more important to be able to carry out basic calculations – addition and subtraction, multiplication and division – without using a calculator, and to know a range of number facts (such as $\frac{1}{4} = 25\%$, or $4^2 = 16$). **Non-calculator questions** give you practice in mental maths and written calculations. Developing these skills helps you in everyday life as well as in the maths class, so don't be tempted to skip them or to take the easy option and use your calculator in these questions.

Answers

Answers are provided at the back of the book to all the calculations. Use the answers to check your work, and if you get a question wrong, go back to it and see if you can work out where you went wrong. If you can't work out your mistake on your own, try asking a friend or classmate. If they can't help, ask your teacher.

Estimation and rounding

1

I can round a number using an appropriate degree of accuracy, having taken into account the context of the problem.

MNU 3-01a

This chapter will show you how to:
- round a number to an appropriate degree of accuracy
- use estimation to check that calculated answers are reasonable
- use rounding to estimate answers to complex problems
- use rounding and estimation to solve real-life problems
- use basic functions on the calculator and round your answers accordingly.

You should already know:
- the place value column headings
- the basic rules of rounding
- why estimating answers is a useful life skill.

Rounding

There are two main uses of rounding. One is to give an answer to a calculation to a sensible degree of accuracy. The other is to enable you to make an estimate of the answer to a problem (see pages 12–14).

When rounding a number to a required degree of accuracy:

- round **down** if the next digit is 0, 1, 2, 3 or 4
- round **up** if the next digit is 5, 6, 7, 8 or 9.

Example 1·1 Round each of the following to the nearest:

	i 1000	ii 100	iii 10	iv 1
a	3472·3			
b	12 546·7			
c	3998·5			

	i	ii	iii	iv
a	3000	3500	3470	3472
b	13 000	12 500	12 550	12 547
c	4000	4000	4000	3999

Example 1·2 Round each of the following to:

i 1 decimal place   ii 2 decimal places   iii 3 decimal places

a 9·3592   b 4·3236   c 5·9987

	i	ii	iii
a	9·4	9·36	9·359
b	4·3	4·32	4·324
c	6·0	6·00	5·999

Number, money and measure

Exercise 1A

1 Round each of the following to the nearest:
   i 1000   ii 100   iii 10
   a 4563   b 3247   c 5992   d 35 234   e 456
   f 9982·4   g 43 263   h 5237·34   i 9043   j 5006

2 Round each of these numbers to:
   i the nearest whole number   ii one decimal place.
   a 4·72   b 3·07   c 2·634   d 1·932   e 0·78   f 0·92
   g 3·99   h 2·64   i 3·18   j 3·475   k 1·45   l 1·863

3 Round each of these numbers to two decimal places.
   a 4·722   b 3·097   c 6·234   d 4·935   e 0·784
   f 0·992   g 3·999   h 2·604   i 3·185   j 6·496

4 Round each of these numbers to three decimal places.
   a 6·3261   b 8·1217   c 1·0023   d 9·8708   e 5·5546
   f 1·9989   g 0·0025   h 5·4211   i 3·7816   j 0·9954

Estimates

UNITED v CITY	
Crowd	41 923
Score	2 – 1
Time of first goal	42 min 13 sec
Price of a pie	95p

Which of the numbers above can be approximated? Which need to be given exactly?

You should have an idea if the answer to a calculation is about the right size or not. There are a number of ways to check the size of your answer. One way is to round numbers off and do a mental calculation to see if an answer is about the right size.

Example 1·3

Estimate answers to these calculations by rounding each number to the greatest place value.

a $\dfrac{21·3 + 48·7}{6·7}$ b $31·2 \times 48·5$ c $359 \div 42$ d $63 \times 0·73$

a Round off the numbers on the top to 20 + 50 = 70. Round off 6·7 to 7. Then 70 ÷ 7 = 10.

b Round off to 30 × 50, which is 3 × 5 × 100 = 1500.

c Round off to 400 ÷ 40, which is 40 ÷ 4 = 10.

d Round off to 60 × 0·7 = 6 × 10 × 0·7 = 6 × 7 = 42.

Estimation and rounding

Exercise 1B

1 Estimate the answer for each of these by rounding each number to the nearest 10.

a 24 × 11 b 53 × 12 c 42 × 18 d 42 × 42 e 21 × 69

2 Estimate the answer for each of these by rounding the larger number to the nearest hundred and the smaller number to the nearest ten.

a 1205 ÷ 31 b 203 ÷ 41 c 1387 ÷ 21 d 1406 ÷ 72 e 1584 ÷ 41

3 Estimate the answer for each of these by rounding each number to the nearest 10.

a $\dfrac{194 + 814}{112 + 90}$ b $\dfrac{213 + 73}{63 - 23}$ c $\dfrac{132 + 88}{78 + 28}$ d $\dfrac{795 + 98}{54 - 21}$

4 Estimate the answer to each of the following by rounding the bigger number to the nearest 10 and the smaller number to one decimal place.

a 72 × 0·56 b 61 × 0·67 c 39 × 0·81 d 42 × 0·17

e 57 × 0·33 f 68 × 0·68 g 38 × 0·19 h 23 × 0·91

i 43 × 0·86 j 28 × 0·75 k 34 × 0·52 l 116 × 0·18

★ 5 Estimate the answer to each of these problems.

a 2968 − 392 b 231 × 18 c 792 ÷ 38 d $\dfrac{36·7 + 23·2}{14·6}$

e 423 × 423 f 157·2 + 38·2 g $\dfrac{102·7 - 43·8}{18·8 - 8·9}$ h $\dfrac{38·9 \times 61·2}{39·6 - 18·4}$

6 Amy bought 6 bottles of pop at 46p per bottle. The shopkeeper asked her for £3·16. Without working out the correct answer, explain why this is wrong.

7 A first class stamp is 60p. I need eight. Will £4·50 be enough to pay for them? Explain your answer clearly.

8 In a shop I bought a 53p comic and a £1·47 model car. The till said £54·47. Why?

9 Which is the best approximation for 50·7 − 39·2?

a 506 − 392 b 51 − 39 c 50 − 39 d 5·06 − 3·92

10 Which is the best approximation for 19·3 × 42·6?

a 20 × 40 b 19 × 42 c 19 × 40 d 19 × 43

11 Which is the best estimate for 54·6 ÷ 10·9?

a 500 ÷ 100 b 54 ÷ 11 c 50 ÷ 11 d 55 ÷ 11

12 Delroy had £10. In his shopping basket he had a magazine costing £2·65, some batteries costing £1·92 and a CD costing £4·99. Without adding up the numbers, how could Delroy be sure he had enough to buy the goods in the basket? Explain a quick way for Delroy to find out if he could afford a 45p bar of chocolate as well.

Number, money and measure

Challenge

1. a. Without working out areas or counting squares, explain why the area of the square shown must be between 36 and 64 grid squares.

   b. Now calculate the area of the square.

   c. Using an 8 × 8 grid, draw a square with an area of exactly 50 grid squares.

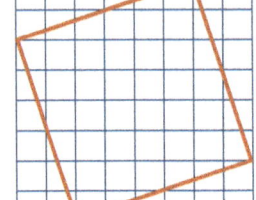

2. 62 ÷ 0·39 can be approximated as 60 ÷ 0·4 = 600 ÷ 4 = 150. Estimate the answer to each of the following divisions.

   a. 62 ÷ 0·56 b. 139 ÷ 0·67 c. 39 ÷ 0·81 d. 42 ÷ 0·17

   e. 57 ÷ 0·33 f. 68 ÷ 0·68 g. 38 ÷ 0·19 h. 178 ÷ 0·91

   i. 269 ÷ 0·86 j. 38 ÷ 0·75 k. 34 ÷ 0·52 l. 116 ÷ 0·18

Using a calculator and rounding

You should have your own calculator, so that you can get used to it. Make sure that you understand how to use the basic functions (×, ÷, +, −) and the brackets keys.

Example 1·4 Use a calculator to work out: a $\dfrac{242 + 118}{88 - 72}$ b $\dfrac{63 \times 224}{32 \times 36}$

The line that separates the top numbers from the bottom numbers indicates that it is the whole of the top divided by the whole of the bottom. You use brackets to do this.

a Key the calculation as (242 + 118) ÷ (88 − 72) = 22·5.

b Key the calculation as (63 × 224) ÷ (32 × 36) = 12·25.

Exercise 1C

1. Using brackets as shown above, work out the value of each of these. Round off your answers to one decimal place.

   a $\dfrac{194 + 866}{122 + 90}$ b $\dfrac{213 + 73}{63 - 19}$ c $\dfrac{132 + 88}{78 - 28}$ d $\dfrac{792 + 88}{54 - 21}$

   e $\dfrac{790 \times 84}{24 \times 28}$ f $\dfrac{642 \times 24}{87 - 15}$ g $\dfrac{107 + 853}{24 \times 16}$ h $\dfrac{57 - 23}{18 - 7\cdot 8}$

2. Without using a calculator, estimate the answer to $\dfrac{231 + 167}{78 - 32}$.

   Now use a calculator to work out the answer to one decimal place. Is it about the same?

3. Use a calculator to work out the following, rounding the answer to one decimal place.

   a 8·3 × (4·2 − 1·9)   b 12·3 ÷ (3·2 + 1·7)   c (3·2 + 1·9) ÷ (5·2 − 2·1)

4. Use a calculator to find the following to two decimal places.

   a 985 divided by 23   b 802 divided by 36

Estimation and rounding

Challenge

A furniture design company is making 3000 round tables of diameter 50 cm. Each table requires a metal strip around its edge. The length of metal needed for one table is found using the calculation $\pi \times 50$.

Two workers use different methods to calculate the total amount of metal needed.

Worker 1 says: 'The metal needed for one table is $\pi \times 50 = 157 \cdot 1$ cm to 1 dp. So the total length of metal needed is $157 \cdot 1 \times 3000 = 471\,300$ cm.'

Worker 2 says: 'The total length of metal needed is $\pi \times 50 \times 3000 = 471\,238 \cdot 9$ cm to 1 dp.'

Both men seem to have carried out the same calculation.

Why are their answers different?

What do you have to be careful about when rounding?

Can you see why Worker 1's method could be costly to the company if repeated regularly?

- By working on this topic I understand how to round numbers appropriately and how to use rounding to estimate and check answers.

- I can round to a required degree of accuracy. ★ Exercise 1A Q2

- I have learnt that rounding numbers in a calculation allows me to make a quick estimate of the answer. ★ Exercise 1B Q5

- I can use the basic functions on a calculator and can explain how and why I have rounded my answer. ★ Exercise 1C Q1

- I know that using a rounded answer in further calculations can lead to inaccurate results. *Challenge* page 15

Number, money and measure

2

I can use a variety of methods to solve number problems in familiar contexts, clearly communicating my processes and solutions.

MNU 3-03a

This chapter will show you how to:
- read a problem carefully to work out the information and the methods you need to solve the problem
- choose the correct method to solve a problem
- check your answers using different methods
- work with whole numbers and decimals
- use written methods for long multiplication and long division
- explain and communicate your methods and answers.

You should already know:
- how to use standard written methods for the four operations
- how to round numbers to 1 decimal place
- how to multiply and divide by 10.

This chapter will help you to analyse number problems and choose the correct strategy to find the required solution. The ability to perform mental and written calculations with numbers is an essential part of numeracy and this chapter will help build confidence in these areas.

Multiplying and dividing by 10, 100 and 1000

Example 2·1   Work out 3·5 × 100.

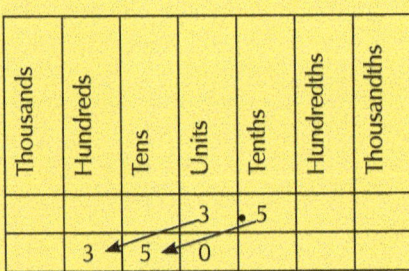

So 3·5 × 100 = 350.

The digits move one place to the left when you multiply by 10, two places to the left when you multiply by 100, and three places to the left when you multiply by 1000.

Number and number processes

Example 2·2   Work out 23 ÷ 1000.

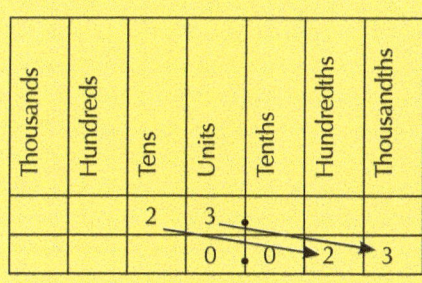

So 23 ÷ 1000 = 0·023.

In the same way, the digits move one place to the right when you divide by 10, two places to the right when you divide by 100, and three places to the right when you divide by 1000.

Exercise 2A

1. Write down the answer to each of these.

    a   3 × 6 = ...   b   6 × 3 = ...   c   6 × 7 = ...   d   4 × 2 = ...

    e   5 × 5 = ...   f   2 × 8 = ...   g   4 × 5 = ...   h   8 × 0 = ...

    i   6 × 10 = ...   j   7 × 1 = ...   k   4 × 7 = ...   l   10 × 10 = ...

    m   9 × 8 = ...   n   9 × 6 = ...   o   9 × 3 = ...   p   5 × 9 = ...

    q   7 × 5 = ...   r   7 × 9 = ...   s   6 × 8 = ...   t   7 × 8 = ...

2. Work out each of these.

    a   6 × 40 = ...   b   5 × 60 = ...   c   5 × 40 = ...   d   60 × 5 = ...

    e   70 × 6 = ...   f   800 × 3 = ...   g   6 × 300 = ...   h   7 × 800 = ...

    i   30 × 9 = ...   j   50 × 40 = ...   k   9 × 800 = ...   l   80 × 90 = ...

3. Without using a calculator work out:

    a   34 × 10   b   89 × 100   c   7 × 100

    d   4 × 1000   e   34 ÷ 10   f   89 ÷ 100

    g   7 ÷ 100   h   4 ÷ 1000   i   58 ÷ 1000

4. Find the missing number in each case.

    a   3 × 10 = ☐   b   3 × ☐ = 300

    c   3 ÷ 10 = ☐   d   3 ÷ ☐ = 0.03

 5. Without using a calculator work out:

    a   4·5 × 10   b   0·6 × 10   c   5·3 × 100

    d   0·03 × 100   e   5·8 × 1000   f   0·7 × 1000

    g   4·5 ÷ 10   h   0·6 ÷ 10   i   5·3 ÷ 100

    j   0·03 ÷ 100   k   5·8 ÷ 1000   l   0·04 ÷ 10

    m   5·01 ÷ 10   n   6·378 × 100   o   0·21 × 1000

Number, money and measure

6 Find the missing number in each case.

a $0.3 \times 10 = \square$
b $0.3 \times \square = 300$
c $0.3 \div 10 = \square$

d $0.3 \div \square = 0.003$
e $\square \div 100 = 0.03$
f $\square \div 10 = 30$

g $\square \times 1000 = 30\,000$
h $\square \times 10 = 300$
i $\square \div 1000 = 0.3$

7 Fill in the missing operation in each case.

a $0.37 \rightarrow \boxed{\times 100} \rightarrow 37$
b $567 \rightarrow \boxed{} \rightarrow 5.67$

c $0.07 \rightarrow \boxed{} \rightarrow 70$
d $650 \rightarrow \boxed{} \rightarrow 65$

e $0.6 \rightarrow \boxed{} \rightarrow 0.006$
f $345 \rightarrow \boxed{} \rightarrow 0.345$

8 Copy, complete and work out the total of this shopping bill:

1000 matches at £0·04 each   = · · · · · ·

100 packets of chocolate buttons at £0·37 each   = · · · · · ·

10 cans of cola at £0·95 each   = · · · · · ·

Whole numbers

Example 2·3

a Find the product of 9 and 6.

b Find the remainder when 347 is divided by 5.

a Product means 'multiply'. So, $9 \times 6 = 54$.

b Using short division gives:

$$5 \overline{)34^47}$$ quotient 69

The remainder is $47 - 45 = 2$.

You know that any multiple of 5 ends in either 0 or 5, so the remainder must be $7 - 5 = 2$.

Number and number processes

Example 2·4

a Find the difference between 453 and 237.
b Work out 43 × 4.
c Work out 7 × 90.
d Work out 400 × 3.

a Set out the problem in columns:

$$\begin{array}{r} 4\overset{4}{\cancel{5}}{}^1 3 \\ -237 \\ \hline 216 \end{array}$$

You have to borrow from the tens column because 7 cannot be taken from 3.

b Using grid multiplication gives:

×	40	3
4	160	12

So, 43 × 4 = 160 + 12 = 172

Using the column method gives:

$$\begin{array}{r} 43 \\ \times 4 \\ \hline 172 \\ {}_{1} \end{array}$$

c 7 × 90 = 7 × 9 × 10
       = 63 × 10
       = 630

d 400 × 3 = 100 × 4 × 3
         = 100 × 12
         = 1200

Exercise 2B

1 Copy each of these grids and fill in the gaps.

a 34 × 7

×	30	4
7		

b 4 × 26

×		4
20		
6		

c 52 × 7

×	50	2
7		

2 Use the grid method as in Question 1, or any other method, to find:

a 58 × 7   b 35 × 8   c 2 × 34   d 19 × 8
e 6 × 32   f 42 × 7   g 56 × 6   h 3 × 33

3 Find the sum and product of: a 7 and 20   b 2 and 50

4 The local video shop is having a sale. DVDs are £4·99 each or five for £20.

a What is the cost of three DVDs?
b What is the cost of ten DVDs?
c What is the greatest number of DVDs you can buy with £37? Explain your answer.

 5 a Three consecutive integers have a sum of 90. What are they?

b Two consecutive integers have a product of 132. What are they?
c Explain why there is more than one answer to this problem:
   Two consecutive integers have a difference of 1. What are they?

6 Two consecutive numbers add up to 13. What are the numbers?

7 Two consecutive numbers add up to 29. What are the numbers?

Number, money and measure

8 Two consecutive numbers add up to 37. What are the numbers?

9 Two consecutive numbers multiply together to give 30. What are the numbers?

10 Two consecutive numbers multiply together to give 90. What are the numbers?

11 Two consecutive numbers multiply together to give 56. What are the numbers?

12 a Use your calculator to find two consecutive odd numbers which multiplied together give an answer of 143.

   b Use the digits 1, 3 and 4 and the multiplication sign × once only to make the largest possible answer.

13 Here is a magic square. Each row, column and diagonal adds up to 15.

8	1	6
3	5	7
4	9	2

Complete these magic squares so that each row, column and diagonal adds up to 15.

4		8
	7	

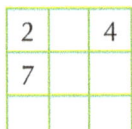

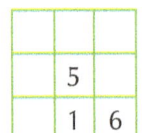

More multiplying and dividing whole numbers

Example 2·5 Work out 36 × 4.

Below are three examples of the ways this calculation can be done. The answer is 144.

Grid method (partitioning)

×	30	6
4	120	24	144

Column method (expanded working)

```
   36
 ×  4
   24   (4 × 6)
  120   (4 × 30)
  144
```

Column method (compacted working)

```
   36
 ×  4
  144
    2
```

Example 2·6 Work out 543 ÷ 8.

Below are three examples of the ways this can be done. The answer is 67, remainder 7.

Repeated subtraction

```
   543
 − 320   (40 × 8)
   223
 − 160   (20 × 8)
    63
 −  48   (6 × 8)
    15
 −   8   (1 × 8)
     7   (67 × 8)
```

Short division

```
      6 7 rem 7
   8)54⁶3
```

Short division (decimal answer)

```
      6 7 · 8 7 5
   8)54⁶3·⁷0⁶0⁴0
```

So 543 ÷ 8 = 67·9 to 1dp

Number and number processes

Exercise 2C For each question, check your answer using an appropriate method.

1. Work out each of the following multiplication problems. Use any method you are happy with.

    a 17 × 3   b 32 × 4   c 19 × 5   d 56 × 6
    e 2 × 346   f 3 × 541   g 7 × 147   h 9 × 213

2. A van does 34 miles to a gallon of petrol. How many miles can it do if the petrol tank holds 8 gallons?

3. The school photocopier can print 82 sheets a minute. If it runs without stopping for 7 minutes, how many sheets will it print? Explain your answer.

4. Each day 7 Jumbo jets fly from London to San Francisco. Each jet can carry up to 348 passengers. How many people can travel from London to San Francisco each day? Explain your answer.

5. Blank CDs cost 45p each. How much will 5 CDs cost? Give your answer in pounds.

6. A daily newspaper sells advertising by the square inch. On Monday, it sells 163 square inches at £9 per square inch. How much money does it get from this advertising?

7. Work out each division giving your answer in:

    i remainder form   ii decimal form rounded to one decimal place.

    a 79 ÷ 7   b 55 ÷ 3   c 124 ÷ 5
    d 112 ÷ 6   e 71 ÷ 4   f 215 ÷ 6

8. The local library has 13 000 books. Each shelf holds 50 books. How many shelves are there?

9. A company has 89 boxes to move by van. The van can carry 7 boxes at a time. How many trips must the van make to move all the boxes? Explain your answer.

10. How many bubble packs of 40 nails can be filled from a carton of 450 nails? Explain your answer.

11. a To raise money, Musselburgh Running Club are going to do a relay race from Musselburgh to Hamilton, which is 79 kilometres. Each runner will run 8 kilometres. How many runners will be needed to cover the distance?

    b Sponsorship will bring in £9 per kilometre. How much money will the club raise?

Number problems

A bus starts at Moodiesburn and makes four stops before reaching Abronhill. At Moodiesburn 23 people get on. At Carrickstone 12 people get off and 14 people get on. At Greenfaulds 15 people get off and 4 people get on. At Seafar 5 people get off and 6 people get on. At Cumbernauld 9 people get off and 8 get on. At Abronhill the rest of the passengers get off. How many people are on the bus?

Number, money and measure

When you solve problems, you need to develop a strategy: that is, a way to go about the problem. You also have to decide which mathematical operation you need to solve it. For example, is it addition, subtraction, multiplication or division or a combination of these? Something else you must do is to read the question fully before starting. The answer to the problem above is one! The driver.

Read the questions below carefully.

Exercise 2D

1 It cost six people £27 to go to the cinema. How much would it cost eight people?

2 Ten pencils cost £4·50. How much would seven pencils cost?

3 30 can be worked out as 33 − 3. Can you find two other ways of working out 30 using the same digit three times?

4 Arrange the numbers 1, 2, 3 and 4 in each of these to make the problem correct.

a ☐ + ☐ = ☐ + ☐   b ☐ × ☐ = ☐☐   c ☐☐ ÷ ☐ = ☐

5 A water tank holds 500 litres. How much has been used if 143·7 litres are left in the tank?

6 Strips of paper are 40 cm long. They are stuck together with a 10 cm overlap.

   a How long would two strips glued together be?
   b How long would four strips glued together be?

7 A can of cola and a chocolate bar together cost £1·50. Two cans of cola and a chocolate bar together cost £2·40. How much would three cans of cola and four chocolate bars cost?

8 The bill in a restaurant comes to £99·20. There are 8 people at the table. They decide to leave a £2 tip each **and** share the bill equally. How much will each person pay? Show your working and explain the reasons for your approach.

9 Diana is buying bedroom furniture. She buys a bed for £599, a chest of drawers for £165 and a chair for £88. The store is doing a half-price sale. How much does Diana pay? Explain your method and your answer.

10 A supermarket claims it will refund twice the difference if a customer finds any item cheaper elsewhere. The supermarket sells a kilogram of sugar for £1·07. A customer finds that his local grocer is selling 1 kg of sugar for 79p. How much of a refund should the customer get?

11 On a trip to Glasgow, the sat nav on Scott's car is showing the following information: distance travelled = 56 miles, distance to destination = 47 miles. The next time Scott looks at the sat nav it shows: distance travelled = 81 miles. What will the sat nav show as the distance to destination?

12 To make a number chain, start with any number.

> When the number is even, divide it by 2.
> When the number is odd, multiply it by 3 and add 1.

If you start with 13, the chain becomes 13, 40, 20, 10, 5, 16, 8, 4, 2, 1, 4, 2, 1, ...
The chain repeats 4, 2, 1, 4, 2, 1. So, stop the chain when it gets to 1.
Start with other numbers below 20. What is the longest chain you can make before you get to 1?

Challenge Using the numbers 1, 2, 3 and 4 and any mathematical signs, make all of the numbers from 1 to 10.

For example: $2 \times 3 - 4 - 1 = 1$, $4 \times 2 - 3 = 5$

Once you have found all the numbers up to 10, can you find totals above 10?

Using a calculator

Example 2.7 Use a calculator to work out: a $\dfrac{215 + 154}{164 - 82}$ b $\dfrac{246 \times 48}{15 \times 64}$

The line that separates the top numbers from the bottom numbers acts as a divide sign (÷).
You must use brackets when entering into the calculator.

a Key the calculation as: $215 + 154 \div 164 - 82 = 4 \cdot 5$

b Key the calculation as: $246 \times 48 \div 15 \times 64 = 12 \cdot 3$

Exercise 2E

1 Without using a calculator, work out the value of each of these by evaluating the top line and bottom line first before dividing.

  a $\dfrac{17 + 8}{7 - 2}$ b $\dfrac{53 - 8}{3 \cdot 5 - 2}$ c $\dfrac{19 \cdot 2 - 1 \cdot 7}{5 \cdot 6 - 3 \cdot 1}$

2 Use a calculator to do the calculations in Question 1. Remember to use brackets. Do you get the same answers? For each part, write down the sequence of keys that you pressed to get the answer.

3 Work out the value of each of these. Round your answers to one decimal place if necessary.

  a $\dfrac{194 + 866}{122 + 90}$ b $\dfrac{213 + 73}{63 - 19}$ c $\dfrac{132 + 88}{78 - 28}$ d $\dfrac{792 + 88}{54 - 21}$

  e $\dfrac{790 \times 84}{24 \times 28}$ f $\dfrac{642 \times 24}{87 - 15}$ g $\dfrac{107 + 853}{24 \times 16}$ h $\dfrac{57 - 23}{18 - 7 \cdot 8}$

 4 Estimate the answer to $\dfrac{432 + 266}{93 - 38}$.

Now use a calculator to work out the answer to one decimal place. Is it about the same?

Number, money and measure
Long multiplication

Example 2·8 Work out 36 × 43.

Below are two examples of the ways this calculation can be done. The answer is 1548.

Grid method (partitioning)

×	30	6	
40	1200	240	1440
3	90	18	108
			1548

Column method (compacted working)

```
   36
 × 43
  108   (3 × 36)
 1440   (40 × 36)
 ----
 1548
```

Example 2·9 Work out 26 × 238.

Grid method

×	200	30	8	
20	4000	600	160	4760
6	1200	180	48	1428
				6188

Column method

```
   238
 ×  26
  1428   (6 × 238)
  4760   (20 × 238)
  ----
  6188
```

Exercise 2F

1. Use the grid method to work out the following long multiplication problems.

   a 15 × 16   b 18 × 22   c 43 × 27   d 62 × 31
   e 17 × 241   f 317 × 24   g 45 × 257   h 406 × 72

2. Use the column method to work out the long multiplication problems in Question 1.

 3. Work out each of the following long multiplication problems. Use any method you are happy with.

   a 17 × 23   b 32 × 42   c 19 × 45   d 46 × 56
   e 12 × 346   f 541 × 32   g 27 × 147   h 213 × 39

Long division

Example 2·10 Work out 543 ÷ 31.

Below are two examples of the ways this can be done. The answer is 17, remainder 16.

Repeated subtraction

```
  543
 -310    (10 × 31)
  233
 -155    (5 × 31)
   78
 - 62    (2 × 31)
   16    (17 × 31)
```

Traditional method

```
      17
 31)543
     31↓
     233
     217
      16
```

Number and number processes

Exercise 2G

1 Use repeated subtraction to work out the following long multiplication problems.

  a 512 ÷ 16
  b 399 ÷ 19
  c 506 ÷ 22
  d 744 ÷ 31
  e 864 ÷ 36
  f 945 ÷ 27
  g 2173 ÷ 53
  h 1472 ÷ 46

2 Use the traditional method to work out the long division problems in Question 1.

★ 3 Work out each of the following long division problems. Use any method you are happy with. Some of the problems will have a remainder.

  a 684 ÷ 19
  b 966 ÷ 23
  c 972 ÷ 36
  d 625 ÷ 25
  e 930 ÷ 38
  f 642 ÷ 24
  g 950 ÷ 33
  h 800 ÷ 42

Exercise 2H

Decide whether each of the following 10 problems involves long multiplication or long division. Then do the appropriate calculation, showing your method clearly. For each question, check your answer using an appropriate method.

1 Each day 17 Jumbo jets fly from London to San Francisco. Each jet can carry up to 348 passengers. How many people can travel from London to San Francisco each day? Show your working and explain your answer.

2 A company has 897 boxes to move by van. The van can carry 23 boxes at a time. How many trips must the van make to move all the boxes?

3 The same van does 32 miles to a gallon of petrol. How many miles can it do if the petrol tank holds 18 gallons? Explain your answer.

★ 4 The school photocopier can print 85 sheets a minute. If it runs without stopping for 45 minutes, how many sheets will it print?

5 The RE department has printed 525 sheets on Buddhism. These are put into folders in sets of 35. How many folders are needed? Explain your answer.

6 a To raise money, Wath Running Club are going to do a relay race from Wath to Edinburgh, running a 384 kilometre route. Each runner will run 24 kilometres. How many runners will be needed to cover the distance? Explain your answer.

  b Sponsorship will bring in £32 per kilometre. How much money will the club raise?

7 Blank CDs are 65p each. How much will a box of 35 CDs cost? Give your answer in pounds.

8 A magazine sells advertising by the square inch. On Monday, it sells 232 square inches at £15 per square inch. How much money does it get from this advertising?

9 A second-hand book store has about 9000 books. On average a shelf holds 47 books. How many shelves are there?

10 How many packs of 30 tea bags can be filled from a carton of 400 tea bags?

Number, money and measure
More decimals

Example 2·11

Without a calculator, work out 13·4 × 0·63.

There are many ways to do this. Three are shown. In all cases you should first estimate the answer:

13·4 × 0·63   13 × 0·6 = 1·3 × 6 = 7·8

Remember also that there are three decimal places in the numbers being multiplied (13·4 × 0·63), so there will be three digits after the decimal point in the answer.

In two methods the decimal points are ignored during the calculation and put back into the answer.

Column method
```
    134
  ×  63
   ----
    402
   8040
   ----
   8442
```

Grid method 1

×	100	30	4	Total
60	6000	1800	240	8040
3	300	90	12	402
			Total	8442

Grid method 2

×	10	3	0·4	Total
0·6	6	1·8	0·24	8·04
0·03	0·3	0·09	0·012	0·402
			Total	8·442

By all three methods the answer is 13·4 × 0·63 = 8·442.

Example 2·12

Work out: 4·32 ÷ 1·2

First estimate the answer: 4·32 ÷ 1·2   4 ÷ 1 = 4

Write without the decimal points, i.e. 432 ÷ 12.

Repeated subtraction
```
    432
  − 360    (30 × 12)
   ----
     72
  −  72    (6 × 12)
   ----
      0    (36 × 12)
```
The answer is 4·32 ÷ 1·2 = 3·6.

Traditional method
```
       36
   12)432
       36↓
       --
       72
       72
       --
        0
```
Putting the decimal points back in gives: 4·32 ÷ 1·2 = 3·6.

For adding and subtracting decimals see chapter 8 (MNU 3-07a) on pages 64–66.

Number and number processes

Exercise 2I

1. Without using a calculator, work out the answers to the following. Use any method you are happy with.

   a 73 × 9·4 b 5·82 × 4·5 c 12·3 × 2·7 d 1·24 × 10·3
   e 2·78 × 0·51 f 12·6 × 0·15 g 2·63 × 6·5 h 0·68 × 0·42

2. A rectangle is 2·46 m by 0·67 m. What is the area of the rectangle?

3. Without using a calculator, work out the following. Use any method you are happy with.

   a 3·36 ÷ 1·4 b 1·56 ÷ 2·4 c 5·688 ÷ 3·6 d 20·28 ÷ 5·2
   e 22·23 ÷ 6·5 f 2·89 ÷ 3·4 g 5·75 ÷ 23 h 2·304 ÷ 0·24

4. A rectangle has an area of 3·915 cm². The length is 2·7 cm. Calculate the breadth by dividing the area by the length.

More number problems

Example 2·13

A box contains 12 identical toy cars:

Each toy has 4 lights.
Each toy weighs 200 g.
The box weighs 150 g.

a How many lights are there altogether?
b If 3 cars are removed, what is the total weight of the box and its contents?

a 12 cars with 4 lights each = 12 × 4 = 48 lights.
b If 3 cars are removed, there are 9 left.
The cars weigh 9 × 200 g = 1800 g.
The box weighs 150 g.
So the total weight = 1800 + 150 = 1950 g.

Exercise 2J

1. Find two odd numbers that sum to 48.

2. The product of 2 and 3 is 6, because 2 × 3 = 6. Work out the product of 6 and 7.

3. A cupboard space is 70 cm high. Tins are 15 cm high. How many layers of tins will fit in the cupboard?

Number, money and measure

4 Here is a rule for the number grids.
   Use the rule to fill in the missing numbers.

   This number is the difference of the numbers on the bottom line.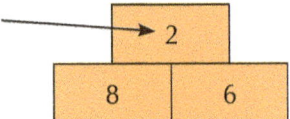

   a 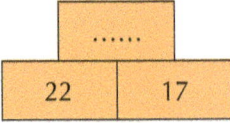   b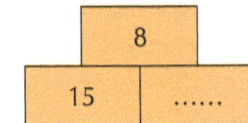

   c 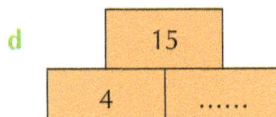   d

   a: top ......, bottom 22, 17
   b: top ......, bottom 2, −3
   c: top 8, bottom 15, ...
   d: top 15, bottom 4, ...

5 Yoghurts are sold individually for 65p. They are also sold in multipacks of 6 for £3·50. Which is cheaper? Explain your answer.

6 The total age of two brothers is 110 years. The difference in their ages is 4 years. How old is the younger brother? Show your working and explain your method.

7 There are 5 blue, 3 red and 2 white counters in a bag. The counters are numbered from 1 to 10. Each counter weighs 6 g.

   Match each statement to the correct calculation:

The weight of the odd numbered counters	10 × 6
The total weight of the counters	(5 + 3) × 6
The weight of the counters that are blue or red	5 × 6

8 A photocopying company charges 5p per sheet. How many sheets can be photocopied for £3?

9 Olivia is twice as old as Jack. The sum of their ages is 36 years. How old are they?

10 A class of 25 pupils scored a total of 3125 points in an online maths competition. What was the average score per pupil?

11 Autosales Garage is selling petrol at 143·9 pence per litre. Work out the cost of buying:
   a 5 litres
   b 60 litres

12 Eric works in a fast food restaurant. His pay is £8·52 per hour. What is his pay for working 45 hours?

Number and number processes

Challenge

Here is a magic square.
All the rows and columns add up to 34.
Complete the magic square

There are lots of ways of making 34 in this magic square using patterns of 4 numbers.

For example, 9 + 7 + 4 + 14 = 34 or 3 + 8 + 9 + 14 = 34.

How many ways can you find to make 34?

	2	3	
5			
9	7	6	12
4	14		1

- By working on this topic I can choose the most suitable method to solve number problems and explain my choices, methods and answers.

- I understand the effect on place value when multiplying and dividing by powers of 10.
  ★ Exercise 2A Q5

- I understand the meaning of product, sum, difference and consecutive. ★ Exercise 2B Q5

- I can choose the correct method (multiply or divide) to solve number problems.
  ★ Exercise 2C Q9

- I can choose the correct method (add, subtract, multiply or divide) to solve number problems.
  ★ Exercise 2D Q9

- I can use my calculator and check that the answer is sensible using estimation.
  ★ Exercise 2E Q4

- I can explain the box and column methods for long multiplication and can carry out accurate written calculations. ★ Exercise 2F Q3

- I can explain repeated subtraction and the traditional method for long division and can carry out accurate written calculations. ★ Exercise 2G Q3

- I can apply my knowledge of written methods for long multiplication and long division.
  ★ Exercise 2H Q4

- I can extend my knowledge of long multiplication and long division to decimal numbers.
  ★ Exercise 2I Q2

- I can apply my skills with number to solve real-life problems. ★ Exercise 2J Q5

Number, money and measure

3

I can continue to recall number facts quickly and use them accurately when making calculations.

MNU 3-03b

This chapter will show you how to:
- use mental strategies to calculate and check solutions to number problems
- use your knowledge of addition and subtraction number bonds to extend the range of calculations you can carry out
- use your knowledge of multiplication and division number bonds to extend the range of calculations you can carry out
- use the relationships between addition and subtraction, and multiplication and division, to carry out calculations and solve problems.

You should already know:
- how to break down a calculation into simpler steps
- how to use the four operations with whole numbers and decimals
- odd and even numbers
- the meaning of place value in decimal numbers.

This chapter will help you to use the number facts you know and choose appropriate strategies to help you work out new facts and to check your calculations.

Quick methods and estimation

A quick way to check whether an answer is right is to estimate the answer.

Example 3·1   Work out:

a 30×400 b $3000 \div 15$

a $30 \times 400 = 3 \times 10 \times 4 \times 100 = 12\,000$

b $3000 \div 15 = \dfrac{3000}{15} = \dfrac{\overset{2}{\cancel{3000}}}{\cancel{15}} = 200$

Example 3·2   Estimate the answer to each of these:

a 31×53 b $3127 \div 60$

a $31 \times 53 \approx 30 \times 50 = 1500$ (Correct answer is 1643.)

b $\dfrac{3127}{60} \approx \dfrac{3000}{60} = \dfrac{\overset{5}{\cancel{3000}}}{\cancel{60}} = 50$ (Correct answer is 52·1 to 1 dp.)

Number and number processes

Exercise 3A

1 Work out the following.

a 2 × 10   b 20 × 10   c 7 × 100   d 70 × 20
e 30 × 40   f 300 × 70   g 20 × 40   h 50 × 60
i 70 × 20   j 60 × 30   k 40 × 80   l 30 × 3000
m 600 × 70   n 200 × 200   o 80 × 30   p 50 × 50

2 Three answers are given for each calculation. Which one is correct?

a 20 × 90   180, 1800, 18 000   b 300 × 40   120, 1200, 12 000
c 50 × 400   2000, 20 000, 200 000   d 7 × 60   420, 4200, 42 000

3 Work out the following.

a 1200 ÷ 40   b 200 ÷ 40   c 1400 ÷ 20   d 1400 ÷ 70
e 1500 ÷ 50   f 1600 ÷ 100   g 1800 ÷ 30   h 4000 ÷ 40
i 2000 ÷ 40   j 2400 ÷ 60   k 1800 ÷ 60   l 1500 ÷ 30
m 4000 ÷ 20   n 2800 ÷ 40   o 3500 ÷ 70   p 4800 ÷ 80

★ 4 For each of these: i estimate   ii calculate the answer.

a 23 × 11   b 43 × 12   c 72 × 18   d 32 × 42
e 31 × 69   f 18 × 38   g 48 × 58   h 72 × 22
i 63 × 33   j 39 × 81   k 54 × 21   l 36 × 49

5 For each of these: i estimate   ii calculate the answer (rounding to 1 dp).

a 1178 ÷ 32   b 207 ÷ 38   c 1412 ÷ 22   d 1378 ÷ 68
e 1534 ÷ 48   f 1578 ÷ 98   g 1824 ÷ 32   h 1998 ÷ 37
i 1998 ÷ 41   j 2376 ÷ 62   k 3742 ÷ 52   l 1965 ÷ 75

6 For each of these: i estimate   ii calculate the answer (rounding to 1 dp).

a $\dfrac{194+816}{122+90}$ b $\dfrac{213+73}{63-13}$ c $\dfrac{132+88}{78-28}$ d $\dfrac{792+88}{54-21}$

7 a On squared paper, draw a shape that is ten times bigger in both directions than the shape on the right.

How many little squares are there in the larger shape?

b Repeat part **a** with the shape on the right.

c Without drawing, work out how many squares there would be in a shape that is 100 times bigger in both directions than the shape in part **a**.

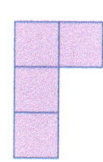

Number, money and measure

Challenge Multiplications and divisions which have lots of zeros can cause difficulties if done using a calculator.

Use mental methods to work out the following.

The first two questions are done for you.

1. Use mental methods to work out $30\,000 \times 4\,000\,000$.

$$30\,000 \times 4\,000\,000 = 3 \times 4 \times 10\,000 \times 1\,000\,000$$
$$= 12 \times 10\,000\,000\,000$$
$$= 120\,000\,000\,000$$

2. Estimate the value of $\dfrac{8\,125\,751}{412\,075}$.

$$\dfrac{8\,125\,751}{412\,075} \approx \dfrac{8\,000\,000}{400\,000} = \dfrac{8\,0\cancel{00}\,\cancel{000}}{4\cancel{00}\,\cancel{000}} = 20$$

3. Use mental methods to work out $7\,000\,000\,000 \times 6\,000\,000$.

4. Use mental methods to estimate the value of $7\,129\,836 \times 29\,451$.

5. Estimate the value of $\dfrac{2\,017\,832}{510\,965}$.

Mental checks

Example 3·3 Explain why these calculations must be wrong.

a $25 \times 63 = 1573$ b $21 \times 49 = 102$

a The last digit should be 5, because the product of the last digits is 15. That is, $25 \times 63 = \ldots 5$

b The answer is roughly $20 \times 50 = 1000$.

Example 3·4 Estimate answers to these calculations.

a $\dfrac{31\cdot3 + 58\cdot7}{9\cdot3}$ b $21\cdot2 \times 47\cdot5$ c $284 \div 43$

a Round off the numbers on the top to $30 + 60 = 90$. Round off $9\cdot3$ to 9. Then $90 \div 9 = 10$.

b Round off to 20×50, which is $2 \times 5 \times 100 = 1000$.

c Round off to $280 \div 40$, which is $28 \div 4 = 7$.

Number and number processes

Example 3·5 By using the inverse operation, check if each calculation is correct.

a 330 ÷ 6 = 55   b 240 − 79 = 169

a By the inverse operation, 330 = 6 × 55. This is true and can be checked mentally:
6 × 50 = 300, 6 × 5 = 30, 300 + 30 = 330.

b By the inverse operation, 169 + 79 must end in 8 as 9 + 9 = 18, so 240 − 79 = 169 cannot be correct.

Exercise 3B

1 Explain why these calculations must be wrong.

a 14 × 43 = 1080   b 61 × 83 = 723   c $\dfrac{23·5 + 81·2}{9·9} = 20·04$

d 440 ÷ 8 = 45   e 290 − 37 = 257   f 424·3 + 15·03 = 439·6

2 Estimate the answer to each of these problems.

a 4788 − 691   b 321 × 29   c 691 + 142   d $\dfrac{56·7 + 19·5}{16·2}$

e 507 × 507   f 187·3 ÷ 47·6   g $\dfrac{20·3 + 31·4}{24·9 - 19·8}$ h $\dfrac{92·1 \times 21·7}{82·7 - 21·7}$

3 David had £10. In his shopping basket he had a comic costing £2·75, some pens costing £1·82 and a DVD costing £4·95. Without adding up the numbers, how could David be sure he had enough to buy the goods in the basket? Explain a quick way for David to find out if he could afford a 35p bag of crisps as well.

★ 4 Sarah bought 8 bottles of pop at 56p per bottle. The shopkeeper asked her for £5·48. Without working out the correct answer, explain why this is wrong.

5 A first class stamp is 60p. I need four. Will £2 be enough to pay for them? Explain your answer clearly.

6 In a shop Isla bought a 60p juice and a £1·95 magazine. The till said £61·95. Why?

7 Estimate the value the arrow is pointing at in each of these.

a  b   c

Extending knowledge of number facts

Example 3·6 Use the fact that 6 + 7 = 13 to write down the answer to the following.

a 600 + 700   b 0·13 − 0·07   c 126 + 17

a The digits 6 and 7 have the same place value so they can be added together. Six hundred add seven hundred equals thirteen hundred or 600 + 700 = 1300.

b Thirteen hundredths subtract seven hundredths equals six hundredths or 0·13 − 0·07 = 0·06.

c Partition into hundreds, tens and units and add elements with the same place value: 100 + 20 + 6 + 10 + 7 = 100 + 30 + 13 = 143.

33

Number, money and measure

Exercise 3C

1. Using the fact that 3 + 5 = 8, write down the answers to the following.
   a 300 + 500
   b 0·3 + 0·5
   c £8 million subtract £5 million

2. Using the fact that 9 + 6 = 15, write down the answers to the following.
   a 60 kg + 90 kg
   b 0·09 cm + 0·06 cm
   c 15 000 − 9000

3. Use 14 + 6 = 20 to match the calculations and answers.

140 + 60	60
1·4 + 0·6	2·0
14 000 + 6000	14 000
0·020 − 0·006	200
200 − 140	1·020
0·14 + 0·06	202·0
20 000 − 6000	600
0·6 + 201·4	20 000
2000 − 1400	0·20
1·014 + 0·006	0·014

★ 4. Use basic number facts to help answer the following.
   a 70 000 + 60 000
   b 1·5 − 0·6
   c 400 + 700
   d 160 − 80
   e 0·8 + 0·7
   f 120 − 70
   g 6000 + 5000
   h 0·003 + 0·008
   i 1·09 + 0·07
   j 20 040 + 80
   k 1130 − 50
   l 4·14 − 0·08
   m £1·3 million − £0·9 million
   n 6·009 kg + 1·008 kg
   o 101·2 mm − 0·6 mm

5. You should be able to answer the following questions mentally.

   a Dundee United bought a striker for £70 000 and sold him two years later for £120 000. How much profit did they make?

   b Grace won the 100 m sprint by 0·4 seconds. If her time was 11·8 seconds, what was the time for the second place runner?

   c One breakfast cereal contains 1·5 g of fibre per serving while another contains 0·9 g. What is the difference between the fibre quantities?

   d Audrey buys a new coat online. It has been reduced from £170 by £90. How much did she pay?

Multiplying decimals

Number and number processes

This section will give you more practice on multiplying decimals.

Example 3·7

Find:

a 0·2 × 3   b 0·2 × 0·3   c 40 × 0·8

a 2 × 3 = 6. There is one decimal place in the multiplication, so 0·2 × 3 = 0·6.

b 2 × 3 = 6. There are two decimal places in the multiplication, so there are two in the answer. So, 0·2 × 0·3 = 0·06.

c Rewrite the problem as an equivalent product, that is
   40 × 0·8 = 4 × 10 × 0·8 = 4 × 8 = 32.

Example 3·8

A sheet of card is 0·5 mm thick. How thick is a pack of card containing 80 sheets?

This is a multiplication problem:

0·5 × 80 = 0·5 × 10 × 8 = 5 × 8 = 40 mm

Exercise 3D

1 Without using a calculator, write down the answers to the following.

a 0·2 × 4   b 0·3 × 2   c 0·6 × 7   d 0·7 × 5
e 0·2 × 8   f 0·8 × 3   g 0·9 × 1   h 0·4 × 4
i 0·8 × 7   j 0·5 × 9   k 0·9 × 6   l 0·7 × 9

2 Without using a calculator, write down the answers to the following.

a 0·2 × 0·4   b 0·3 × 0·2   c 0·6 × 0·7   d 0·7 × 0·5
e 0·2 × 0·8   f 0·8 × 0·3   g 0·9 × 0·1   h 0·4 × 0·4
i 0·8 × 0·7   j 0·5 × 0·9   k 0·9 × 0·6   l 0·7 × 0·9

★ **3** Without using a calculator, work out the following.

a 30 × 0·8   b 0·6 × 20   c 0·6 × 50   d 0·2 × 60
e 0·3 × 40   f 0·4 × 50   g 0·7 × 20   h 0·2 × 90
i 0·5 × 80   j 70 × 0·6   k 30 × 0·1   l 80 × 0·6

4 Without using a calculator, work out the following.

a 0·04 × 0·2 b 0·4 × 0·08 c 0·07 × 0·04 d 0·009 × 0·6
e 0·5 × 0·008 f 0·06 × 0·05 g 0·01 × 0·07 h 0·07 × 0·07

5 Without using a calculator, work out the following.

a 300 × 0·8   b 0·06 × 400 c 0·6 × 500   d 0·02 × 600
e 0·005 × 8000 f 300 × 0·01 g 600 × 0·006 h 0·04 × 8000

6 Screws cost 0·9p. A company orders 3000 screws. How much will this cost?

7 A grain of sand weighs 0·6 milligrams. How much will 700 grains weigh?

Number, money and measure

Challenge

1. Given that 46 × 34 = 1564, write down the answer to:
   a 4·6 × 34   b 4·6 × 3·4   c 1564 ÷ 3·4   d 15·64 ÷ 0·034

2. Given that 57 × 32 = 1824, write down the answer to:
   a 5·7 × 0·032   b 0·57 × 32000   c 5700 × 0·32   d 0·0057 × 32

3. Given that 2·8 × 0·55 = 1·54, write down the answer to:
   a 28 × 55   b 154 ÷ 55   c 15·4 ÷ 0·028   d 0·028 × 5500

Working backwards

Example 3·9   A gardener charges £8 per hour. Write down a formula, in words, for the total charge when the gardener is hired for several hours. Work out the cost of hiring the gardener for 6 hours.

The formula is:

　　The charge is equal to the number of hours worked multiplied by eight pounds

If the gardener is hired for 6 hours, the charge = 6 × £8
　　　　　　　　　　　　　　　　　　　　　　= £48

So, the charge is £48.

Example 3·10   I think of a number, add 3 and then double it. The answer is 16. What is the number?

Working this flowchart backwards gives:

So, 16 ÷ 2 = 8
　　 8 − 3 = 5

The answer is 5.

Exercise 3E

1. I think of a number, double it and add 1. The answer is 19.

   Work the flow diagram backwards to find the number.

2. I think of a number, multiply it by 3 and subtract 5. The answer is 25.

   a Copy and complete the flow diagram.
   b Work backwards to find the number.

Number and number processes

3 I think of a number, divide it by 2 and add 5. The answer is 11.

   a Copy and complete the flow diagram.
   b Work backwards to find the number.

4 I think of a number, multiply by 2 and add 7. The answer is 15.
   Work backwards to find the number.

5 I think of a number, divide it by 3 and subtract 2. The answer is 5.
   Work backwards to find the number.

★ 6 The diagram shows how to change a temperature in °F to °C.

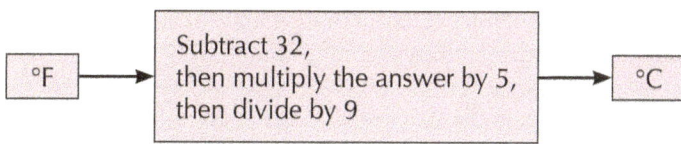

   a What is 41°F in °C?   b What is 68°C in °F?

Number problems

Look at the recipe. This is for four people. How much of each ingredient is needed to make a chocolate cake for eight people?

Example 3·11 Give an example to show that any even number multiplied by any other even number always gives an even number.

Two examples are:

   $2 \times 4 = 8$ $10 \times 6 = 60$

Both 8 and 60 are even numbers.

Example 3·12 Take any three consecutive numbers. Add the first number to the third number and divide the answer by 2. What do you notice?

Take, for example, 1, 2, 3 and 7, 8, 9. These give:

   $1 + 3 = 4$ $7 + 9 = 16$
   $4 \div 2 = 2$ $16 \div 2 = 8$

Whichever three consecutive numbers you choose, you should always get the middle number.

Number, money and measure

Exercise 3F

1. Copy and complete each of the following number problems, filling in the missing digits.

   a. 3☐
       + ☐7
       ———
         4 9

   b. 4☐
       + ☐3
       ———
         7 6

   c. 1 3☐
       +   ☐7
       ———
         ☐4 9

   d. 8☐
       – ☐2
       ———
         4 5

   e. 2 3 8
       – ☐☐☐
       ———
         1 1 7

   f.   ☐3 5
       ×   4
       ———
        5☐☐

2. Write down an example to show that when you add two odd numbers, the answer is always an even number.

3. Write down an example to show that when you add two even numbers, the answer is always an even number.

4. Write down an example to show that when you add an odd number to an even number, the answer is always an odd number.

5. Write down an example to show that when you multiply together two odd numbers, the answer is always an odd number.

6. Write down an example to show that when you multiply an odd number by an even number, the answer is always an even number.

7. By finding the cost of 1 litre, work out which bottle is the best value for money.

8. Which is the better value for money?

   a. 6 litres for £12 or 3 litres for £9.

   b. 4 kg for £10 or 8 kg for £18.

   c. 200 g for £4 or 300 g for £5.

   d. Six chocolate bars for £1·50 or four chocolate bars for 90p.

9. A recipe uses 450 g of meat and makes a meal for five people. How many grams of meat would be needed to make a meal for 15 people?

Number and number processes

Challenge

Write down a three-digit number using three different digits. Reverse the digits and write down this new number. You should now have two different three-digit numbers. Subtract the smaller number from the bigger number.

Your answer will have either two or three digits. If it has two digits (for example 99), rewrite it with three (099).

Now reverse the digits of your answer and write down that number. Add this number to your previous answer. Your final answer should be a four-digit number.

Either repeat using different numbers, or compare your answer with someone else's. Write down what you notice.

What happens if you do not use different digits at the start?

- By working on this topic I can explain how to use number facts I know to work out and check answers.

- I can use mental methods to estimate solutions to problems. ★ Exercise 3A Q4

- I can use my understanding of number properties to explain why some answers must be incorrect. ★ Exercise 3B Q4

- I can extend basic addition and subtraction facts to solve related problems. ★ Exercise 3C Q4

- I can use my understanding of place value to multiply by decimal numbers. ★ Exercise 3D Q3

- I can explain how to work backwards through a problem. ★ Exercise 3E Q6

- I understand the properties of odd and even numbers and how they affect answers to calculations. ★ Exercise 3F Q6

Number, money and measure

4

I can use my understanding of numbers less than zero to solve simple problems in context.

MNU 3-04a

This chapter will show you how to:
- use a number line to add and subtract negative numbers
- use the symbols > (greater than) and < (less than)
- add and subtract negative numbers in real-life problems.

You should already know:
- how to use a number line for numbers greater than zero
- how to add and subtract positive numbers.

The number line and inequalities

Temperature 32 °C
Latitude 17° South
Time 09 30 h GMT

Temperature −13 °C
Latitude 84° North
Time 23 24 h GMT

Look at the two pictures. What are the differences between the temperatures, the latitudes and the times?

All numbers have a sign. Positive numbers (numbers greater than zero) have a + sign in front of them although we do not always write it. Negative numbers (numbers less than zero) have a − sign in front of them. We *always* write the negative sign. Zero is not positive or negative.

The positions of positive and negative numbers can be put on a number line, as below.

This is very useful, as it helps us to compare positive and negative numbers and also to add and subtract them. When comparing the size of numbers we use the symbols > (is greater than) and < (is less than). The smaller (pointed) end of the symbol always points at the smaller number.

Example 4·1

Which is bigger, −7 or −3?

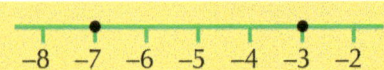

Because −3 is further to the right on the line, it is the larger number. We can write −7 < −3 which is read as −7 is less than −3.

Number and number processes

Example 4·2 Work out the answers to: a 3 − 2 − 5   b −3 − 5 + 4 − 2

a Starting at zero and 'jumping' along the number line gives an answer of −4.

b −3 − 5 + 4 − 2 = −6

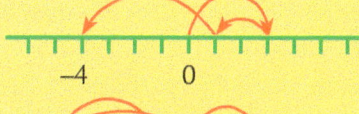

Example 4·3 −3 < 6 means 'negative 3 is less than 6'.

−4 > −7 means 'negative 4 is greater than negative 7'.

State whether these are true or false:

a 7 > 9   b −2 > −1   c −2 > −5

Marking each of the pairs on a number line, we can see that a is false, b is false and c is true.

Exercise 4A

1 a Circle each whole number greater than 0 and less than 7.

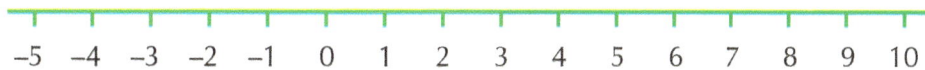

b Circle each number greater than −4 and less than 2.

2 State whether each of these is true (T) or false (F).

a 9 > 7   b 9 < 16   c 5 < −6   d −6 > −4   e −3 < −2

3 Put <, > or = into each ☐ to make a true sentence.

a 5 ☐ 9   b 4 ☐ 15   c 59 ☐ 48   d 13 ☐ 12

e Two hundred and two ☐ 202   f Two thousand and two ☐ 202

★ **4** Put the correct sign, > or <, between each pair of numbers.

a −5 ... 4   b −7 ... −10   c 3 ... −3   d −12 ... −2

5 Find the number that is halfway between each pair of numbers.

a −8, −2   b −6, +4   c −9, −1

★ **6** Work out the answer to each of these.

a 6 − 9   b 2 − 7   c 1 − 3   d −8 + 8
e −6 + 9   f −7 − 3   g −2 + 3   h −14 + 7
i −2 − 3 + 4   j −1 + 1 − 2   k −3 + 4 − 7   l −102 + 103 − 5

Number, money and measure

7 Find the missing number to make each of these true.

a $+2 - 6 = \square$ b $+4 + \square = +7$ c $-4 + \square = 0$

d $+5 - \square = -1$ e $+3 + 4 = \square$ f $\square - 5 = +7$

g $\square - 5 = +2$ h $+6 - \square = 0$ i $\square + 5 = -2$

j $+2 - 2 = \square$ k $\square - 2 = -4$ l $-2 - 4 = \square$

8 a A fish is 2 m below the surface of the water. A fish eagle is 15 m above the water. How many metres must the bird descend to get the fish?

  b Alf has £25 in the bank. He writes a cheque for £35. How much has he got in the bank now?

Challenge

1 A maths test consists of 20 questions. Three points are given for a correct answer and two points are deducted if an answer is wrong or not attempted.

Work out the scores for the following people:

  a Aisha gets 12 right and 8 wrong.
  b Bill gets 10 right and 10 wrong.
  c Charles gets 8 right and 12 wrong.
  d Dilash gets 9 right and 11 wrong.

What times table are all your answers in?

2 What happens when there are four points for a correct answer and minus two for a wrong answer?

A computer spreadsheet is useful for this activity.

Negative numbers in real life

The questions in the following Exercise will give you practice at solving real-life problems which involve negative numbers.

Exercise 4B

1 Write down the highest and lowest temperature in each group.

  a 4°C, −2°C, 0°C   b −8°C, −6°C, −10°C   c −20°C, −19°C, −5°C

2 In each of the following write down the difference in temperature.

  a −2°C and 5°C   b −10°C and −22°C   c 4°C and −5°C

Number and number processes

3 On Monday the temperature at noon was 5°C. Over the next few days the following temperature changes were recorded:

Mon to Tues   down 3°C

Tues to Weds   up 1°C

Weds to Thurs   down 6°C

Thurs to Fri   up 2°C

What was the temperature on Friday?

4 Part of a bank account statement is shown. Copy the table and complete the Balance column.

Deposits (£)	Withdrawals (£)	Balance (£)
100		100
	120	−20
30		10
	60	−50
	40	
10		
	30	

5 On these bank statements some deposits and withdrawals are missing. Copy the statements, filling in the missing entries.

a

Deposits (£)	Withdrawals (£)	Balance (£)
100		100
	60	40
		110
		−50
		−10
		20
		0

b

Deposits (£)	Withdrawals (£)	Balance (£)
500		500
	200	300
		−100
		50
		0
		−75
		−10

Number, money and measure

6. In golf the par for a hole is the number of golf strokes it should take to complete the hole. If a player takes one stroke fewer than the par for a hole this is called a birdie (par − 1), and two strokes fewer than par is called an eagle (par − 2). If a player takes one stroke more than par this is called a bogey (par + 1), and two more than par is called a double bogey (par + 2).

   Here is a golfer's scorecard:

Hole	1	2	3	4	5	6	7	8	9
Par	4	4	3	4	5	4	4	4	3
Score	3	5	4	4	3	5	6	3	3

   a The player birdied the first hole. Describe how he played the other 8 holes using the words birdie, bogey, etc.

   b After two holes he was level par. What was his score against par after:

   i 5 holes   ii 9 holes

7. At a golf championship three players – Donald, Woods and McIlroy – are playing together. After five holes the scoreboard shows their scores as:

   Donald −2

   Woods +1

   McIlroy −3

   Over the next three holes the players get the following scores:

   Donald: birdie, par, birdie

   Woods: eagle, birdie, par

   McIlroy: par, double bogey, birdie

   What will the scoreboard show now?

8. Two important Greek mathematicians were Pythagoras, thought to have been born in the year 570 BC, and Archimedes, born in 287 BC.

   a How many years ago was each mathematician born? (Note: there is no year zero.)

   b How many years apart were they born?

   c They both died at the age of 75. In what year did each of them die?

   d Sketch a time line showing their birth years along with the year 1 AD and the current year.

9. The Mariana Trench in the Pacific Ocean has a maximum depth of 10·9 km below sea level. Mount Everest is 8·8 km above sea level.

   a How much higher than the base of the trench is the summit of Mount Everest?

   b If Mount Everest was set in the deepest part of the trench, how far would its summit be below sea level?

Number and number processes

Adding negative numbers

Look at the following pattern for addition:

$3 + 2 = 5$

$3 + 1 = 4$

$3 + 0 = 3$

$3 + (-1) = 2$ (this is the same as $3 - 1 = 2$)

$3 + (-2) = 1$ (this is the same as $3 - 2 = 1$)

As the number being added decreases so does the answer. When you add a negative number you are actually doing a subtraction.

This can also be seen when you change the order of the addition.

$3 + (-1) = (-1) + 3$

$ = 2$ (which is $3 - 1$)

Adding a negative number is the same as subtracting a positive number.

Note that when calculations involve negative numbers, sometimes brackets are put around the negative number so that the negative sign doesn't get confused with the sign for subtraction.

Example 4·4

Work out the answers to:

a $5 + (-2)$ b $20 + (-3)$ c $(-3) + (-4)$

a $5 + -2 = 5 - 2$
$ = 3$

b $20 + -3 = 20 - 3$
$ = 17$

c $(-3) + (-4) = (-3) - 4$
$ = -7$

Example 4·5

Work out the answers to:

a $3 + 4 + (-5)$ b $12 + (-7) - 2$

a $3 + 4 + (-5) = 3 + 4 - 5$
$ = 2$

b $12 + (-7) - 2 = 12 - 7 - 2$
$ = 3$

Number, money and measure

Exercise 4C

1 Copy and complete the patterns.

   a   4 + 3 = 7   **b**   9 + 3 = 12   **c**   5 + 1 = 6   **d**   (−4) + 3 = −1
          4 + 2 = 6   9 + 2 = 11   5 + 0 = 5   (−4) + 2 = −2
          4 + 1 = 5   9 + 1 =   5 + (−1) = 4   (−4) + 1 = −3
          4 + 0 =   9 + 0 =   5 + (−2) =   (−4) + 0 =
          4 + (−1) =   9 + (−1) =   5 + (−3) =   (−4) + (−1) =
          4 + (−2) =   9 + (−2) =   5 + (−4) =   (−4) + (−2) =

2 Copy and complete.

   a   6 + (−2)   **b**   8 + (−6)   **c**   20 + (−7)   **d**   16 + (−5)   **e**   (−6) + (−7)
       = 6 − 2   = 8 − 6   = 20 −   =   =
       =   =   =   =   =

3 Use the number line below to help work out the answers to these.

   a   3 − 5   **b**   8 + (−2)   **c**   4 + (−5)   **d**   3 + (−3)
   e   (−2) + (−3)   **f**   2 − 10   **g**   (−4) + 10   **h**   0 − 9
   i   10 + (−5)   **j**   (−8) + (−6)   **k**   12 + (−10)   **l**   6 + (−6)
   m   9 + (−10)   **n**   15 + (−25)   **o**   0 + (−8)   **p**   (−1) + (−7)

−15 −14 −13 −12 −11 −10 −9 −8 −7 −6 −5 −4 −3 −2 −1 0 1 2 3 4 5 6 7 8 9 10 11 12 13 14 15

★ 4 Calculate.

   a   20 + (−5)   **b**   (−5) + (−20)   **c**   60 − 100   **d**   (−10) + (−20)
   e   20 + (−16)   **f**   (−30) + 30   **g**   15 + (−40)   **h**   120 − 240
   i   13 + (−8)   **j**   (−100) + (−20)   **k**   16 + (−25)   **l**   (−9) + (−19)

5 Calculate.

   a   5 + 6 + (−8)   **b**   (−1) − 1 + (−1)   **c**   20 + (−14) + (−6)   **d**   (−16) + 3 + (−10)

6 Find the total of the following lists of numbers

   a   5, −4, 10, −7, −9, 3   **b**   −12, 20, 5, −8, −15, 30

7 In each magic square, all the rows, columns and diagonals add up to the same total. Copy and complete the squares.

a

		−8
	−5	−3
−2		

b

0		
−5		
−4		−6

c

3		1
	0	
		−3

Number and number processes

Challenge In this 4 × 4 magic square all of the rows, columns and diagonals add to –6.

Copy and complete the square.

–9			5
2			–4
		6	–1
	1	–5	

Subtracting negative numbers

Look at the following pattern for subtraction:

$3 - 2 = 1$

$3 - 1 = 2$

$3 - 0 = 3$

$3 - (-1) = 4$ (this is the same as $3 + 1 = 4$)

$3 - (-2) = 5$ (this is the same as $3 + 2 = 5$)

You can see that as the number being subtracted decreases the answer increases. When you subtract a negative number you actually add.

Subtracting a negative number is the same as adding a positive number.

Example 4·6 Work out the answers to:

a $6 - (-2)$ b $14 - (-3)$ c $(-7) - (-5)$

a $6 - (-2) = 6 + 2$
$= 8$

b $14 - (-3) = 14 + 3$
$= 17$

c $(-7) - (-5) = (-7) + 5$
$= -2$

Example 4·7 Work out the answers to:

a $5 + 6 - (-3)$ b $15 - (-4) - 2$

a $5 + 6 - (-3) = 5 + 6 + 3$
$= 14$

b $15 - (-4) - 2 = 15 + 4 - 2$
$= 17$

Number, money and measure

Exercise 4D

1 Copy and complete the patterns.

a 4 − 3 = 1
   4 − 2 = 2
   4 − 1 = 3
   4 − 0 =
   4 − (−1) =
   4 − (−2) =

b 9 − 3 = 6
   9 − 2 = 7
   9 − 1 = 8
   9 − 0 =
   9 − (−1) =
   9 − (−2) =

c (−5) − 2 = −7
   (−5) − 1 = −6
   (−5) − 0 = −5
   (−5) − (−1) =
   (−5) − (−2) =
   (−5) − (−3) =

d (−11) − 1 = −12
   (−11) − 0 = −11
   (−11) − (−1) = −10
   (−11) − (−2) =
   (−11) − (−3) =
   (−11) − (−4) =

2 Copy and complete.

a 6 − (−2)
   = 6 + 2
   =

b 10 − (−6)
   = 10 + 6
   =

c 20 − (−8)
   = 20 +
   =

d −14 − (−5)
   =
   =

e (−6) − (−7)
   =
   =

3 Use the number line below to help work out the answers to these.

a 3 − (−8)
b 7 − (−2)
c (−6) − (−5)
d 13 − (−2)
e (−2) − (−3)
f (−2) − (−10)
g (−4) − (−10)
h (−10) − (−9)
i 7 − (−5)
j (−8) − (−6)
k (−9) − (−8)
l (−6) − (−6)
m (−9) − (−1)
n (−15) − (−25)
o 0 − (−8)
p (−3) − (−7)

−15 −14 −13 −12 −11 −10 −9 −8 −7 −6 −5 −4 −3 −2 −1 0 1 2 3 4 5 6 7 8 9 10 11 12 13 14 15

4 Calculate.

a 20 − (−8)
b (−20) − (−15)
c 30 − (−100)
d (−10) − (−10)
e 22 − (−18)
f 30 − (−30)
g −15 − (−40)
h −300 − (−240)
i 16 − (−8)
j (−120) − (−70)
k 16 − (−25)
l (−8) − (−19)

5 Calculate.

a 7 + 6 − (−8)
b (−2) − 1 − (−1)
c 30 − (−12) + (−9)
d (−10) + 3 − (−10)

6 Choose a number from each list and subtract one from the other. Repeat for at least four pairs of numbers. What are the biggest and smallest answers you can find?

A	B
4	−3
−7	9
−5	−8
3	−5
8	−1

48

Number and number processes

7 Calculate.

- a 6 − 9
- b (−8) + 5
- c 7 + (−5)
- d 8 − (−2)
- e (−6) + (−5)
- f (−4) − (−2)
- g 12 − 30
- h (−6) − (−10)
- i 4 + (−11)
- j (−8) − 6
- k 20 − (−20)
- l 1 + (−5)
- m 0 − (−7)
- n (−15) + (−1)
- o (−7) − (−7)
- p (−3) − 18
- q 9 − (−10)
- r (−50) + (−50)
- s 2 − 100
- t (−2) − (−100)

8 Copy and work out the missing numbers.

- a 7 + ☐ = 6
- b 10 − ☐ = 12
- c ☐ + (−5) = 7
- d (−8) − ☐ = 4

- By working on this topic I can explain how to use negative numbers in calculations and when solving real-life problems.

- I understand what the symbols < and > mean and can apply them to negative numbers. ★ Exercise 4A Q4

- I can explain how the number line extends below zero and can work out problems that have answers below zero. ★ Exercise 4A Q6

- I can work with negative numbers in real-life situations. ★ Exercise 4B Q5

- I know that adding a negative number is the same as subtracting a positive number and I can accurately calculate such problems. ★ Exercise 4C Q4

- I know that subtracting a negative number is the same as adding a positive number and I can accurately calculate such problems. ★ Exercise 4D Q4

Number, money and measure

5

I have investigated strategies for identifying common multiples and common factors, explaining my ideas to others, and can apply my understanding to solve related problems.

MTH 3-05a

This chapter will show you how to:
- determine common multiples
- determine common factors
- use the term lowest common multiple (LCM) and know how to find the LCM of a set of numbers
- use the term highest common factor (HCF) and know how to find the HCF of a set of numbers.

You should already know:
- how to check numbers for divisibility
- how to identify multiples of a number using a range of methods
- how to identify multiples of fractions and decimals
- how to create a sequence of multiples
- how to identify factors of a number using your knowledge of multiplication and division.

Common multiples

You should already know that the **multiples** of a number are all the numbers which **it fits into** exactly.

For example, the multiples of 7 are 7, 14, 21, 28, … (all the numbers in the 7 times table).

Example 5·1

Which of the following are multiples of 3?

a 24   b 35   c 58   d 102

If each number can be divided by 3 exactly, it must be a multiple of 3.

a 24 ÷ 3 = 8, so 24 is a multiple of 3.
b 35 ÷ 3 = 11 r 2, so 35 is not a multiple of 3.
c 58 ÷ 3 = 19 r 1, so 58 is not a multiple of 3.
d 102 ÷ 3 = 34, so 34 is a multiple of 3.

Lowest common multiple

Any pair of numbers has many common multiples. The lowest of these is called the **lowest common multiple** (LCM). This can be found by listing the first few multiples of both numbers until you see the first number common to both lists.

Multiples, factors and primes

Example 5·2 Find the LCM of 6 and 8.

Write out the first few multiples of each number:
  6 12 18 **24** 30 36 42 **48** 54 …
  8 16 **24** 32 40 **48** 56 …

You can see which are the common multiples, the lowest of which is 24.

So, **24** is the LCM of 6 and 8.

Notice that once you have found the lowest common multiple of 24, you get another common multiple every 24 after that. This will be useful for finding the LCM of three numbers.

Exercise 5A

1 Write down the numbers in the row below that are multiples of:

  a 2   b 3   c 5   d 10

  | 10 | 4 | 23 | 18 | 69 | 81 | 8 | 65 | 33 | 72 | 100 |

2 Write down the first ten multiples of the following numbers.

  a 4   b 5   c 8   d 9   e 10

3 Which of the following are multiples of 3?

  a 42   b 75   c 356   d 786

4 Use your answers to Question 2 to help to find the LCM of the following pairs.

  a 5 and 8   b 4 and 10   c 4 and 9   d 8 and 10

 5 Find the LCM of the following pairs.

  a 5 and 9   b 5 and 25   c 3 and 8   d 4 and 6
  e 8 and 12   f 12 and 15 g 9 and 21   h 7 and 11

[*Hint:* Write out the multiples of each number.]

6 John has given the following clues about his birthday:
  • the day he was born is a multiple of 7 bigger than 25
  • the month he was born is a multiple of 3 and 4.

  What is the date of his birthday?

 7 The number 3 bus takes 18 minutes to complete its route and return to the station. The number 7 bus takes 12 minutes to complete its route and return to the station.

  a If both buses leave the station at 9 am, when will they next be at the station at the same time?

  b If both buses run from 9 am to 3 pm, how many times will they arrive at the station at the same time?

Number, money and measure

LCM of three or more numbers

You can find the LCM of three numbers using the fact that common multiples appear at regular intervals. Start by finding the LCM of the biggest two numbers first as this will give you the smallest number you need to check for a common multiple of all three numbers.

Example 5·3

Find the LCM of:

a 4, 6 and 8

b 3, 5 and 7

a Find the LCM of 6 and 8 first:

Multiples of 6:   6   12   18   **24**   30   36 …

Multiples of 8:   8   16   **24**   32 …

The LCM of 6 and 8 is 24.

Now check if it is a multiple of 4:

24 ÷ 4 = 6 so 24 is a multiple of 4.

The LCM of 4, 6 and 8 is 24.

Note: If 24 wasn't a multiple of 4 you would try 48, 72, 96, … until you found one.

b Find the LCM of 5 and 7 first:

Multiples of 5:   5   10   15   20   25   30   **35**   40 …

Multiples of 7:   7   14   21   28   **35**   42   49 …

The LCM of 5 and 7 is 35.

Now check whether this number is a multiple of 3:

35 ÷ 3 = 11 r 2 so 35 is not a multiple of 3.

Common multiples appear at regular intervals so the next common multiple of 5 and 7 will be after another 35. Focus on the 35 times table until you find a multiple of 3:

70 ÷ 3 = 23 r 1 so 70 is not a multiple of 3.

105 ÷ 3 = 35 so 105 is a multiple of 3.

The LCM of 3, 5 and 7 is 105.

Exercise 5B

 1 Find the lowest common multiple (LCM) of the following.

a 2, 3 and 5   b 3, 4 and 8   c 4, 7 and 8

d 5, 6 and 7   e 2, 3 and 7   f 5, 7 and 12

2 A bell rings every 3 minutes, a horn blasts every 4 minutes and a siren blares every 5 minutes. At 5 pm they all sound at the same time. When will they next sound at the same time?

 3 Danny, Lara and Louise are competing in a long distance race which involves running several laps around a running track. Danny takes 5 minutes to complete a lap, while Lara takes 8 minutes and Louise takes 3 minutes. They all start at the same time from the starting line. How many minutes will it take until they are all back at the starting line at the same time?

Multiples, factors and primes

4 Jim pays his gas bill every 9 weeks, his electricity bill every 8 weeks and his phone bill every 6 weeks. They all came out of his bank account on 2 May. After how many weeks will they next all come out at the same time?

Common factors

You should already know that the **factors** of a number are all the numbers which **fit into it** exactly.

For example, the factors of 8 are 1, 2, 4 and 8.

Every number has a limited number of factors.

To find the factors of a number we find them in **factor pairs**. For example, 2 and 4 are a factor pair of 8 because 2 × 4 = 8.

Example 5·4

Is 4 a factor of 64?

64 ÷ 4 = 16

Yes, 4 is a factor of 64 as it fits in exactly. The answer to the division calculation, 16, is also a factor. 4 and 16 are a factor pair of 64.

Example 5·5

Write down all the factors of:

a 24   b 56

a **24**   b **56**

   1 × 24   1 × 56
   2 × 12   2 × 28
   3 × 8   4 × 14
   4 × 6   7 × 8

   The factors of 24 are   The factors of 56 are
   1, 2, 3, 4, 6, 8, 12 and 24.   1, 2, 4, 7, 8, 14, 28 and 56.

Working systematically minimises the amount of work you have to do as you know that you have finished if the next number to check is already in your list.

Highest common factor

Some numbers share common factors. The biggest of these is called the **highest common factor** (HCF).

The highest common factor of two or more numbers is found by following these three steps.

- List the factors of each number.
- Look for and list the common factors.
- Look for the highest common factor in this list.

53

Number, money and measure

Example 5·6 Find the HCF of 18 and 42.

18	42
1 × 18	1 × 42
2 × 9	2 × 21
3 × 6	3 × 14
	6 × 7

The common factors of 18 and 42 are 1, 2, 3 and 6.

The HCF of 18 and 42 is 6.

Exercise 5C

1 Is 3 a factor of:

   a 65   b 81   c 75   d 154

2 7 is a factor of 378. What is its factor partner?

3 Write out all the factors of the following.

   a 15   b 20   c 32   d 12   e 25

4 Use your answers to Question 3 to help find the HCF of the following pairs.

   a 15 and 20   b 15 and 25   c 12 and 20   d 20 and 32

 5 Find the HCF of the following pairs.

   a 15 and 18   b 12 and 32   c 12 and 22   d 8 and 12

   e 2 and 18   f 8 and 18   g 18 and 27   h 7 and 11

6 Sally is 48. She has two children at primary school.

   Both children have ages which are currently factors of Sally's age.

   a How old are her two children?

   b Will either of their ages still be factors of Sally's age in exactly 1 year's time?

7 Mr White is deciding how to lay out the tables in his restaurant. He has eight tables which he can lay out in four different ways.

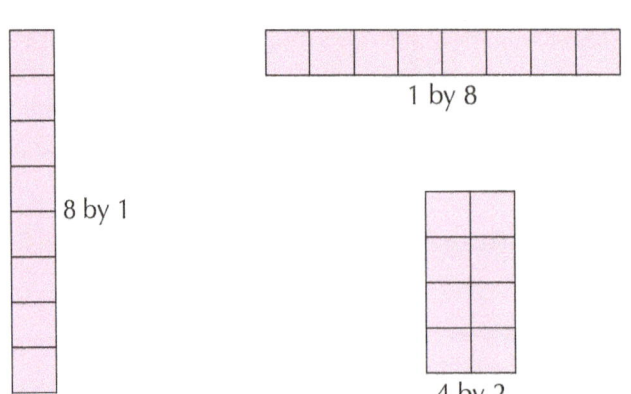

List the different ways that you can set out 18 tables.

Multiples, factors and primes

Exercise 5D

1. a   Two numbers have an LCM of 24 and an HCF of 2. What are they?
   b   Two numbers have an LCM of 18 and an HCF of 3. What are they?
   c   Two numbers have an LCM of 60 and an HCF of 5. What are they?

2. Copy and complete the table.

x	y	Product	HCF	LCM
4	14	56	2	28
9	21			
12	21			
18	24			

Describe any relationships that you can see in the table.

 3   Below is a Venn diagram looking at multiples and factors. Copy the diagram and write the following numbers in the correct circles. If a number doesn't belong in any circle then write it outside the circles.

3, 6, 12, 15, 32, 24, 17, 1, 36, 16, 27, 18, 42, 4, 2, 9

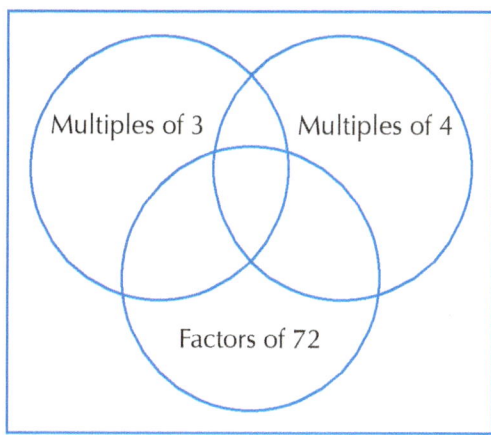

4   Three consecutive pages in a book have a product of 210. What are their numbers?

5. a   What is the smallest number with only three factors?
   b   What is the smallest number with only six factors?

 6   Use these clues to work out the mystery number.

- It is a number less than 50.
- It is a multiple of 3.
- It is a factor of 168.
- It is a multiple of 4.
- It is a factor of 72.
- It is not a factor of 36.

Number, money and measure

Challenge

1. You can also find the HCF of algebraic expressions. Find the HCF of the following expressions:

    a $4ab$ and $10bc$
    b $15m$ and $6mn$
    c $7xyz$ and $3wxz$

2. a What is the HCF and the LCM of: i 5, 7   ii 3, 4   iii 2, 11
   b Two numbers, x and y, have an HCF of 1. What is the LCM of x and y?

3. a What is the HCF and LCM of: i 5, 10   ii 3, 18   iii 4, 20
   b Two numbers, x and y (where y is bigger than x), have an HCF of x. What is the LCM of x and y?

- By working on this topic I can explain how to use common multiples and common factors to solve associated problems.

- I can find the LCM of two numbers. ★ Exercise 5A Q5

- I can find the LCM of three numbers and can explain why I find the LCM of the highest two first. ★ Exercise 5B Q1

- I can recognise and solve real-life problems using LCMs. ★ Exercise 5B Q3

- I can find the HCF of two numbers. ★ Exercise 5C Q5

- I can use my knowledge of multiples and factors to solve problems. ★ Exercise 5D Q3

Multiples, factors and primes

6

I can apply my understanding of factors to investigate and identify when a number is prime.

MTH 3-05b

This chapter will show you how to:
- identify prime numbers
- find prime factors of a number using different methods
- express a number as a product of prime numbers using a suitable method
- find the lowest common multiple of two numbers using their prime factors.

You should already know:
- how to find the factors of a number
- what the term product means.

Prime factors

A **prime number** is any number which has exactly two **factors**. It is only divisible by itself and 1. The first 10 prime numbers are: 2, 3, 5, 7, 11, 13, 17, 19, 23, 29. You need to know these.

The **prime factors** of a number are the prime numbers which, when multiplied together, give that number.

Every whole number can be made by multiplying primes together.

What are the prime factors of 120 and 210?

Example 6·1   Find the prime factors of 18.

Using a prime factor tree, split 18 into 3 × 6 and 6 into 3 × 2 and then stop because 2 and 3 are prime numbers.
So, 18 = 2 × 3 × 3.

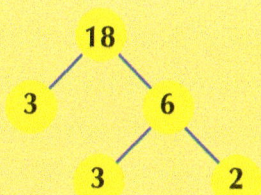

Note: You can start with any factor pairs but must end up with prime numbers at the end of each branch.

Number, money and measure

Example 6·2 Write 24 as the product of its prime factors.

Using the division method:
```
2 | 24
2 | 12
2 |  6
3 |  3
     1   ← Stop when you reach 1.
```

The prime factors of 24 are 2, 2, 2, 3.

So, $24 = 2 \times 2 \times 2 \times 3$.

Exercise 6A

1. Using a prime factor tree, work out the prime factors of the following.
   - a 8
   - b 10
   - c 16
   - d 20
   - e 28
   - f 34
   - g 35
   - h 52
   - i 60
   - j 180

2. Using the division method work out the prime factors of the following.
   - a 42
   - b 75
   - c 140
   - d 250
   - e 480

★ 3. Using your preferred method, work out the prime factors of the following.
   - a 200
   - b 32
   - c 72
   - d 105
   - e 220

4. Write the numbers in Question 3 as a product of primes.

Example 6·3 Use prime factors to find the highest common factor (HCF) and lowest common multiple (LCM) of 24 and 54.

$24 = 2 \times 2 \times 2 \times 3$, $54 = 2 \times 3 \times 3 \times 3$

You can see that 2×3 is common to both lists of prime factors.

Put these in the centre, overlapping, part of the diagram.

Put the other prime factors in the outside of the diagram.

The product of the centre numbers, $2 \times 3 = 6$, is the HCF.

The product of all the numbers, $2 \times 2 \times 2 \times 3 \times 3 \times 3 = 216$, is the LCM.

Exercise 6B

1. Using the diagrams below, work out the HCF and LCM of the following pairs.

   a) 30 and 72
   b) 50 and 90
   c) 48 and 84

2. The prime factors of 120 are 2, 2, 2, 3, 5. The prime factors of 150 are 2, 3, 5, 5.

   Put these numbers into a diagram like those in Question 1.

   Use the diagram to work out the HCF and LCM of 120 and 150.

Multiples, factors and primes

3 The prime factors of 210 are 2, 3, 5, 7. The prime factors of 90 are 2, 3, 3, 5.

Put these numbers into a diagram like those in Question 1.

Use the diagram to work out the HCF and LCM of 210 and 90.

★ **4** Use prime factors to work out the HCF and LCM of the following pairs.

    a 200 and 175   b 56 and 360   c 42 and 105

Challenge

1 **Prime numbers less than 100**

The prime numbers less than 100 are 2, 3, 5, 7, 11, 13, 17, 19, 23, 29, 31, 37, 41, 43, 47, 53, 59, 61, 67, 71, 73, 79, 83, 89 and 97.

  a How many prime numbers less than 100 are 1 more than a multiple of 6 (for example, 13)?

  b How many prime numbers less than 100 are 1 less than a multiple of 6 (for example, 11)?

  c How many prime numbers less than 100 are neither 1 less nor 1 more than a multiple of 6?

  d What do the answers to parts **a**, **b** and **c** suggest about prime numbers greater than 3?

2 **Prime puzzles**

The prime numbers up to 20 are 2, 3, 5, 7, 11, 13, 17 and 19.

  a Joe said: 'Two prime numbers can never add together to make another prime number.' Is Joe correct? Explain your answer.

  b Can you make any prime number as the sum of three prime numbers? Explain your answer.

3 **Perfect squares**

You can tell if a number is a perfect square and find its square root from prime factors. For example, $144 = 2 \times 2 \times 2 \times 2 \times 3 \times 3$. You can split the factors into two equal groups: $2 \times 2 \times 3$ and $2 \times 2 \times 3$. Multiply one of the groups to find the square root: $2 \times 2 \times 3 = 12$.

Using this method, calculate the square root of:

  a 225   b 256   c 169   d 441

- By working on this topic I have learnt the definition of a prime number and can identify prime numbers.
- I can find the prime factors of a number using a suitable method. ★ Exercise 6A Q3
- I can use prime factors to find the LCM and HCF of a pair of numbers. ★ Exercise 6B Q4

Number, money and measure

7

Having explored the notation and vocabulary associated with whole number powers and the advantages of writing numbers in this form, I can evaluate powers of whole numbers mentally or using technology.

MTH 3-06a

This chapter will show you how to:
- recognise and use index notation
- calculate whole number powers of a value
- write an expression in index form
- use mental strategies and calculator functions to solve calculations involving powers
- simplify more complex expressions in index form.

You should already know:
- the order of operations
- that the order you work in when multiplying numbers together doesn't matter.

Powers and index form

When multiplying a number by itself several times it is often more useful to simplify it by writing it in **index form**.

For example, $2 \times 2 \times 2$ can be written as 2^3. This is read as '2 to the **power** of 3'.

The 2 is called the **base** and the 3 is called the **index** or **power**.

The index tells you how many of the base you have to multiply together.

You can use your calculator to work with powers. Make sure you know which buttons to use to calculate the values of numbers expressed as powers, and how to express whole numbers as powers. If you're not sure how to do this, ask a friend or your teacher.

Example 7·1   Write the following in index form.

a 3×3
b $4 \times 4 \times 4 \times 4 \times 4$
c $g \times g \times g$
d $5 \times 4 \times 4 \times 5 \times 4$

a There are two 3s being multiplied, so $3 \times 3 = 3^2$. This can be read as '3 squared'.
b There are five 4s being multiplied, so $4 \times 4 \times 4 \times 4 \times 4 = 4^5$.
c There are three gs being multiplied, so $g \times g \times g = g^3$. This can be read as '$g$ cubed'.
d Rearranging the order (because multiplication is commutative – that is, the order doesn't matter), $5 \times 4 \times 4 \times 5 \times 4 = 4 \times 4 \times 4 \times 5 \times 5 = 4^3 \times 5^2$.

Powers and roots

Example 7·2 Calculate the following.

a 4^3 b 7^2 c 10^4 d $4^3 - 3^2$

a $4^3 = 4 \times 4 \times 4 = 64$
b $7^2 = 7 \times 7 = 49$
c $10^4 = 10 \times 10 \times 10 \times 10 = 10\,000$
d $4^3 - 3^2 = 4 \times 4 \times 4 - 3 \times 3 = 64 - 9 = 55$

You are expected to be able to **mentally** calculate the squares of numbers from 1 to 10 and the cubes of numbers from 1 to 5. Other powers may require you to do some working.

Exercise 7A

1 Write the following in index form.

a $3 \times 3 \times 3 \times 3$ b $5 \times 5 \times 5$ c $7 \times 7 \times 7 \times 7 \times 7$
d 2×2 e $1 \times 1 \times 1 \times 1$ f 10×10
g $4 \times 4 \times 4$ h $3 \times 3 \times 3 \times 3 \times 3 \times 3$ i $2 \times 2 \times 2 \times 2$
j $4 \times 4 \times 4 \times 4 \times 4 \times 4 \times 4$ k $f \times f \times f$ l $m \times m \times m \times m \times m$

2 Calculate, showing working where necessary.

a 2^3 b 3^2 c 5^3 d 10^6 e 2^5 f 1^5
g 3^3 h 8^2 i 2^4 j 6^3 k 4^2 l 3^4

3 Write the following in index form.

a $3 \times 4 \times 3 \times 3$ b $5 \times 5 \times 7 \times 7 \times 5$
c $2 \times 3 \times 3 \times 3 \times 2 \times 2 \times 3$ d $5 \times 6 \times 7 \times 7 \times 5 \times 6 \times 6$
e $4 \times 3 \times 9 \times 3 \times 4 \times 9 \times 3$ f $10 \times 10 \times 7 \times 7 \times 3 \times 10 \times 7$
g $m \times m \times n \times n \times m \times n \times m$ h $t \times t \times w \times v \times w$
i $a \times b \times b \times b \times c \times a \times a \times a \times c$

4 Calculate

a $5^2 + 4^2$ b $5^3 - 2^2$ c $3^2 + 2^4$ d $5^3 - 3^2$
e $1^2 + 2^2 + 3^2$ f $2^6 - 5^2$ g $10^4 - 3^3 - 8^2$ h $5^2 + 7^3 - 5^3$

5 The area of a square is given by the formula area $= l^2$, where l is the length of the sides.

Calculate the area, in cm², of a square with length of side

a 5 cm   b 7 cm   c 9 cm   d 12 cm

6 The volume of a cube is given by the formula volume $= l^3$, where l is the length of the sides.

Calculate the volume, in cm³, of a cube with length of side

a 3 cm   b 6 cm   c 10 cm   d 4 cm

61

Number, money and measure

Exercise 7 B — Use a calculator for these questions.

1. Calculate the following.

   a 5^6 b 3^5 c 6^5 d 7^4 e 3^7 f 5^5 g 2^9

2. Check your answers to Exercise 7A Question 4.

3. Calculate the following. Give your answers to two decimal places.

   [*Hint*: Put brackets around the numerator and denominator when you enter the expression in your calculator.]

   a $\dfrac{5^4 - 2^2}{2^3}$ b $\dfrac{3^5 + 5^4}{3^2 + 7^3}$ c $\dfrac{2^5 + 4^4}{4^2}$

Example 7·3

Express the following numbers in terms of powers.

a 4   b 8   c 27   d 25

a $4 = 2 \times 2 = 2^2$

b $8 = 2 \times 2 \times 2 = 2^3$

c $27 = 3 \times 3 \times 3 = 3^3$

d $25 = 5 \times 5 = 5^2$

Exercise 7 C

1. Express the following numbers in terms of 3^n.

   a 9   b 81   c 243

2. Express the following numbers in terms of powers.

   a 36   b 64   c 100   d 49

3. Express the following numbers in terms of   i 2^n ii 4^n

   a 16   b 64   c 256

4. What do you notice about your answers in Question 3?

Challenge

Sometimes expressions involving multiplication and division of terms in index form can be simplified. You can only do this when the terms have the same base.

For example:

- $4^2 \times 4^3 = (4 \times 4) \times (4 \times 4 \times 4) = 4^5$

- $6^8 \div 6^5 = \dfrac{6 \times 6 \times 6 \times \cancel{6} \times \cancel{6} \times \cancel{6} \times \cancel{6} \times \cancel{6}}{\cancel{6} \times \cancel{6} \times \cancel{6} \times \cancel{6} \times \cancel{6}} = 6^3$

- $\dfrac{3^2 \times 3^4}{3^3} = \dfrac{(3 \times 3) \times (3 \times \cancel{3} \times \cancel{3} \times \cancel{3})}{\cancel{3} \times \cancel{3} \times \cancel{3}} = 3^3$

1. Simplify the following.

   a $5^3 \times 5^6$ b $6^7 \times 6^5$ c $4^3 \times 4^4$ d $8^3 \times 8^6$

   e $9^2 \times 9^4 \times 9^3$ f $10^4 \div 10^2$ g $4^7 \div 4^3$ h $6^7 \div 6^4$

Continued

Continued

2 Write in simplest form.

a $\dfrac{5^4 \times 5^3}{5^5}$ b $\dfrac{2^7 \times 2^6}{2^5}$ c $\dfrac{4^3 \times 4^2}{4^4}$ d $\dfrac{7^2 \times 7^5 \times 7^3}{7^6}$

3 Look at your answers to Question 1. Can you spot a quick way of working out the final answer to each part without having to write it out in full?

Copy and complete the following sentences:
- To multiply terms with the same base you ____ the powers.
- To divide terms with the same base you ____ the powers.

4 Using the above, try simplifying the following.

a $m^5 \times m^4$ b $t^8 \times t^3$ c $y^8 \div y^5$

d $m^{12} \div m^7$ e $\dfrac{w^4 \times w^5}{w^6}$ f $\dfrac{f^4 \times f^5 \times f^2}{f^7}$

- By working on this topic I can confidently use the terms **base**, **index** and **power** to describe an expression in index form.

- I can express the product of terms with the same base in index form. ★ Exercise 7A Q1

- I can evaluate a term in index form, both mentally and with working where necessary. ★ Exercise 7A Q2

- Using the fact that multiplication is commutative, I can rewrite the product of several terms as a product of terms in index form. ★ Exercise 7A Q3

- I know how to enter a term in index form on a calculator and can evaluate expressions in index form using a calculator. ★ Exercise 7B Q2

Number, money and measure

8

I can solve problems by carrying out calculations with a wide range of fractions, decimal fractions and percentages, using my answers to make comparisons and informed choices for real-life situations.

MNU 3-07a

This chapter will show you how to:
- solve problems involving the four basic operations with decimal fractions
- find a fraction of a quantity
- multiply by a fraction
- write one quantity as a percentage of another
- find a percentage of a quantity with and without a calculator
- find a percentage increase or decrease
- solve money and other real-life problems using your knowledge of percentages, fractions and decimals.

You should already know:
- the place value column headings
- how to add and subtract decimals in familiar contexts
- how to multiply and divide decimal fractions by whole numbers
- how to change improper fractions to mixed numbers
- how to multiply and divide decimal fractions by 10, 100 and 1000
- how to find a unit fraction of a quantity
- how to change between fractions, decimal fractions and percentages.

Adding and subtracting decimals

Example 8·1 Work out:

a $64·06 + 178·9 + 98·27$ b $20 - 8·72 - 6·5$

The numbers need to be lined up in columns with their decimal points in line. Blank places are filled with zeros. Part b needs to be done in two stages.

```
a    64·06           b    ¹⁹⁹¹           ⁰¹
    178·90              2̶0̶·0̶0           1̷1·28
  + 98·27              -  8·72          - 6·50
    ──────              ──────           ──────
    341·23              11·28            4·78
     ²²¹¹
```

Fractions, decimal fractions and percentages

Example 8·2

In a science lesson a student adds 0·45 kg of water and 0·72 kg of salt to a beaker that weighs 0·092 kg. He then pours out 0·6 kg of the mixture. What is the total mass of the beaker and mixture remaining?

This has to be set up as an addition and subtraction problem, that is:

0·092 + 0·45 + 0·72 − 0·6

The problem has to be done in two stages:

```
  0·092          ⁰ ¹
  0·450        1·262
+ 0·720      − 0·600
  ─────        ─────
  1·262        0·662
  ¹ ¹
```

So the final mass is 0·662 kg or 662 grams.

Exercise 8A

1 Work out each of the following.

  a 8·3 + 4·6   b 8·3 − 4·9   c 5·1 + 2·6 + 1·4
  d 9·6 + 6·5 + 2·2 e 8·3 + 6·9 − 2·1   f 6·7 + 3 − 5·7
  g 4·5 − 1·2 − 2·3 h 8·2 − 2·9 − 2·7   i 4·5 cm + 2·1 cm + 8·6 cm
  j 7·3 m − 3·7 m − 2·5 m

★ 2 Work out the following.

  a 7·05 + 2·9 + 7 + 0·64   b 8·7 + 9 + 14·02 + 1·035
  c 11·423 + 15·72 − 12·98   d 42·7 + 67·3 − 35·27
  e 19·87 + 2·8 − 13·46 + 12·873 + 8·9 f 12 − 5·096 + 3·21
  g 23·907 + 8 − 9·25   h 4·32 + 65·098 + 172·3
  i 7·25 + 19·3 − 12·06 − 0·008
  j 21·35 + 6·72 − 12·36 − 9·476 + 16·406 − 7·64

3 In an experiment a beaker of water has a mass of 1·104 kg. The beaker alone weighs 0·125 kg. What is the mass of water in the beaker?

4 There are 1000 grams in a kilogram. Calculate the mass of the following shopping baskets. (Work in kilograms.)

  a 3·2 kg of apples, 454 g of jam, 750 g of lentils, 1·2 kg of flour
  b 1·3 kg of sugar, 320 g of strawberries, 0·65 kg of rice

5 Calculate the following. First make sure all units are in km. Remember: to change from metres to kilometres you divide by 1000 (because there are 1000 m in 1km).

  a 7·45 km + 843 m + 68 m   b 3·896 km + 723 m + 92 m
  c 8·76 km + 463 m − 892 m   d 16 km − 435 m − 689 m
  e 7·8 km + 5·043 km − 989 m

6 A rectangle is 2·35 m by 43 cm. What is its perimeter (in metres)?

Number, money and measure

7 A piece of string is 5 m long. Pieces of length 84 cm, 1·23 m and 49 cm are cut from it. How much string is left (in metres)?

8 A large container of oil contains 20 litres. Over five days the following amounts are poured from the container:

   2·34 litres, 1·07 litres, 0·94 litres, 3·47 litres, 1·2 litres

   How much oil is left in the container?

Multiplying decimals

This section will give you more practice on multiplying integers and decimals.

Example 8·3

Write down the answer to each of the following, using the fact that 27 × 4 = 108

a 27 × 0·4   b 2·7 × 4   c 2·7 × 0·4

a There is one decimal place in the multiplication 27 × 0·4, so there is one decimal place in the answer. Therefore you have:

$$27 \times 0.4 = 10.8$$

b There is one decimal place in the multiplication 2·7 × 4, so there is one decimal place in the answer. Therefore you have:

$$2.7 \times 4 = 10.8$$

c There are two decimal places in the multiplication 2·7 × 0·4, so there are two decimal places in the answer. Therefore you have:

$$2.7 \times 0.4 = 1.08$$

Example 8·4

Calculate mentally.

a 0·3 × 0·05   b 0·07 × 0·003

a There are three decimal places in the multiplication, so there are three in the answer. Therefore, using the fact that 3 × 5 = 15:

$$0.3 \times 0.05 = 0.015$$

b There are five decimal places in the multiplication, so there are five in the answer. Therefore, using the fact that 7 × 3 = 21:

$$0.07 \times 0.003 = 0.00021$$

Exercise 8B

Do not use a calculator to answer any of these questions.

1 Write down the answers to each of the following, using the fact that 83 × 24 = 1992.

   a 8·3 × 24   b 83 × 2·4   c 8·3 × 2·4   d 0·83 × 0·24

2 Write down the answers to each of the following, using the fact that 25 × 32 = 800.

   a 2·5 × 32   b 250 × 3·2   c 2·5 × 3·2   d 2·5 × 0·32

Fractions, decimal fractions and percentages

3 Write down the answers to the following.

a 0·2 × 3   b 0·4 × 2   c 0·6 × 6   d 0·7 × 2
e 0·2 × 4   f 0·8 × 4   g 0·6 × 1   h 0·3 × 3
i 0·7 × 8   j 0·5 × 8   k 0·9 × 3   l 0·6 × 9

4 Write down the answers to the following.

a 0·2 × 0·3   b 0·4 × 0·2   c 0·6 × 0·6   d 0·7 × 0·2
e 0·2 × 0·4   f 0·8 × 0·4   g 0·6 × 0·1   h 0·3 × 0·3
i 0·7 × 0·8   j 0·5 × 0·8   k 0·9 × 0·3   l 0·6 × 0·9

★ **5** Work out the following.

a 0·02 × 0·4   b 0·8 × 0·04   c 0·07 × 0·08   d 0·006 × 0·9
e 0·8 × 0·005 f 0·06 × 0·03 g 0·01 × 0·02   h 0·07 × 0·07

Challenge

1 Work out each of the following.

a 0·7 × 0·7

b 0·3 × 0·3

c 0·7 × 0·7 − 0·3 × 0·3

d 0·7 − 0·3

2 What do you notice about your answers to questions **1c** and **1d**?

3 Repeat using two other decimals which add up to 1 (for example, 0·8 and 0·2, or 0·87 and 0·13).

See if you can guess the answers.

Example 8·5 Calculate mentally.

a 900 × 0·4   b 50 × 0·04

a Rewrite as an equivalent product. That is:

900 × 0·4 = 90 × 10 × 0·4 = 90 × 4 = 360
or 900 × 0·4 = 9 × 100 × 0·4 = 9 × 40 = 360

b As in part **a**, giving: 50 × 0·04 = 5 × 10 × 0·04 = 5 × 0·4 = 2

Number, money and measure

Example 8·6   Without using a calculator, work out 134 × 0·6.

There are several ways to do this. Three are shown (a column method and two grid methods).

Remember that there is one decimal place in the multiplication, so there will be one in the answer.

In the first two methods, the decimal points are ignored in the multiplication and then placed in the answer.

Column method

```
   134
 ×   6
  ────
   804
    2 2
```

Grid method 1

	100	30	4	Total
6	600	180	24	804

Grid method 2

	100	30	4	Total
0·6	60	18	2·4	80·4

By all three methods the answer is 80·4.

Exercise 8C

Do not use a calculator to answer any of these questions.

1 Work out each of the following, by rewriting as an equivalent product.

- **a** 10 × 0·5
- **b** 0·7 × 10
- **c** 0·3 × 100
- **d** 0·6 × 10
- **e** 10 × 0·7
- **f** 0·8 × 100
- **g** 100 × 0·1
- **h** 0·4 × 100
- **i** 40 × 0·6
- **j** 0·4 × 20
- **k** 0·5 × 200
- **l** 0·7 × 50
- **m** 40 × 0·1
- **n** 0·8 × 300
- **o** 400 × 0·3
- **p** 0·4 × 500

2 Work out each of the following, using one of the methods shown in Example 8·6.

- **a** 2·6 × 5
- **b** 3·4 × 6
- **c** 4·91 × 4
- **d** 6·12 × 5
- **e** 31·5 × 7
- **f** 22·4 × 8
- **g** 14·6 × 6
- **h** 19·1 × 4

 3 Screws cost 0·6p. A company orders 2000 screws. How much will this cost?

 4 A grain of sand weighs 0·6 milligrams. How much would 500 grains weigh?

Fractions, decimal fractions and percentages

Dividing decimals

This section will give you more practice in dividing integers (whole numbers) and decimals.

Example 8·7 Work out each of these.

a 0·8 ÷ 2   b 0·12 ÷ 3   c 27·5 ÷ 5   d 17·4 ÷ 4

a There is one decimal place in the division 0·8 ÷ 2, so there is one decimal place in the answer. Therefore, using the fact that 8 ÷ 2 = 4:

$$0·8 ÷ 2 = 0·4$$

b There are two decimal places in the division 0·12 ÷ 3, so there are two decimal places in the answer. Therefore, using the fact that 12 ÷ 3 = 4:

$$0·12 ÷ 3 = 0·04$$

Note: You should be able to calculate answers to **a** and **b** mentally.

c $5 \overline{)27·^25}$ = 05·5   remember to line up the decimal points

d $4 \overline{)17·^14^20}$ = 04·3 5   to deal with the remainder of 2, write a zero after the 4 (this works because 17·4 is the same as 17·40)

Exercise 8D

Do not use a calculator to answer any of these questions.

1 Work out each of the following mentally.
   a 0·6 ÷ 3   b 0·9 ÷ 3   c 2·4 ÷ 4   d 3·5 ÷ 5
   e 2·1 ÷ 7   f 4·8 ÷ 8   g 5·4 ÷ 9   h 6·4 ÷ 8

★ 2 Work out each of the following mentally.
   a 0·36 ÷ 2   b 0·45 ÷ 5   c 0·16 ÷ 4   d 0·45 ÷ 9
   e 0·24 ÷ 6   f 0·81 ÷ 9   g 0·63 ÷ 7   h 0·56 ÷ 8

3 Work out each of the following.
   a 2·42 ÷ 2   b 3·25 ÷ 5   c 6·44 ÷ 4   d 8·55 ÷ 9
   e 5·22 ÷ 6   f 8·01 ÷ 9   g 4·27 ÷ 7   h 2·56 ÷ 8

4 Work out each of the following.
   a 18·02 ÷ 2   b 37·95 ÷ 5   c 23·04 ÷ 4   d 49·14 ÷ 9
   e 18·36 ÷ 6   f 99·18 ÷ 9   g 16·52 ÷ 7   h 10·24 ÷ 8

5 The perimeter of a square is 4·32 cm. Work out the length of one side.

6 Six cupcakes cost £5·22. How much does one cake cost?

Number, money and measure

Challenge — Making equivalent division calculations

You can divide by a decimal number by making an equivalent division calculation. The divisions can be made easier by multiplying both parts by 10 until you divide by a whole number instead of a decimal.

For example:

a 4·5 ÷ 0·3 = 45 ÷ 3 = 15

b 56 ÷ 0·07 = 560 ÷ 0·7 = 5600 ÷ 7 = 800

c 12·6 ÷ 0·04 = 126 ÷ 0·4 = 1260 ÷ 4 = 315

1 Calculate the following.

    a 12 ÷ 0·4   b 48 ÷ 0·06   c 2·4 ÷ 0·8   d 72 ÷ 0·6   e 3·4 ÷ 0·2

    f 1·6 ÷ 0·08   g 4·2 ÷ 0·03   h 18·3 ÷ 0·3   i 0·32 ÷ 0·008   j 4·72 ÷ 0·04

2 A bookshelf is 1·5 m wide. How many books can you fit on the shelf if each book is 0·03 m wide?

3 Mrs Smith wants to stock up on pencils for her classroom. She has £10 to spend. How many pencils can she buy if they cost £0·08 each?

Fractions of quantities

This section is going to help you to revise the rules for working with fractions using the unitary method.

Example 8·8 Find: a $\frac{2}{7}$ of £28 b $\frac{3}{5}$ of 45 sweets

a First, find $\frac{1}{7}$ of £28: 28 ÷ 7 = 4. So, $\frac{2}{7}$ of £28 = 2 × 4 = £8.

b First, find $\frac{1}{5}$ of 45 sweets: 45 ÷ 5 = 9. So, $\frac{3}{5}$ of 45 sweets = 3 × 9 = 27 sweets.

Exercise 8E

 1 Find each of these.

    a $\frac{2}{3}$ of £27   b $\frac{3}{5}$ of 75 kg   c $\frac{2}{3}$ of 18 metres

    d $\frac{4}{9}$ of £18   e $\frac{3}{10}$ of £46   f $\frac{5}{8}$ of 840 houses

    g $\frac{3}{7}$ of 21 litres   h $\frac{2}{5}$ of 45 minutes   i $\frac{5}{6}$ of £63

    j $\frac{3}{8}$ of 1600 loaves   k $\frac{4}{7}$ of 35 km   l $\frac{7}{10}$ of 600 crows

    m $\frac{2}{9}$ of £1·26   n $\frac{4}{9}$ of 540 children   o $\frac{7}{12}$ of 144 miles

Fractions, decimal fractions and percentages

2. A bag of rice weighed 1300 g. $\frac{2}{5}$ of it was used to make a meal. How much was left?

3. Mrs Smith weighed 96 kg. She lost $\frac{3}{8}$ of her weight due to a diet. How much did she weigh after the diet?

4. A petrol tank holds 52 litres. $\frac{3}{4}$ is used on a journey. How many litres are left?

5. A sweets machine produces 1400 sweets a minute. $\frac{2}{7}$ of them are red. How many red sweets will the machine produce in an hour?

6. A maths textbook has 448 pages. $\frac{3}{28}$ of the pages are the answers. How many pages of answers are there?

Multiplying fractions

In the last exercise you found a fraction of a quantity. You may not have realised it, but this involved multiplying the two values together. In maths 'of' is often used as another word for multiply.

To multiply two fractions together, you multiply the numerators and multiply the denominators.

Example 8·9 Calculate the following.

a $\frac{3}{7} \times \frac{2}{5}$ b $5 \times \frac{2}{3}$ c $\frac{3}{7}$ of 8   d $\frac{3}{4} \times \frac{2}{5} \times \frac{3}{7}$

a $\frac{3}{7} \times \frac{2}{5} = \frac{3 \times 2}{7 \times 5} = \frac{6}{35}$

b Change the 5 to $\frac{5}{1}$. This gives $\frac{5}{1} \times \frac{2}{3} \times \frac{5 \times 2}{1 \times 3} = \frac{10}{3} = 3\frac{1}{3}$

(Changing from an improper fraction to a mixed number is covered in on pages 88–89.)

c $\frac{3}{7}$ of 8 $= \frac{3}{7} \times \frac{8}{1} = \frac{24}{7} = 3\frac{3}{7}$

d $\frac{3}{4} \times \frac{2}{5} \times \frac{3}{7} = \frac{3 \times 2 \times 3}{4 \times 5 \times 7} = \frac{18}{140} = \frac{9}{70}$ or $\frac{3}{4} \times \frac{2}{5} \times \frac{3}{7} = \frac{3 \times {}^1\!\!\!\not{2} \times 3}{{}^2\!\!\!\not{4} \times 5 \times 7} = \frac{9}{70}$

Example 8·10 A bar of chocolate weighs $\frac{3}{5}$ of a kilogram. How much does $\frac{1}{7}$ of the bar weigh?

$\frac{3}{5} \times \frac{1}{7} = \frac{3}{35}$ kg

Number, money and measure

Example 8·11 A jug of milk contains $\frac{3}{4}$ of a litre. Mike drinks $\frac{1}{5}$ of it. How many litres of milk does he drink?

$$\frac{1}{5} \text{ of } \frac{3}{4} = \frac{1}{5} \times \frac{3}{4} = \frac{3}{20}$$

Mike drinks $\frac{3}{20}$ of a litre.

It is worth noting that when you multiply a number by a fraction less than 1, your answer should be smaller than the original number. (Think: 7 halves is less than 7 wholes!) You can use this knowledge to quickly check that your answer is sensible.

Exercise 8F

1 Calculate the following, simplifying your answer where possible.

a $\frac{2}{3} \times \frac{3}{5}$ b $\frac{3}{4} \times \frac{5}{6}$ c $\frac{5}{7} \times \frac{3}{10}$ d $\frac{3}{4} \times \frac{1}{2}$ e $\frac{4}{9} \times \frac{2}{7}$

f $\frac{3}{5}$ of $\frac{2}{3}$ g $\frac{4}{7}$ of $\frac{3}{4}$ h $\frac{7}{8} \times \frac{6}{7}$ i $\frac{2}{5} \times \frac{7}{9}$ j $\frac{1}{3}$ of $\frac{4}{5}$

2 Work out each of these.

a $3 \times \frac{4}{5}$ b $\frac{4}{7} \times 8$ c $7 \times \frac{2}{3}$ d $\frac{6}{7} \times 3$ e $2 \times \frac{5}{6}$

f $12 \times \frac{3}{4}$ g $\frac{3}{7} \times 49$ h $\frac{5}{6}$ of 20   i $\frac{5}{7}$ of 18   j $\frac{3}{10}$ of 17

3 Calculate the following, giving your answer in simplest form where possible.

a $\frac{1}{2} \times \frac{1}{3} \times \frac{2}{5}$ b $\frac{3}{4} \times \frac{2}{5} \times \frac{1}{2}$ c $\frac{5}{6} \times \frac{3}{4} \times \frac{7}{8}$ d $\frac{1}{4} \times \frac{2}{3} \times \frac{5}{6}$

e $\frac{3}{7} \times \frac{7}{8} \times \frac{1}{2}$ f $\frac{2}{3} \times \frac{4}{5} \times \frac{1}{7}$ g $\frac{4}{7} \times \frac{1}{2} \times 14$ h $\frac{2}{3} \times \frac{5}{7} \times 6$

4 Farmer Jenks feeds each of his horses $\frac{2}{5}$ of a bale of hay every week. How many bales does he need to feed 12 horses each week?

5 A snail can crawl $\frac{5}{6}$ of a metre every hour. How far could it travel in 8 hours?

6 $\frac{1}{4}$ of Class 4B are girls. $\frac{2}{3}$ of the girls have blonde hair. What fraction of the class are blonde girls?

7 Jenny spends $\frac{3}{4}$ of an hour at the gym. She spends $\frac{1}{5}$ of her time on the treadmill. What fraction of an hour does she spend on the treadmill?

8 Mr Smith spends $\frac{1}{2}$ of his working week at his desk. He spends $\frac{1}{7}$ of this time on the phone to clients. What fraction of his time at work does he spend on the phone to clients?

9 A recipe for shortbread requires $\frac{1}{2}$ a cup of sugar. How much sugar do you need if you are making only $\frac{3}{4}$ of the amount indicated in the recipe?

Fractions, decimal fractions and percentages

Expressing one quantity as a percentage of another

Example 8·12 Without using a calculator, find:

a 18 as a percentage of 25.   b 39 as a percentage of 300.

a Write as a fraction $\frac{18}{25}$. Multiply the top and bottom by 4, which gives $\frac{72}{100}$.
So 18 is 72% of 25.

b Write as a fraction $\frac{39}{300}$. Cancel the top and bottom by 3, which gives $\frac{13}{100}$.
So 39 is 13% of 300.

Example 8·13 a What percentage of 80 is 38?   b What percentage of 64 is 14?

a Write as a fraction $\frac{38}{80}$. Divide through to give the decimal 0·475. Then multiply by 100 to give $47\frac{1}{2}$%. Or, simply multiply $\frac{38}{80}$ by 100.

b Write as a fraction $\frac{14}{64}$. Divide through to give the decimal 0·21875. Then multiply by 100 giving 22% (rounded off from 21·875). Or, simply multiply $\frac{14}{64}$ by 100.

Example 8·14 Ashram scored 39 out of 50 in a physics test, 56 out of 70 in a chemistry test and 69 out of 90 in a biology test. In which science did he do best?

Write each score as a fraction then convert to a percentage.

Physics: $\frac{39}{50}$ = 78%

Chemistry: $\frac{56}{70}$ = 80%

Biology: $\frac{69}{90}$ = 77% (rounded off)

So chemistry was the best mark.

Exercise 8G

1 Without using a calculator, express the first quantity as a percentage of the second.

a 32 out of 50   b 17 out of 20   c 2 out of 5   d 16 out of 25
e 12 out of 20   f 3 out of 10   g 64 out of 100   h 18 out of 25
i 33 out of 50   j 8 out of 50   k 2 out of 25   l 6 out of 25
m 3 out of 20   n 3 out of 5   o 15 out of 50   p 48 out of 100

2 In some tests, Trevor scored 39 out of 50 in maths, 16 out of 20 in English and 19 out of 25 in science. Convert each of these scores to a percentage. In which test did Trevor do best?

3 Mr Wilson pays £50 a month to cover his electricity, gas and oil bills. Electricity costs £24. Gas costs £18 and the rest is for oil. What percentage of the total does each fuel cost?

Number, money and measure

 4 My phone bill last month was £25. Of this, I spent £7 on text messages, £12 on long-distance calls and the rest on local calls. What percentage of the bill did I spend on each of type of call?

5 Use a calculator to work out what percentage the first quantity is of the second (round off to the nearest percent if necessary).

    a 33 out of 60   **b** 18 out of 80   **c** 25 out of 75   **d** 26 out of 65

    e 5 out of 125   **f** 84 out of 150   **g** 62 out of 350   **h** 48 out of 129

6 In a maths exam worth 80 marks, 11 marks are allocated to number, 34 marks are allocated to algebra, 23 marks are allocated to geometry and 12 marks are allocated to statistics. Work out the percentage allocated to each topic (round the answers off to the nearest percent) and add these up. Why is the total more than 100%?

Challenge

Copy the cross-number puzzle. Work out each percentage. Then use the puzzle to fill in the missing numbers in the clues.

Across
- **1** 54 out of 200
- **3** 33 out of 50
- **5** 13 out of 25
- **8** 99 out of 100
- **10** …… out of 200
- **12** 27 out of 50
- **14** 38 out of 200

Down
- **2** …… out of 400
- **4** 134 out of 200
- **6** …… out of 300
- **7** 100 out of 400
- **9** …… out of 20
- **11** 110 out of 1000
- **13** 23 out of 50

Percentages of quantities

This section will show you how to find a percentage of a quantity with and without a calculator.

Using 10% and 1%

When finding a percentage of a quantity it is useful to split it up into groups of 10% and 1%. For example, 23% can be split up into two lots of 10% plus three lots of 1%. This is useful because you can find 10% and 1% quickly:

- To find 10% of a quantity, divide by 10 (10% = $\frac{1}{10}$)
- To find 1% of a quantity, divide by 100 (1% = $\frac{1}{100}$).

It is also useful to know that you can find 5% quickly by halving 10%.

Fractions, decimal fractions and percentages

Example 8·15 Without a calculator find:

  a 30% of 560   b 7% of 362   c 46% of 423

  a 10% of 560 = 56, so 30% of 560 = 56 × 3 = 168
  b 1% of 362 = 3·62, so 7% of 362 = 3·62 × 7 = 25·34
  c 10% of 423 = 42·3,   *or* 10% of 423 = 42·3
     so 40% = 42·3 × 4 = 169·2   so, 5% of 423 is 21·15
     1% of 423 = 4·23,   46% of 423 = 169·2 + 21·15 + 4·23
     so 6% of 423 = 4·23 × 6 = 25·38   = 194·58
     46% of 423 = 169·2 + 25·38
                = 194·58

Example 8·16 Use a calculator to work out:

  a 6% of £190   b 63% of 75 eggs

  a (6 ÷ 100) × 190 = £11·40
  b (63 ÷ 100) × 75 = 47·25 = 47 eggs

Example 8·17 Which is greater, 42% of 560 or 62% of 390? You may use a calculator.

(42 ÷ 100) × 560 = 235·2
(62 ÷ 100) × 390 = 241·8
62% of 390 is greater.

Exercise 8H

1 Calculate:

  a 10% of 240   b 5% of 70   c 60% of 150   d 40% of 32
  e 15% of 540   f 45% of 320   g 35% of 460   h 75% of 280
  i 10% of 45   j 20% of 95   k 30% of 45   l 65% of 160

2 Work out each of the following.

  a 12% of 320   b 49% of 45   c 31% of 260   d 18% of 68
  e 11% of 12   f 28% of 280   g 52% of 36   h 99% of 206

3 Work out each of these. Round your answers appropriately.

  a 13% of £560   b 46% of 64 books   c 73% of 190 chairs
  d 34% of £212   e 64% of 996 pupils   f 57% of 120 buses
  g 37% of 109 plants   h 78% of 345 bottles   i 62% of 365 days
  j 93% of 2564 people k 54% of 456 fish   l 45% of £45
  m 65% of 366 eggs   n 7% of £684   o 9% of 568 chickens

Number, money and measure

4 Which is bigger:
   a 45% of 68 or 34% of 92?
   b 22% of £86 or 82% of £26?
   c 28% of 79 or 69% of 31?
   d 32% of 435 or 43% of 325?

5 A bus garage holds 50 buses. 34% are single-deckers, the rest are double-deckers.
   a How many single-deckers are there?
   b What percentage are double-deckers?
   c How many double-deckers are there?

6 A Jumbo Jet carries 400 passengers. On one trip, 52% of the passengers were British, 17% were American, 12% were French and the rest were German.
   a How many people of each nationality were on the plane?
   b What percentage were German?

Percentage increase and decrease

SPORTY SHOES
$\frac{1}{3}$ off all trainers

SHOES-FOR-YOU
30% off all trainers

Which shop gives the better value?

Example 8·18

a A clothes shop has a sale and reduces its prices by 20%. How much is the sale price of each of the following?

   i   A jacket originally costing £45.
   ii A dress originally costing £125.

   i   20% of 45 is 2 × 10% of 45 = 2 × 4·5 = 9. So, the jacket costs:
       £45 − £9 = £36
   ii 20% of 125 is 2 × 10% of 125 = 2 × 12·50 = 25. So, the dress costs:
       £125 − £25 = £100

b A company gives all its workers a 5% pay rise. How much is the new wage of each of these workers?

   i   Joan who gets £240 per week
   ii Jack who gets £6·60 per hour

   i   10% of 240 = 24
       So 5% of 240 = $\frac{1}{2}$ of 24 = 12.
       So, Joan gets £240 + £12 = £252 per week.
   ii 10% of 6·60 = 0·66
       So 5% of 6·60 = $\frac{1}{2}$ of 0·66 = 0·33.
       So, Jack gets £6·60 + £0·33 = £6·93 per hour.

Fractions, decimal fractions and percentages

Exercise 81

Do not use a calculator for these questions.

1. A bat colony has 40 bats. Over the breeding season, the population increases by 30%.
   a How many new bats were born?
   b How many bats are there in the colony after the breeding season?

2. In a wood there are 20 000 midges. During the evening, bats eat 40% of the midges.
   a How many midges were eaten by the bats?
   b How many midges were left after the bats had eaten?
   c What percentage of midges remain?

3. Work out the final amount when:
   a £45 is increased by 10%
   b £48 is decreased by 10%
   c £120 is increased by 20%
   d £90 is decreased by 20%
   e £65 is increased by 5%
   f £110 is decreased by 5%
   g £250 is increased by 25%
   h £300 is decreased by 25%
   i £6·80 is increased by 10%
   j £5·40 is decreased by 10%

4. a In a sale, all prices are reduced by 15%. Give the new price of an item that previously cost:
   i £18   ii £26   iii £50   iv £70

   b An electrical store increases its prices by 5%. Give the new price of an item that previously cost:
   i £200   ii £130   iii £380   iv £100

You may use a calculator for the rest of this Exercise.

5. A petri dish contains 2400 bacteria. These increase overnight by 23%.
   a How many extra bacteria are there?
   b How many bacteria are there the next morning?

6. A rabbit colony has 220 rabbits. As a result of disease, 45% die.
   a How many rabbits die from disease?
   b How many rabbits are left after the disease?
   c What percentage of the rabbits remain?

7. Work out the final price in euros when:
   a € 65 is increased by 12%
   b € 65 is decreased by 14%
   c € 126 is increased by 22%
   d € 530 is decreased by 28%
   e € 95 is increased by 32%
   f € 32 is decreased by 31%
   g € 207 is increased by 55%
   h € 421 is decreased by 18%
   i € 6·85 is increased by 40%
   j € 5·40 is decreased by 28%

Number, money and measure

 8 a In a sale all prices are reduced by $12\frac{1}{2}$%. Give the new price of items that previously cost:

    i £23·50   ii £66   iii £56·80   iv £124

b An electrical company increases its prices by $7\frac{1}{2}$%. Give the new price of items that previously cost:

    i £250   ii £180   iii £284   iv £199

Challenge

1 Write down 20% of 100.

2 Write down 20% of 120.

3 If £100 is increased by 20%, how much do you have?

4 If £120 is decreased by 20%, how much do you have?

5 Draw a poster to explain why a 20% increase followed by a 20% decrease does not return you to the value you started with.

Real-life problems

Percentages occur everyday in many situations.

Example 8·19

Which costs less: a pair of £45 shoes offered at $\frac{1}{3}$ off or a pair of £35 jeans offered at 20% off?

$\frac{1}{3}$ of £45 = £15

So cost of shoes = £45 − £15 = £30

10% of £35 = £3.50

so 20% of £35 = £3.50 × 2 = £7.00

So cost of jeans = £35 − £7.00 = £28

The pair of jeans costs less (by £2).

Example 8·20

In class 1A 12 out of the 25 pupils are girls. In class 1B 45% of the pupils are girls. Which class has the higher percentage of girls?

The fraction of pupils in 1A which are girls is $\frac{12}{25}$. Multiplying top and bottom by 4 gives $\frac{48}{100}$.

So the percentage of girls in 1A is 48%.

So class 1A has a higher percentage of girls (by 3%).

Fractions, decimal fractions and percentages

Exercise 8J

1. Two TV shops are offering these deals: TV Mania is offering $\frac{1}{3}$ off marked prices and TV Bargains is offering 20% off marked prices.

   Isla wants to buy a certain type of TV. In TV Mania its marked price is £300, while in TV Bargains its marked price is £280.

   Which shop offers the better deal after their discounts are applied?

2. Mike is looking to buy gravel for his driveway. Two companies offer the following promotions:
   - The Gravel King – 4 kg bag with 25% extra free for £10.
   - Jimmy Rubble – 4 kg bag with $\frac{1}{5}$ extra free for £10.

   Which shop offers the better bargain?

3. a Which is the better deal, 20% off or $\frac{1}{4}$ off?

   b Which is the better deal, 30% off or $\frac{1}{3}$ off?

4. In his survey of 20 cars, Vince found that 6 were red. In his survey, Alex found that 35% of the cars were red. Whose survey had the lower percentage of red cars?

5. Michelle and her husband Jarrod both lost weight after dieting. Jarrod originally weighed 80 kg and lost 10% of his weight. Michelle originally weighed 72 kg and lost $\frac{1}{8}$ of her weight. Who lost more weight?

6. James bought a car for £5000 while David bought a car for £4800. After a year, James's car had lost 20% of its value while David's car had lost $\frac{1}{12}$ of its value. Whose car was more valuable and by how much?

7. Which is bigger: $\frac{2}{5}$ of £30 or 30% of £40?

8. Peter wants to buy a pair of trainers. In one shop they are priced at £50. In another shop they are priced at $\frac{1}{5}$ off the marked price of £60. He decides to buy them in the first shop. Is he correct to do so? You must give a reason.

Number, money and measure

- By working on this topic I can explain how to solve calculations and real-life problems using fractions, decimals and percentages.

- Using my knowledge of place value I can accurately add and subtract decimal fractions in a variety of contexts. ★ Exercise 8A Q2

- By counting the number of decimal places in a question I know where to put the decimal point in my answer when multiplying and dividing with decimals. ★ Exercise 8B Q5 ★ Exercise 8D Q2

- I can multiply decimals by rewriting a calculation as an equivalent product. ★ Exercise 8C Q1

- I can find simple fractions of a quantity by first finding the unit fraction and then multiplying by the numerator. ★ Exercise 8E Q1

- I can multiply fractions by multiplying the numerators and the denominators and know how this links to finding a fraction of a quantity. ★ Exercise 8F Q4

- I can write one quantity as a percentage of another and can use my results to compare fractions of different amounts. ★ Exercise 8G Q2

- By using combinations of 10% and 1% I can find a percentage of a quantity without using a calculator. ★ Exercise 8H Q2

- I can find a percentage of a quantity using a calculator. ★ Exercise 8H Q3

- I can find a percentage increase or decrease with or without a calculator and can solve problems related to this. ★ Exercise 8I Q5

- I can apply my knowledge of percentages and fractions to solve a variety of real-life problems. ★ Exercise 8J Q8

Fractions, decimal fractions and percentages

9

By applying my knowledge of equivalent fractions and common multiples, I can add and subtract commonly used fractions.

MTH 3-07b

This chapter will show you how to:
- add and subtract fractions with the same denominator
- use common multiples to create equivalent fractions in order to add and subtract fractions with different denominators.

You should already know:
- how to find the lowest common multiple of two or more numbers
- how to find equivalent fractions
- how to simplify fractions
- how to find a fraction of a quantity
- how to change mixed numbers to improper fractions.

Adding and subtracting fractions

Look at the following fraction additions.

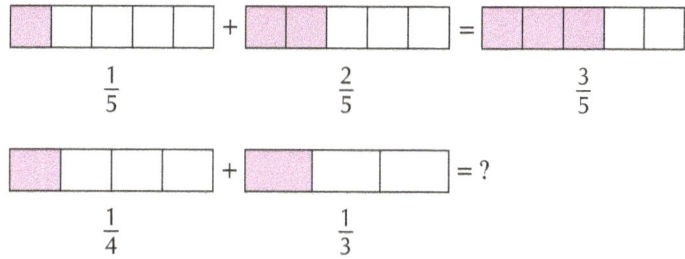

The first addition is simple because both shapes are cut into pieces which are the same size – the fractions have the same **denominator**. The second addition is more difficult because both shapes have been cut into different size pieces (the fractions have different denominators), making the answer less obvious. In general, to add or subtract with fractions, you need to have the same denominator. You will learn how to solve both types of problem in this chapter.

Number, money and measure

Adding and subtracting fractions with a common denominator

Example 9·1 Add together:

a $\dfrac{1}{5}+\dfrac{2}{5}$ b $\dfrac{3}{8}+\dfrac{1}{8}$ c $\dfrac{1}{3}+\dfrac{1}{3}+\dfrac{2}{3}$

In each addition, the denominators are the same. So, the **numerators** are just added.

a $\dfrac{1}{5}+\dfrac{2}{5}=\dfrac{3}{5}$

b $\dfrac{3}{8}+\dfrac{1}{8}=\dfrac{4}{8}=\dfrac{1}{2}$ (The answer here has been cancelled down.)

c $\dfrac{1}{3}+\dfrac{1}{3}+\dfrac{2}{3}=\dfrac{4}{3}=1\dfrac{1}{3}$ (This answer is an improper fraction, so it should be written as a mixed number.)

Example 9·2 Subtract:

a $\dfrac{5}{6}-\dfrac{1}{6}$ b $\dfrac{5}{8}-\dfrac{3}{8}$ c $3-\dfrac{1}{5}$

In each subtraction, the denominators are the same.
So, the numerators are just subtracted.

a $\dfrac{5}{6}-\dfrac{1}{6}=\dfrac{4}{6}=\dfrac{2}{3}$

b $\dfrac{5}{8}-\dfrac{3}{8}=\dfrac{2}{8}=\dfrac{1}{4}$

Note that both of these answers have been cancelled down.

c $3-\dfrac{1}{5}=2\dfrac{5}{5}-\dfrac{1}{5}=2\dfrac{4}{5}$

Exercise 9A

1 Add each of the following pairs of fractions. Cancel down or write as a mixed number where necessary.

a $\dfrac{1}{3}+\dfrac{1}{3}$ b $\dfrac{2}{5}+\dfrac{1}{5}$ c $\dfrac{1}{7}+\dfrac{2}{7}$ d $\dfrac{1}{4}+\dfrac{1}{4}$

e $\dfrac{1}{5}+\dfrac{3}{5}$ f $\dfrac{3}{8}+\dfrac{3}{8}$ g $\dfrac{5}{6}+\dfrac{5}{6}$ h $\dfrac{3}{4}+\dfrac{3}{4}$

2 Subtract each of the following pairs of fractions. Cancel down where necessary.

a $\dfrac{2}{3}-\dfrac{1}{3}$ b $\dfrac{2}{5}-\dfrac{1}{5}$ c $\dfrac{2}{7}-\dfrac{1}{7}$ d $\dfrac{3}{4}-\dfrac{1}{4}$

e $\dfrac{3}{5}-\dfrac{1}{5}$ f $\dfrac{5}{8}-\dfrac{1}{8}$ g $\dfrac{5}{6}-\dfrac{1}{6}$ h $\dfrac{5}{9}-\dfrac{2}{9}$

Fractions, decimal fractions and percentages

3 Work out each of the following.

a $\dfrac{3}{7}+\dfrac{2}{7}$ b $\dfrac{3}{11}+\dfrac{4}{11}$ c $\dfrac{3}{5}+\dfrac{1}{5}$ d $\dfrac{7}{13}+\dfrac{3}{13}$

e $\dfrac{7}{11}-\dfrac{2}{11}$ f $\dfrac{4}{9}-\dfrac{2}{9}$ g $\dfrac{3}{7}-\dfrac{2}{7}$ h $\dfrac{3}{5}-\dfrac{1}{5}$

★ 4 Work out each of the following. Cancel down to lowest terms.

a $\dfrac{3}{8}+\dfrac{1}{8}$ b $\dfrac{1}{12}+\dfrac{5}{12}$ c $\dfrac{3}{10}+\dfrac{1}{10}$ d $\dfrac{7}{15}+\dfrac{2}{15}$

e $\dfrac{4}{9}-\dfrac{1}{9}$ f $\dfrac{3}{10}-\dfrac{1}{10}$ g $\dfrac{5}{6}-\dfrac{1}{6}$ h $\dfrac{7}{12}-\dfrac{5}{12}$

5 Add the following fractions. Convert to mixed numbers or cancel down to lowest terms.

a $\dfrac{1}{3}+\dfrac{1}{3}$ b $\dfrac{5}{6}+\dfrac{5}{6}$ c $\dfrac{3}{10}+\dfrac{3}{10}$ d $\dfrac{1}{9}+\dfrac{2}{9}$

e $\dfrac{4}{15}+\dfrac{13}{15}$ f $\dfrac{7}{9}+\dfrac{5}{9}$ g $\dfrac{5}{12}+\dfrac{1}{12}$ h $\dfrac{3}{7}+\dfrac{5}{7}+\dfrac{2}{7}$

6 Calculate the following. Convert to mixed numbers.

a $7-\dfrac{2}{7}$ b $1-\dfrac{1}{6}$ c $5-\dfrac{3}{10}$ d $3-\dfrac{2}{9}$

e $1-\dfrac{2}{15}$ f $2-\dfrac{4}{9}$ g $6-\dfrac{5}{12}$ h $5-\dfrac{2}{3}$

7 Kevin, Dave and Richard cut a cake into eighths. Kevin ate 1 piece, while Dave and Richard ate 3 pieces each. What fraction of the cake was left?

8 Laura lives $\dfrac{5}{6}$ km from the shops. How far is it to walk from her house to the shops and back?

9 A bread recipe needs $\dfrac{4}{5}$ kg of wholemeal flour and $\dfrac{3}{5}$ kg of white flour. How much flour is needed altogether?

Adding and subtracting fractions with different denominators

When denominators are not the same, they must be made the same before the numerators are added or subtracted. To do this, first you need to find the lowest common multiple (LCM) of the denominators. Once you have done this, you must change each fraction to its equivalent fraction which has this LCM as its denominator.

Number, money and measure

Example 9·3 Work out:

a $\dfrac{2}{3}+\dfrac{1}{4}$ b $\dfrac{8}{9}-\dfrac{5}{6}$ c $\dfrac{3}{4}+\dfrac{5}{8}$

a $\dfrac{2}{3}+\dfrac{1}{4}$ convert to equivalent fractions (LCM of 3 and 4 = 12)

$=\dfrac{8}{12}+\dfrac{3}{12}$ $\dfrac{2}{3}\times\dfrac{4}{4}=\dfrac{2\times 4}{3\times 4}=\dfrac{8}{12}$ $\dfrac{1}{4}\times\dfrac{3}{3}=\dfrac{1\times 3}{4\times 3}=\dfrac{3}{12}$

$=\dfrac{11}{12}$ add fractions

b $\dfrac{8}{9}-\dfrac{5}{6}$ convert to equivalent fractions (LCM of 9 and 6 = 18)

$=\dfrac{16}{18}-\dfrac{15}{18}$ $\dfrac{8}{9}\times\dfrac{2}{2}=\dfrac{8\times 2}{9\times 2}=\dfrac{16}{18}$ $\dfrac{5}{6}\times\dfrac{3}{3}=\dfrac{5\times 3}{6\times 3}=\dfrac{15}{18}$

$=\dfrac{1}{18}$ subtract fractions

c $\dfrac{3}{4}+\dfrac{5}{8}$ convert to equivalent fractions (LCM of 4 and 8 = 8)

$=\dfrac{6}{8}+\dfrac{5}{8}$ $\dfrac{3}{4}\times\dfrac{2}{2}=\dfrac{3\times 2}{4\times 2}=\dfrac{6}{8}$

$=\dfrac{11}{8}$ add fractions

$=1\dfrac{3}{8}$ rewrite as mixed number

Note that only one of the fractions needed to be changed as one was already written in eighths.

Exercise 9B

1 Find the lowest common multiple of the following pairs of numbers.

a (3, 4) b (5, 6) c (3, 5) d (2, 3)
e (4, 5) f (2, 4) g (6, 9) h (4, 6)

2 Add the following fractions. Use your answers in Question 1 to help.

a $\dfrac{2}{3}+\dfrac{1}{4}$ b $\dfrac{2}{5}+\dfrac{1}{6}$ c $\dfrac{1}{3}+\dfrac{2}{5}$ d $\dfrac{1}{3}+\dfrac{1}{2}$

e $\dfrac{1}{5}+\dfrac{1}{4}$ f $\dfrac{1}{2}+\dfrac{1}{4}$ g $\dfrac{5}{6}+\dfrac{1}{9}$ h $\dfrac{1}{6}+\dfrac{1}{4}$

3 Subtract the following fractions.

a $\dfrac{1}{3}-\dfrac{1}{4}$ b $\dfrac{2}{5}-\dfrac{1}{6}$ c $\dfrac{2}{5}-\dfrac{1}{3}$ d $\dfrac{1}{2}-\dfrac{1}{3}$

e $\dfrac{2}{5}-\dfrac{1}{4}$ f $\dfrac{1}{2}-\dfrac{1}{4}$ g $\dfrac{5}{6}-\dfrac{1}{9}$ h $\dfrac{5}{6}-\dfrac{3}{4}$

Fractions, decimal fractions and percentages

4 Convert the following fractions to equivalent fractions with a common denominator, and then work out the answer, cancelling down or writing as a mixed number if appropriate:

a $\dfrac{1}{3} + \dfrac{1}{4}$ b $\dfrac{1}{6} + \dfrac{1}{3}$ c $\dfrac{3}{10} + \dfrac{1}{4}$ d $\dfrac{1}{8} + \dfrac{5}{6}$

e $\dfrac{4}{15} + \dfrac{3}{10}$ f $\dfrac{7}{8} + \dfrac{5}{6}$ g $\dfrac{7}{12} + \dfrac{1}{4}$ h $\dfrac{3}{4} + \dfrac{1}{3} + \dfrac{1}{2}$

i $\dfrac{2}{3} - \dfrac{1}{8}$ j $\dfrac{5}{6} - \dfrac{1}{3}$ k $\dfrac{3}{10} - \dfrac{1}{4}$ l $\dfrac{8}{9} - \dfrac{1}{6}$

m $\dfrac{4}{15} - \dfrac{1}{10}$ n $\dfrac{7}{8} - \dfrac{5}{6}$ o $\dfrac{7}{12} - \dfrac{1}{4}$ p $\dfrac{3}{4} + \dfrac{1}{3} - \dfrac{1}{2}$

5 Convert the following fractions to equivalent fractions with a common denominator. Then work out the answer. Cancel down or write as a mixed number if appropriate.

a $\dfrac{1}{8} + \dfrac{3}{5}$ b $\dfrac{5}{7} + \dfrac{1}{4}$ c $\dfrac{3}{14} + \dfrac{3}{8}$ d $\dfrac{5}{9} + \dfrac{5}{6}$

e $\dfrac{5}{12} + \dfrac{3}{10}$ f $\dfrac{3}{8} + \dfrac{1}{6}$ g $\dfrac{7}{12} + \dfrac{1}{8}$ h $\dfrac{3}{4} + \dfrac{1}{3} + \dfrac{5}{12}$

i $\dfrac{3}{5} - \dfrac{1}{8}$ j $\dfrac{5}{7} - \dfrac{1}{4}$ k $\dfrac{1}{4} - \dfrac{3}{14}$ l $\dfrac{5}{9} - \dfrac{1}{6}$

m $\dfrac{5}{12} - \dfrac{1}{10}$ n $\dfrac{7}{8} - \dfrac{5}{14}$ o $\dfrac{7}{12} - \dfrac{1}{8}$ p $\dfrac{3}{4} + \dfrac{3}{14} - \dfrac{3}{8}$

6 A magazine has $\dfrac{1}{3}$ of its pages for advertising, $\dfrac{1}{12}$ for letters and the rest for articles.

a What fraction of the pages is for articles?

b If the magazine has 120 pages, how many are used for articles?

7 A survey of pupils showed that $\dfrac{1}{5}$ of them walked to school, $\dfrac{2}{3}$ came by bus and the rest came by car.

a What fraction came by car?

b If there were 900 pupils in the school, how many came by car?

8 A farmer plants $\dfrac{2}{7}$ of his land with wheat and $\dfrac{3}{8}$ with maize. The rest is used for cattle.

a What fraction of the land is used to grow crops?

b What fraction is used for cattle?

9 John gives $\dfrac{1}{2}$ of his pocket money to Sarah and $\dfrac{1}{3}$ of it to Dave.

a What fraction of his pocket money has he given away?

b What fraction does he have left?

10 In the fridge there is a carton of milk containing $\dfrac{2}{5}$ of a litre. There is also another carton of milk containing $\dfrac{3}{4}$ of a litre. Harry drinks $\dfrac{1}{3}$ of a litre of milk from one of the cartons. How much milk is left in the fridge altogether?

Number, money and measure

Challenge

1 **Ancient Egyptians**

The ancient Egyptians only used unit fractions, that is, fractions with a numerator of 1. So they would write $\frac{5}{8}$ as $\frac{1}{2} + \frac{1}{8}$.

a Write the following as the sum of two unit fractions.

   i $\frac{3}{8}$ ii $\frac{3}{4}$

   iii $\frac{7}{12}$ iv $\frac{2}{3}$

b Write the following as the sum of three unit fractions.

   i $\frac{7}{8}$ ii $\frac{5}{6}$

   iii $\frac{5}{8}$ iv $\frac{23}{24}$

2 **More ancient Egyptians**

The ancient Egyptians thought that 360 was a magical number because it had lots of factors.

a Write down all the factors of 360.

b Write down the following fractions as equivalent fractions with a denominator of 360.

$$\frac{1}{2}, \frac{1}{3}, \frac{1}{4}, \frac{1}{5}, \frac{1}{6}, \frac{1}{8}, \frac{1}{9}, \frac{1}{10}, \frac{1}{12}$$

c Use these results to work out:

   i $\frac{1}{2} + \frac{1}{3}$ ii $\frac{1}{2} + \frac{1}{6}$ iii $\frac{1}{2} + \frac{1}{5}$ iv $\frac{1}{3} + \frac{1}{4}$

   v $\frac{1}{6} + \frac{1}{8}$ vi $\frac{1}{3} - \frac{1}{5}$ vii $\frac{1}{4} - \frac{1}{6}$ viii $\frac{1}{8} - \frac{1}{10}$

   ix $\frac{1}{3} + \frac{1}{4} + \frac{1}{5}$ x $\frac{1}{6} + \frac{1}{12} - \frac{1}{4}$

Cancel down your answers to their simplest form.

- By working on this topic I know that in order to add and subtract fractions they must have the same denominator.
- I know that when adding and subtracting fractions with a common denominator I only have to focus on the numerators. ★ Exercise 9A Q4
- I can confidently add and subtract fractions with different denominators by using my knowledge of LCMs and equivalent fractions. ★ Exercise 9B Q4
- I can solve a variety of problems which involve adding and subtracting fractions.
  ★ Exercise 9B Q10

Fractions, decimal fractions and percentages

10

Having used practical, pictorial and written methods to develop my understanding, I can convert between whole or mixed numbers and fractions.

MTH 3-07c

This chapter will show you how to:
- convert mixed numbers to improper fractions
- convert improper fractions to mixed numbers.

You should already know:
- how to simplify fractions
- how to interpret a pictorial representation of a fraction
- that a mixed number is made up of a whole number part and a fraction part
- that an improper fraction is a fraction where the numerator is bigger than the denominator.

Converting mixed numbers to improper fractions

The following picture shows the **mixed number** $3\frac{2}{5}$. Can you write $3\frac{2}{5}$ as an **improper** fraction to show how many fifths there are altogether?

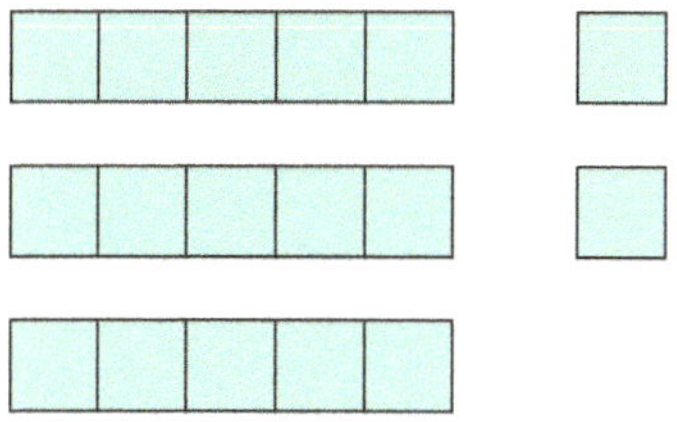

Each whole shape has five pieces (fifths), giving a total of 17 fifths. $3\frac{2}{5}$ as an improper fraction is written as $\frac{17}{5}$. Fractions like $\frac{17}{5}$ are called improper fractions or top-heavy fractions because the numerator (top) is bigger than the denominator (bottom).

Example 10·1

a How many sevenths are in 4 whole ones?

b How many fifths are in $2\frac{3}{5}$?

a There are 7 sevenths in one whole, so there are 4 × 7 = 28 sevenths in 4 whole ones.
b There are 5 fifths in one whole so there are 2 × 5 = 10 fifths in 2 whole ones, giving a total of 10 + 3 = 13 fifths in $2\frac{3}{5}$.

Number, money and measure

Example 10·2 Change the following mixed numbers to improper fractions.

a $2\frac{1}{3}$ b $5\frac{3}{8}$

a There are 3 thirds in 1 whole so there are 2 × 3 = 6 thirds in 2 wholes, giving a total of 6 + 1 = 7 thirds in $2\frac{1}{3}$.

Therefore, $2\frac{1}{3} = \frac{7}{3}$.

b There are 8 eighths in 1 whole so there are 5 × 8 = 40 eighths in 5 wholes, giving a total of 40 + 3 = 43 eighths in $5\frac{3}{8}$.

Therefore, $5\frac{3}{8} = \frac{43}{8}$.

Exercise 10A

1.   a How many sixths are in 1?   b How many sixths are in 3?
      c How many eighths are in 1?   d How many eighths are in 4?

2.   a How many sixths are in $3\frac{5}{6}$?   b How many eighths are in $4\frac{1}{2}$?
      c How many tenths are in $2\frac{3}{10}$?   d How many ninths are in $5\frac{7}{9}$?

★ 3 Convert each of these mixed numbers to an improper fraction.

a $1\frac{1}{4}$ b $2\frac{1}{2}$ c $3\frac{1}{6}$ d $4\frac{2}{7}$ e $5\frac{1}{8}$ f $2\frac{3}{5}$

g $1\frac{7}{9}$ h $3\frac{3}{4}$ i $4\frac{2}{5}$ j $2\frac{3}{11}$ k $4\frac{5}{8}$ l $3\frac{2}{9}$

m $1\frac{3}{4}$ n $1\frac{4}{5}$ o $2\frac{2}{3}$ p $2\frac{3}{4}$ q $3\frac{1}{6}$ r $3\frac{2}{3}$

4. Match the improper fractions to mixed numbers.

$\frac{5}{2}$ $\frac{7}{4}$ $\frac{7}{5}$ $\frac{11}{5}$

$2\frac{1}{2}$ $1\frac{2}{5}$ $1\frac{3}{4}$ $2\frac{1}{5}$

Converting improper fractions to mixed numbers

The picture shows 5 pizza halves. How many whole pizzas can you make? How many are left?

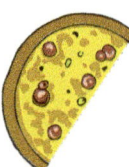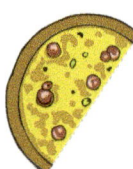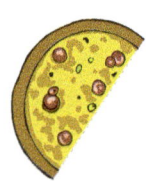

You need 2 halves to make a whole pizza. So with 5 halves you can make 2 whole pizzas, with 1 piece left over. This could be found by doing the following calculation: 5 ÷ 2 = 2 r 1. That is, you can make 2 wholes with 1 half left over.

So 5 halves or $\frac{5}{2} = 2\frac{1}{2}$.

Fractions, decimal fractions and percentages

Example 10·3 Write these improper (top-heavy) fractions as mixed numbers.

a $\dfrac{3}{2}$ b $\dfrac{9}{5}$ c $\dfrac{17}{5}$

a $\dfrac{3}{2} = 3 \div 2 = 1 \text{ r } 1 = 1\dfrac{1}{2}$

b $\dfrac{9}{5} = 9 \div 5 = 1 \text{ r } 4 = 1\dfrac{4}{5}$

c $\dfrac{17}{5} = 17 \div 5 = 3 \text{ r } 2$, so $\dfrac{17}{5} = 3\dfrac{2}{5}$

Example 10·4 Write the following as mixed numbers.

a $\dfrac{48}{15}$ b The fraction of a kilometre given by 3150 metres

a $48 \div 15 = 3 \text{ r } 3$, so $\dfrac{48}{15} = 3\dfrac{3}{15}$, which can cancel to $3\dfrac{1}{5}$. (Note: It is usually easier to cancel after the fraction has been written as a mixed number rather than before.)

b 1 kilometre is 1000 metres, so the fraction is $\dfrac{3150}{1000} = 3\dfrac{150}{1000} = 3\dfrac{3}{20}$.

Exercise 10B

1 Write each of the following as a mixed number.

a $\dfrac{5}{4}$ b $\dfrac{7}{2}$ c $\dfrac{11}{6}$ d $\dfrac{9}{2}$ e $\dfrac{11}{8}$ f $\dfrac{10}{7}$ g $\dfrac{5}{3}$ h $\dfrac{16}{5}$

★ 2 Change each of these improper fractions into a mixed number.

a $\dfrac{9}{4}$ b $\dfrac{7}{3}$ c $\dfrac{11}{7}$ d $\dfrac{8}{3}$ e $\dfrac{8}{7}$ f $\dfrac{13}{8}$

g $\dfrac{10}{3}$ h $\dfrac{14}{3}$ i $\dfrac{14}{5}$ j $\dfrac{17}{6}$ k $\dfrac{17}{5}$ l $\dfrac{21}{5}$

3 Write each of these fractions as a mixed number. Cancel down if appropriate.

a Seven thirds b Sixteen sevenths c Twelve fifths d Nine halves

e $\dfrac{20}{7}$ f $\dfrac{24}{5}$ g $\dfrac{13}{3}$ h $\dfrac{19}{8}$

i $\dfrac{146}{12}$ j $\dfrac{78}{10}$ k $\dfrac{52}{12}$ l $\dfrac{102}{9}$

4 Write each of the following as a mixed number in its simplest form.

a $\dfrac{14}{12}$ b $\dfrac{15}{9}$ c $\dfrac{24}{21}$ d $\dfrac{35}{20}$ e $\dfrac{28}{20}$ f $\dfrac{70}{50}$

g $\dfrac{28}{24}$ h $\dfrac{26}{12}$ i $\dfrac{44}{24}$ j $\dfrac{32}{10}$ k $\dfrac{36}{24}$ l $\dfrac{75}{35}$

Number, money and measure

Challenge By changing each mixed number to an improper fraction, answer the following questions.

a. Share $3\frac{3}{5}$ kg of sand equally between three bags.

b. John has $7\frac{1}{2}$ chocolate bars. If he shares the chocolate equally between 5 people, how much will each person get?

c. A plank of wood is $5\frac{1}{4}$ metres long. It is cut into 7 equal pieces. How long is each piece?

- By working on this topic I can confidently explain how to convert between mixed numbers and improper fractions.
- I can change a mixed number to an improper fraction. ★ Exercise 10A Q3
- By dividing the numerator by the denominator and focusing on the remainder, I can convert an improper fraction to a mixed number. ★ Exercise 10B Q2

Fractions, decimal fractions and percentages

11

I can show how quantities that are related can be increased or decreased proportionally and apply this to solve problems in everyday contexts.

MNU 3-08a

This chapter will show you how to:
- understand and use proportion
- read and understand ratios
- understand the difference between ratio and proportion
- understand equivalence of ratios
- simplify ratios by dividing all elements by the same number
- use equivalent ratios to solve numerical problems
- divide quantities according to a given ratio.

You should already know:
- how to find a highest common factor
- how to express one quantity as a fraction of another
- how to express one quantity as a percentage of another
- about the equivalence of fractions, percentages and decimals.

Proportion

Look at the fish tank. There are three types of fish – plain, striped and spotted.

Proportion can be expressed using a fraction, percentage or decimal. When thinking about proportion, a useful phrase is 'one **in** every x parts'.

What proportion of the fish are plain? What proportion are striped? What proportion are spotted?

Proportion is a way of comparing the parts of a quantity to the whole quantity.

Example 11·1 What proportion of this metre rule is shaded?

40 cm out of 100 cm are shaded. This is 40% (or 0·4 or $\frac{2}{5}$).

91

Number, money and measure

Example 11·2   Look at the diagram below. What proportion of squares is black?

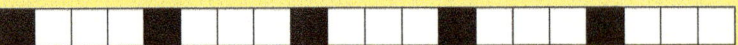

There is 1 black square for every 3 white squares. So, 1 in every 4 squares is black. This is 25% or $\frac{1}{4}$ or 0·25.

Example 11·3   A fruit drink is made by mixing 20 cl of orange juice with 60 cl of pineapple juice. What is the proportion of orange juice in the drink?

Total volume of drink is 20 + 60 = 80 cl.

The proportion of orange is 20 out of 80 = $\frac{20}{80} = \frac{1}{4}$.

Example 11·4   Another fruit drink is made by mixing orange juice and grapefruit juice. The proportion of orange is 40%. 60 cl of orange juice is used. What proportion is grapefruit? How much grapefruit juice is used?

The proportion of grapefruit is 100% − 40% = 60%. Now 40% = 60 cl, so 10% = 15 cl. Hence, 60% = 90 cl of grapefruit juice.

Example 11·5   Five pens cost £3·25. How much do 8 pens cost?

First, work out the cost of 1 pen: £3·25 ÷ 5 = £0·65.

Hence, 8 pens cost 8 × £0·65 = £5·20.

Exercise 11A

1. For each of these metre rules, work out what proportion of the rule is shaded. Write each answer as a percentage, fraction and decimal.

   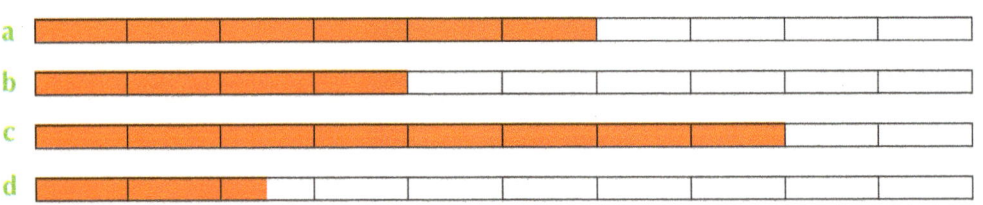

2. For each bag of black and white balls, work out what proportion of the balls is black. Write each answer as a fraction, simplified if possible.

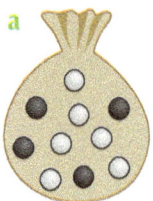
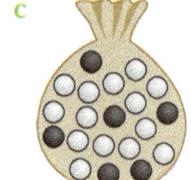

Fractions, decimal fractions and percentages

3 Look at the squares below.

a What proportion of the squares is black? Write your answer as a fraction and as a percentage.

b What proportion is white?

c In the diagram there are 20 squares altogether. 4 are black and 16 are white.
Copy and fill in the table.

Number of black squares	Number of white squares
1	4
2	8
3	
5	
	32
	40

4 A squash drink is made by mixing 15 cl of blackcurrant concentrate with 45 cl of water.

a What is the total volume of the drink?

b What proportion of the drink is blackcurrant concentrate?

c Copy and complete the table. Give proportions as fractions.

Blackcurrant (cl)	Water (cl)	Total volume (cl)	Blackcurrant proportion
20	50	70	$\frac{2}{7}$
10	40		
25	50		
15	60		

5 Three bars of soap cost £1·80. How much would each of the following cost?

a 1 bar   b 12 bars   c 30 bars

6 Five euros is worth £4·15. How many pounds will I get for each of the following numbers of euros?

a 1 euro   b 8 euros   c 600 euros

7 These are the ingredients to make four pancakes.

a How much of each ingredient will be needed to make 12 pancakes?

b How much of each ingredient will be needed to make six pancakes?

> 1 egg
> 3 ounces of plain flour
> 5 fluid ounces of milk

93

Number, money and measure

8 One litre of fruit squash contains 24 cl of fruit concentrate and the rest is water.

  a   What proportion of the drink is fruit concentrate?

  b   What proportion of the drink is water?

9 Answer the following questions but be careful! One of them is a trick question.

  a   3 kg of sugar cost £1·80. How much do 4 kg of sugar cost?

  b   A man can run 10 km in 40 minutes. How long does he take to run 12 km?

  c   In two days my watch loses 20 seconds. How much time does it lose in a week?

  d   It takes me 5 seconds to dial the 10 digit number of a friend who lives 100 km away. How long does it take me to dial the 10 digit number of a friend who lives 200 miles away?

10 a A bottle of shampoo costs £2·62 and contains 30 cl. A different bottle of the same shampoo costs £1·50 and contains 20 cl. Which is the better buy?

  b   A large roll of sticky tape has 25 metres of tape and costs 75p. A small roll of sticky tape has 15 metres of tape and costs 55p. Which roll is better value?

  c   A pad of A4 paper costs £1·10 and has 120 sheets. A thicker pad of A4 paper costs £1·50 and has 150 sheets. Which pad is the better buy?

  d   A small tin of peas contains 250 grams and costs 34p. A large tin costs 70p and contains 454 grams. Which tin is the better buy?

Ratio

In the fish tank in the picture on page 91, there are 2 spotted fish and 8 striped fish. This means that there is 1 spotted fish for every 4 striped fish.

- Copy and complete each of these statements:
  - There is 1 spotted fish for every …… plain fish.
  - There are 4 striped fish for every …… plain fish.

- Now use the four bags of balls in Question 2 of Exercise 11A.
  - In bag **a** there are 2 black balls for every …… white balls.
  - In bag **b** there is 1 black ball for every …… white balls.
  - In bag **c** there are 3 black balls for every …… white balls.
  - In bag **d** there are 2 black balls for every …… white balls.

Comparisons like this are called **ratios**.

The fish have been breeding!

What is the ratio of striped fish to spotted fish?

What is the ratio of plain fish to spotted fish?

What is the ratio of plain fish to striped fish?

If five more plain fish are added to the tank, how many more striped fish would have to be added to keep the ratio of plain to striped the same?

When thinking about ratio, a useful phrase is 'one part **for** every x parts'. Whereas proportion compares the number of parts to the whole, ratio compares the number of parts to each other.

Fractions, decimal fractions and percentages

Example 11·6 In the fish tank on page 94 there are 3 spotted fish and 9 striped fish. We say the ratio of spotted fish to striped fish is 3 to 9. This is written as 3 : 9. The ratio of striped fish to spotted fish is 9 to 3 or 9 : 3. The order is important.

Example 11·7 Look at the diagram below.

a What is the ratio of black to white?

b What proportion of squares is black?

a There is 1 black square to 3 white squares. The ratio is 1 to 3 or 1 : 3.

b The proportion of squares that are black is $\frac{1}{4}$.

Example 11·8 In a class of 20 children, 13 are girls. What is the ratio of girls to boys?

First, find the number of boys. 20 − 13 = 7 so there are 13 girls and 7 boys. The ratio of girls to boys is 13 : 7. And the ratio of boys to girls is 7 : 13. Remember, the order is important.

Exercise 11B

1 Look at the diagram below.

a What is the ratio of black to white?

b What is the ratio of white to black?

2 For each of these metre rules, what is the ratio of the shaded part to the unshaded part?

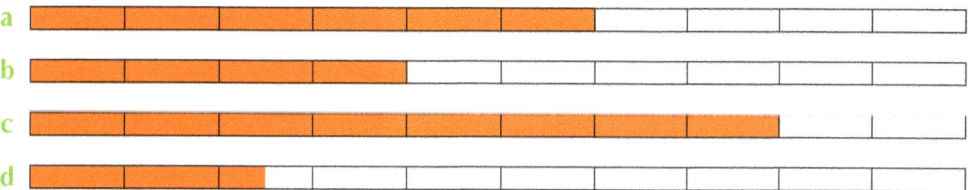

3 For each bag of black and white balls, what is the ratio of black to white balls?

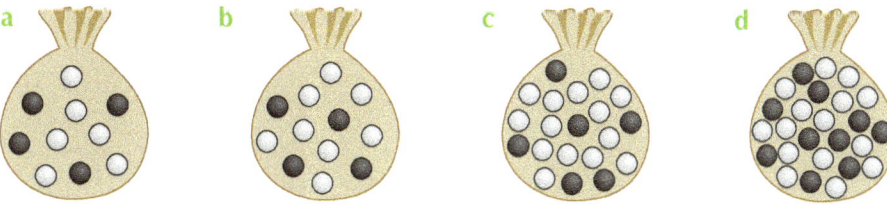

4 In a class of 15 pupils, 8 are boys. What is the ratio of boys to girls?

5 The ratio of British cars to foreign cars in the staff car park is 1 : 4. Explain why the proportion of British cars is 20% and not 25%.

95

Number, money and measure
Simplifying ratios

Some ratios can be simplified, making them easier to work with and understand. To do this, look at the individual elements in a ratio and see if they have any common factors. Then divide by the highest common factor. Sometimes we need to multiply to simplify, for example:

$$\frac{1}{3} : 5$$

$$\frac{1}{3} \times 3 : 5 \times 3$$

$$1 : 15$$

Ratios are always given as whole numbers.

Example 11·9   Reduce the following ratios to their simplest form: a   4 : 6   b   5 : 25

a   The highest common factor of 4 and 6 is 2.
So, divide both values by 2, giving 4 : 6 = 2 : 3.

$$\div 2 \begin{pmatrix} 4 : 6 \\ 2 : 3 \end{pmatrix} \div 2$$

b   The highest common factor of 5 and 25 is 5. So, divide both values by 5, giving 5 : 25 = 1 : 5.

Example 11·10   A fruit drink is made by mixing 20 cl of orange juice with 60 cl of pineapple juice. What is the ratio of orange juice to pineapple juice?

Orange : pineapple = 20 : 60 = 1 : 3 (divide by 20; the highest common factor is 20)

$$\div 20 \begin{pmatrix} 20 : 60 \\ 1 : 3 \end{pmatrix} \div 20$$

Example 11·11   A builder makes concrete by mixing sand, cement and gravel in the ratio 9 : 3 : 6. Simplify this ratio.

All three quantities can be divided by 3 to become 3 : 1 : 2.

Exercise 11C

1   Reduce each of the following ratios to its simplest form.

    a   4 : 8   b   8 : 4   c   2 : 10   d   12 : 9
    e   5 : 20   f   8 : 10   g   6 : 4   h   10 : 15
    i   2 : 14   j   14 : 4   k   6 : 10   l   25 : 30

2   Write down the ratio of shaded : unshaded for each of these metre rules.

Fractions, decimal fractions and percentages

3 There are 300 lights on a Christmas tree. 120 are white, 60 are blue, 45 are green and the rest are yellow.

    a Write down the proportion of each colour as a percentage.

    b Write down each of the following ratios in its simplest form.

        i white : blue   **ii** blue : green

        iii green : yellow   **iv** white : blue : green : yellow

4 In each of the following, find the ratio of the length AB to the length CD. Remember to simplify the ratio where possible.

    a AB = 2 cm, CD = 4 cm

    b AB = 2 cm, CD = 5 cm

    c AB = 6 cm, CD = 3 cm

    d AB = 3 cm, CD = 5 cm

    e AB = 6 cm, CD = 4 cm

5 Express each of the following ratios in its simplest form.

    a 10 mm : 25 mm   **b** 2 mm : 20 mm   **c** 36 cm : 45 cm

    d 200 cm : 40 cm   **e** 500 m : 2000 m   **f** 96 mm : 40 mm

Using equivalent ratios

Sometimes problems can be solved using **equivalent ratios**. This is done by multiplying (or dividing) the elements in the ratio by the same amount. You can use a table to find equivalent ratios.

Example 11·12   Find ratios equivalent to the ratio 4 : 3.

Draw a table and multiply the elements in the ratio by the same amount.

First element	Second element
4	3
8	6
12	9
40	30

Multiplying both elements by 2: a ratio equivalent to 4 : 3 is 8 : 6.
Multiplying both elements by 3: another ratio equivalent to 4 : 3 is 12 : 9
Multiplying both elements by 10: this gives the equivalent ratio 40 : 30.

Number, money and measure

Example 11·13 A batch of pink paint is made by mixing white and red in the ratio 4 : 1. If 12 pots of white paint are to be used, how much red paint is needed?

White	Red
×3 ⟶ 4 / 12	1 / 3 ⟵ ×3

The white element has been multiplied by 3 so the red element must be multiplied by 3 also. So 3 pots of red are needed.

Example 11·14 Another fruit drink is made by mixing orange juice and grapefruit juice in the ratio 2 : 5. 60 cl of orange juice are used. How much grapefruit juice is needed?

The problem is 60 : ? = 2 : 5. You will see that, instead of cancelling, you need to multiply by 30. So, 2 : 5 = 60 : 150.

Orange	Grapefruit
×30 ⟶ 2 / 60	5 / 150 ⵠ ×30

So, 150 cl of grapefruit juice will be needed.

Exercise 11D

1 Copy and complete the tables to find equivalent ratios.

a

First element	Second element
5	3
15	

b

First element	Second element
5	2
	12

c

First element	Second element	Third element
4	3	1
20		

2 A fruit drink is made by mixing orange juice and grapefruit juice in the ratio 3 : 2.

   a Amy uses 30 cl of orange juice. How much grapefruit juice should she use?

   b Struan uses 12 cl of grapefruit juice. How much orange juice should he use?

★ **3** The ratio of black cars to silver cars is 3 : 5.

   a How many silver cars are there if the number of black cars is:

      i 12   ii 30   iii 36   iv 51   v 120

   b How many black cars are there if the number of silver cars is:

      i 10   ii 25   iii 35   iv 55   v 300

Fractions, decimal fractions and percentages

4. Jessica makes a batch of chocolate brownies by mixing chocolate and sugar in the ratio 4 : 5. If she uses 200 grams of chocolate, how much sugar will she need?

5. Concrete is to be made using sand, gravel and cement in the ratio 5 : 3 : 1

    a. If 6 bags of gravel are used, how many bags of sand and cement are needed?

    b. If 6 bags of cement are used, how many bags of gravel and sand are needed?

Dividing quantities in a given ratio

Example 11·15 Divide £50 in the ratio 3 : 7.

There are 3 + 7 = 10 portions. This gives £50 ÷ 10 = £5 per portion. So, one share of the £50 is 3 × £5 = £15, and the other share is 7 × £5 = £35.

You can use a table to help solve this type of problem. Add an extra column giving the total of the elements:

First element	Second element	Total
3	7	10
		£50

× 5

The total has been multiplied by £5, so the elements must also be multiplied by £5, giving:

First element	Second element	Total
3	7	10
£15	£35	£50

Example 11·16 Divide £150 in the ratio 1 : 5.

There are 1 + 5 = 6 portions. This gives £150 ÷ 6 = £25 per portion. So one share of the £150 is 1 × 25 = £25, and the other share is 5 × £25 = £125.

Or, using a table to set out your working:

First element	Second element	Total
1	5	6
£25	£125	£150

× 25

Example 11·17 A fruit juice drink is made by mixing orange juice, lemon juice and grapefruit juice in the ratio 5 : 2 : 3. Find the volume of each type of fruit juice needed to make 20 litres of fruit juice drink.

Orange	Lemon	Grapefruit	Total
5	2	3	10
10 litres	4 litres	6 litres	20 litres

× 2

The total has been multiplied by 2 litres so all elements must be multiplied by 2 litres.

99

Number, money and measure

Exercise 11E

1. Divide £100 in the ratio:
   a 2 : 3   b 1 : 9   c 7 : 3   d 1 : 3   e 11 : 9

★ 2. There are 350 pupils in a primary school. The ratio of girls to boys is 3 : 2. How many boys and girls are there in the school?

3. Freda has 120 CDs. The ratio of pop CDs to dance CDs is 5 : 7. How many of each type of CD are there?

4. James is saving 50p coins and £1 coins. He has 75 coins. The ratio of 50p coins to £1 coins is 7 : 8. How much money does he have altogether?

5. Mr Smith has 24 calculators in a box. The ratio of ordinary calculators to scientific calculators is 5 : 1. How many of each type of calculator does he have?

6. An exam consists of three parts. A mental test, a non-calculator paper and a calculator paper. The ratio of marks for each is 1 : 3 : 4. The whole exam is worth 120 marks. How many marks is each part of the exam worth?

7. To make jam, Josh uses strawberries to preserving sugar in the ratio 3 cups : 1 cup.
   a How many cups of each will he need to make 20 cups of jam altogether?
   b If he has 12 cups of strawberries, how many cups of sugar will he need?
   c If he has $2\frac{1}{2}$ cups of sugar, how many cups of strawberries will he need?

8. a There are 15 bottles on the wall. The ratio of green bottles to brown bottles is 1 : 4. How many green bottles are there on the wall?
   b One green bottle accidentally falls. What is the ratio of green to brown bottles now?

9. a Forty-nine trains pass through Barnsley station each day. They go to Huddersfield or Leeds in the ratio 3 : 4. How many trains go to Huddersfield?
   b One day, due to driver shortages, six of the Huddersfield trains are cancelled and three of the Leeds trains are cancelled. What is the ratio of Huddersfield trains to Leeds trains that day?

Challenge

Uncle Fred has decided to give his nephew and niece, Jack and Jill, £100 between them. He decides to split the £100 in the ratio of their ages. Jack is 4 and Jill is 6.

a How much do each get?

b The following year he does the same thing with another £100. How much do each get now?

c He continues to give them £100 shared in the ratio of their ages for another 8 years. How much will each get each year?

d After the 10 years, how much of the £1000 given in total will Jack have? How much will Jill have?

Fractions, decimal fractions and percentages

- By working on this topic I can explain how to use ratio and proportion to solve problems.
- I understand the meaning of proportion and can express it as a fraction, percentage or decimal.
  ★ Exercise 11A Q1
- I can use the correct notation to express the ratio between two quantities. ★ Exercise 11B Q3
- I can use my knowledge of common factors to simplify ratios. ★ Exercise 11C Q1
- I can apply my understanding of equivalent ratios to solve problems. ★ Exercise 11D Q3
- I can divide a quantity in a given ratio. ★ Exercise 11E Q2

Number, money and measure

12

When considering how to spend my money, I can source, compare and contrast different contracts and services, discuss their advantages and disadvantages, and explain which offer best value to me.

MNU 3-09a

This chapter will show you how to:
- make the correct calculations so that you can make sensible well-informed decisions
- choose the best options when faced with different types of financial services and justify your decisions.

You should already know:
- how to work out the best deals in shops
- to look out for hidden costs
- how to work out what you can afford to buy from a fixed amount of money
- to make sensible comparisons between products.

Services and contracts

When you buy services and contracts, there are usually a range of different models or providers to choose from. By shopping around you can find the closest match for your needs at the best price. In this chapter you will be presented with a number of different situations where you have to find the 'best buy'. If you have access to the internet, newspapers or magazines you could research these scenarios to find the most up-to-date offers.

Example 12·1

Two shops are selling the same football strip. Which shop offers better value?
- Sportsrec is offering the kit at £34·80 less 15% in its sale.
- Gymtec has advertised the strip at £29·99.

To work out the better offer, first calculate the sale price at Sportsrec:

$$\frac{15}{100} \times £34·80 = £5·22$$

Sale price is £34·80 − £5·22 = £29·58

Sportrec's sale price of £29·58 is cheaper than Gymtec's price of £29·99, so Sportsrec offers the better value.

Money

Example 12.2 Which jar of jam offers the better value?

The larger jar is four times bigger than the smaller jar.

So four small jars would cost 4 × £0·89 = £3·56.

One large jar costs £3·59.

So the smaller jar is better value.

Exercise 12A

1. Leisureways sell Kayenno trainers for £69·99 but give 10% discount for Running Club members. All Sports sell the same trainers for £75 but are having a sale for which everything is reduced by 15%. In which shop are the trainers cheaper for Running Club members?

2. A supermarket sells crisps in different sized packets. An ordinary bag contains 30 g and costs 28p. A large bag contains 100 g and costs 90p. A jumbo bag contains 250 g and costs £2·30. Which bag is the best value? You must show all your working.

3. A cash-and-carry sells crisps in boxes. A 12-packet box costs £3·00. An 18-packet box costs £5·00. A 30-packet box costs £8·00. Which box gives the best value?

4. You are trying to choose between two prepaid music download cards. The first offers 16 songs for £24 and the second offers 12 songs for £20. Which is the better buy and why?

5. You are trying to decide which electricity supply firm to go with. You have worked out that you use about 500 units of electricity every month. The two companies you are considering add a **standing charge** every month, which is a fixed sum of money added to your bill to pay for maintenance.

East Elect	North Energy
5p per unit plus a standing charge of £20 per month	2p per unit plus a standing charge of £40 per month

   a. How much would each company charge you for 1 month of electricity use? Calculate the cost, then say which you should choose.

   b. You find out that you get a 10% discount if you pay by **direct debit** (a fixed amount paid automatically from your bank account every month). Recalculate your monthly charge with this new discount.

   c. How much will you spend over a year?

Number, money and measure

6 Frances is shopping online for a new bookcase. She has found three suppliers but needs some help choosing the best option. Which company should she pick and why?

Before doing any calculations, find out the current rate of **value added tax** or VAT (which is an extra cost added to your bill).

Bookworm	ShelveCo	Leafy Wood
£250 plus VAT plus £18·75 delivery	£330 all inclusive	£275 inc. VAT plus 5% delivery and handling

7 Jenny needs to hire a marquee for a birthday party. She has two quotes to choose from:

- Tent Hire Company: £20 set-up fee plus £30/day
- Partyhut: £60 set-up fee plus £22/day

a How much will each firm charge for:

  i 2 days   ii 3 days   iii a week

  Remember to show your calculations.

b Jenny has decided to hire the marquee for 4 days to give her plenty of time to get set up. Which firm should she use?

8 Three different mobile phone companies offer these contracts:

Company 1

£15/month, 100 minutes of free calls, unlimited free texts and 400 MB web access
---

Company 2

£5/month, 20 minutes of free calls, 50 free texts and 100 MB of web access
---

Company 3

£30/month, 500 minutes of free calls, 50 free texts and 200 MB of web access
---

All calls cost 10p/minute over your allocation, texts are 10p each and each company charges £1 for every extra 100 MB over your web access allowance.

Using the three contracts described above, choose which contract would suit you best. Decide how many calls and texts you make and how much web time you use every month. Make sure you show all your calculations.

Money

9 Suppliers of broadband internet access charge according to the following features: speed, downloads, contract length and monthly cost. Here are the details for three companies:

Company	Speed (MB)	Downloads (GB)	Contract length (months)	Monthly cost (£)
Blue	up to 15	2	18	8
Speedynet	up to 24	20	12	12
Zoomweb	up to 10	unlimited	6	5

   a Which package would you choose?

   b Which company would you advise each of the following people to choose? Why?

   　i　Jack likes to download movies and watch them with his friends.

   　ii　Jill is a student who enjoys online gaming but finds money a bit tight.

   　iii　Mrs Henderson is a grandmother who likes to email and can't be bothered changing contracts very often.

   　iv　The Watson family have three computers and two teenage children who love social networking.

   c What are the advantages and disadvantages of each company?

10 You want to get fit so you decide to join a gym. These are the terms for two gyms:

   - Jim's Gym: £40 joining fee and £8·50/month membership
   - Strong-Arm Gym: £2·50 per visit plus £3/month membership

   a Over the next 6 months you visit the gym 30 times. Which gym would be cheaper?

   b If you had visited the gym 60 times, which would be cheaper?

   c You prefer the Strong-Arm Gym. If you go twice per week to the gym, how many visits, over a year, will it be before it costs you £100? (Assume 4 weeks per month.)

11 Mr and Mrs Rubble have just returned from their honeymoon and want to get hard copies of their 360 digital photographs. Mrs Rubble has found three possible solutions.

Local Superstore	Online Foties	DIY at home
prints are 18p each	£5 for every 50 photos plus 20% VAT plus £15 delivery	Ink £17·50 Paper £22·99

   a Which should she choose?

   b What are the advantages and disadvantages of each?

> **Challenge** Research broadband charges online. Can you find a package that suits your needs? Now ask a friend how they use the internet – can you find a deal that meets their needs?

Number, money and measure
Borrowing money

Sometimes you need to borrow money to buy items that you need or want. It could be a loan to buy a car, a mortgage to buy a house or it could be a credit or store card to do your weekly shopping. Borrowing money or using credit will usually cost extra, because the lender will charge **interest**, but you can still shop around to find the best deal.

Example 12·3

You want to borrow £3000 to help buy a new car. You have been offered two different loans to choose from.

- Bank loan offer 1: repay £110 per month over 3 years
- Bank loan offer 2: repay £70 per month over 5 years

Which is the better offer?

Offer 2 looks cheaper because the monthly repayment is lower. But over 5 years you would pay £70 × 60 = £4200. Notice that it would cost you £1200 to borrow the original £3000.

Offer 1 would cost £110 × 36 = £3960 over 3 years. So if you pay back the loan faster it will cost you less, as taking this loan would cost you £960.

So offer 1 is the better deal in the long run but you would need to think carefully – can you afford £110 per month or will you have to accept paying more in the long run?

Exercise 12B

1. Here are two quotes for a loan of £2500:
   - TSBank: 1 year loan at 6%
   - CO-OP Union: repayments £110 per month for 24 months

   Which is the better deal?

2. Tim is offered two loans to buy a new kitchen. Which is the better deal?

Bank	Credit Union
£5500 loan	£5000 loan
Repay £200/month for 36 months	Repay £190/month for 36 months

3. You are saving up for a new bicycle that costs £150. You can either save £20 per week or borrow the money from a friend and repay £14 per week for 12 weeks. What would you do and why?

 4. Mr and Mrs Morrison are buying a house with a mortgage. They need to borrow £156 000. They are offered the following deals. Write a letter to the Morrisons advising which offer they should take, describing the advantages and disadvantages of each offer.

- Offer 1: mortgage repayment £750 per month for 20 years; you are not allowed to pay off the mortgage early; a penalty fee of £3000 will be charged if you move home.
- Offer 2: mortgage repayment £900 per month for 18 years; no penalties if you repay early or move home.
- Offer 3: mortgage repayment £575 per month for 26 years plus £2000 set-up fee; no penalties if you repay early or move home.

Money

5 You want to buy a new video game that costs £60 but you don't have any of the money yet. Which of these options should you take? What are the advantages and disadvantages of each option?

    a Save £5 a week out of your pocket money.

    b Get the money from your mother next month on the promise that you will wash the car, for free, every weekend for the next 10 weeks (you normally get £8 for washing the car).

    c Borrow the money from a friend straightaway and pay back £2 a week for a year.

    d Go without.

6 A car that costs £5995 can be bought on credit by paying a 25% deposit and then paying 24 monthly payments of £199.

    a How much will the car cost on credit?

    b How much extra do you pay using credit compared with paying the cash price?

7 An insurance policy for a car costs £335. It can be bought by paying a 20% deposit and then making six payments of £55·25. How much does the policy cost using this scheme?

Challenge   You can find interest calculators online. Try searching for one, choose the amount of money you want to borrow (the principal), then change the interest rate and see how much it will cost you to borrow the money.

- By working on this topic I can compare different offers and can explain my choices.

- By making calculations I have learnt how to choose between different contracts and services so I can pick the best option. ★ Exercise 12A Q6

- I have learnt how to calculate the cost of borrowing money so I can pick the best option. ★ Exercise 12B Q4

Number, money and measure

13

I can budget effectively, making use of technology and other methods, to manage money and plan for future expenses.

MNU 3-09b

This chapter will show you how to:
- plan your personal spending and budget, using technology where appropriate, for the basic activities of everyday life
- calculate how much money you might need for extra expenses
- allocate money to a budget so that you have enough money for all your needs, or so that you can make sensible decisions about different spending options
- convert pounds sterling into another currency and vice-versa using an exchange rate.

You should already know:
- the meaning of these terms: account, statement, balance, overdraft, interest, credit, debit, ATM, PIN, budget
- how to use a bank account
- how to read and understand bank statements
- which type of bank account is suitable for different purposes
- how to draw up a budget to help you keep track of your income, spending and saving
- the advantages and disadvantages of saving and borrowing money.

Budgeting

How much money do you spend and save every week? What about your family? Keeping a **budget** will help you keep track of your money and save for the future. If you want to buy yourself the latest gadget, plan a family holiday or run a successful business you need to know where your money is going and where you can save it. By keeping a budget you can see if you are getting the best deals available or if you need to make changes to the way you use money.

Many families and businesses use a spreadsheet to keep track of their weekly, monthly or annual **income** and **expenditure**. Can you use a spreadsheet to track your spending?

The internet can be a useful source of information and guidance about keeping a budget. Or you can talk to staff at a bank, building society, credit union or to an independent financial advisor.

Money

Example 13.1   This is the Jones family's weekly budget:

Income:

- wages £500
- child benefit £35

Expenditure:

- rent £100
- food £70
- council tax and water £40
- travel £40
- gas and electricity £30
- babysitter £50
- TV £10
- leisure £80
- insurance £30
- phones £40
- clothes £40

So their total income is £535 and their outgoings are £530. That leaves them savings of £5 per week, which isn't much if they have an unexpected expense. What would they do if their washing machine broke down or they need to buy new school clothes?

Could they save money by switching to pay-as-you-go phones? Do they really need to spend £80 a week on leisure activities or could they find free activities to do instead? What do you think?

If you find that you always spend all your money, there are some things that you should consider doing to try to save some.

- Can you cut back on luxury items or switch services to get a better deal?
- Often paying by direct debit will be cheaper than using a meter or paying by cheque.
- How much interest are you paying on loans or credit cards? Can you reduce it? The annual percentage rate (APR) will indicate whether you are being charged a lot or a competitive rate.

Some things to watch out for and avoid:

> ! borrowing more than you can pay back
> ! using unofficial money lenders (who often charge a very high interest rate)
> ! taking cash out on credit cards
> ! only repaying the minimum amount on credit cards or taking out high interest loans
> ! using loans to pay for other debts
> ! paying on the 'never never', such as hire purchase — this can result in you paying a lot more than the original cost of the item you bought.

Using these services could cause you to slip into unmanageable debt. If you are, you should take advice!

Number, money and measure

Exercise 13A

You may find a spreadsheet programme helpful for some of these questions.

1. Look again at the Jones family's weekly budget in Example 13·1. Choose three areas of expenditure that the family could cut back on. Write a letter to the Jones family explaining what they should cut back on and tell them how much money they could save in one week, in one month and in a year.

2. Matthew is saving up to buy a £200 TV for his bedroom. He decides to track his money over a week to see if he can reduce his spending and save up more quickly.

   a. Using the information below, calculate his income and expenditure.

Day	Income	Expenditure
Monday	pocket money £30	bus fare £2·50, lunch £1·50
Tuesday		bus fare £2·50, lunch £2·00
Wednesday		juice £0·80, lunch £1·80
Thursday		bus fare £2·50, lunch £2·00, football training £4·00, snacks £2·75
Friday	Gran £10	bus fare £2·50, lunch £1·90, snacks £2·00
Saturday		magazine £2·70, clothes £14·55
Sunday	wash the car £10	

   b. How long will it take him to save up £200?

   c. Where can Matthew save money? Make your suggested changes to his weekly expenditure and recalculate how long it will take Matthew to save up £200.

   d. How can Matthew earn more money? Make some suggestions but be realistic about the time he can spend and how much he could earn.

   e. Think about something that you would like to save up for, for example, a TV, a games console, a present for your parents. Decide how much it will cost. Now track your spending over a week and see how long it will take you to save up.

3. Matthew keeps a record of his spending over 4 weeks.

Week	Income	Expenditure
Week 1	£45	£27·50
Week 2	£23·50	£19·60
Week 3	£35·25	£31·30
Week 4	£42·80	£25·75

   a. What is his total income and expenditure for these 4 weeks?

   b. How much money has he saved?

   c. How long will it take him to save up £200?

Money

4. This is September's bank statement for Mrs McDonald. Can you describe the different payment types? What are the advantages and disadvantages of each?

Date	Payment type	Details	Paid out (£)	Paid in (£)	Balance (£)
					200·00
05-Sep	BACS	Glasgow City Council		1200·00	1400·00
09-Sep	Direct debit	Glasgow CC Council tax	156·00		1244·00
10-Sep	Fixed charge	Account fee	10·00		1234·00
12-Sep	CHQ	Fuel bills	80·00		1154·00
15-Sep	Cashpoint	Argyle St	50·00		1104·00
20-Sep	Debit card	Superstore	150·00		954·00
26-Sep	Standing order	Insuresafe	30·00		924·00
27-Sep	Cashback	Superstore	50·00		874·00

You can search online to find out more about different accounts and payment types from banks, building societies and credit unions.

5. Look again at Mrs McDonald's bank statement in Question 4. Can you think of any other expenses she might have to pay for during the month? Copy the statement headings and the final balance of £874·00, add these extra expenses to the 'paid out' column and calculate her new balance.

6. This is October's bank statement for Mr Laing.

Date	Payment type	Details	Paid out (£)	Paid in (£)	Balance (£)
					200·00
01-Oct	Direct debit	Glasgow CC Council tax	156·00		44·00
01-Oct	Direct debit	Savings acc	50·00		−6·00
04-Oct	Bank Giro Credit	Glasgow City Council		1597·00	1591·00
01-Oct	Direct debit	Mortgage	700·00		891·00
07-Oct	CHQ	Gas	80·00		811·00
07-Oct	CHQ	Electricity	50·00		761·00
08-Oct	Fixed charge	Overdraft	30·00		731·00
10-Oct	Cashpoint	Argyle St	50·00		681·00
15-Oct	Fixed charge	Account fee	25·00		656·00
16-Oct	Debit card	Superstore	150·00		506·00
19-Oct	Cashpoint	Buchanan St	100·00		406·00
20-Oct		Interest (net)		0·05	406·05
21-Oct	Direct debit	Credit card	300·00		106·05
25-Oct	Standing order	Insuresafe	30·00		76·05
29-Oct	CHQ	Deposit		100·00	176·05

a. Did Mr Laing save or lose money in October?

b. There are three ways that he could save money, just by changing how or when he pays certain expenses. Try to find and then describe all three.

Number, money and measure

7. James is planning a night out at the cinema. His mother has given him £30 to spend, some of which he might spend in preparation for the night out. Look at the following list of items. What would you advise James to buy for his night out? Calculate your total spend and be ready to justify your choices.

- cinema pick and mix £6·95
- cinema popcorn £9
- new jeans £22
- can of fizzy pop £1·35
- haircut £10
- cinema drink £3·40
- supermarket popcorn £1·75
- supermarket sweets £3·45
- cinema ticket £12
- new top £17·95
- 3D glasses £3·50
- tissues £0·80
- breath mints £1·60
- bus fare £2
- taxi fare £9·50

8. The Smith family have had a family conference to create a wish list of things they want to buy. Mr Smith has also made a list of their debts.

Family debts
- credit card 1 £1000
- credit card 2 £800
- electricity meter £100/month
- student loans £50/month
- fixed rate mortgage £300/month
- Mr Haste (doorstep lender) £200
- overdue gas bill £200

Wish list
- family holiday to Florida for a week £2000
- fridge freezer £489
- dishwasher £349
- 40" TV £720
- two-seater sofa £550
- games console £150

The Smiths have £1000 a month left over after buying essentials. Take the role of a financial advisor and write a letter to the Smith family outlining how they could reduce their debts. Include calculations to show how long it will take to pay off their debts, and show how long it will be before they can afford the items on their wish list.

9. Your school is planning a ski trip to Austria for 30 pupils and staff. The trip will include lunch and dinner for 7 days, bus hire for 7 days and ski hire for 5 days. For each expense you will be presented with a choice of three possible service providers.

   a. Plan the trip so that the cost is less than £500 per person. Create a table or spreadsheet to keep track of your budget.

   Things to think about:
   - How you will advertise the trip and what are the highlights?
   - You will need a contingency fund. What is a contingency fund? How much will you need in it?
   - Do you think you can find better deals than those listed? Try researching online.

Money

Make a presentation to your fellow pupils explaining your decisions and how you decided between the choices. Include a copy of your table/spreadsheet and any calculations you made.

Expense	Choice 1	Choice 2	Choice 3
Flights	Scot Air £2100 for whole group	Simple Land £80pp rtn	No Frills £30pp each way
Lunch	Deluxe food £6 pp	Packed lunch £1.50pp	Buffet lunch £3pp
Evening meal	Fast burger £2pp	Sit down meal £5pp	Buffet dinner £3pp
Ski hire	Wooden Planks £2pp/day	Supa Ski £30pp/week	Second-hand Ski £16pp/week
Accommodation	2 star hotel £300pp	3 star hotel £350pp	4 star hotel £400pp
Insurance	Basic £15pp	Winter sports £20pp	Cover all £25pp
Bus hire	Air-con Special £100/day	Creaky Buses £500/week	Public transport £10pp week pass

b The ski trip is advertised in August and leaves the following April. How much do you have to save each week? Don't forget to factor in other things you may want to save for, such as Christmas or birthdays.

Challenge

1 **School trip**

Try planning a school trip to a local attraction, for example, a museum, science centre, theatre visit, or sports match. What are the costs that need to be planned for? How can the school ensure it gets value for money and make the trip affordable?

Make a list of all the possible expenses.

Research the costs and calculate the cost per person.

2 **The biggest budget in Great Britain**

Research the UK Government's budget. When is it announced and what does it affect? Write a short report to summarise your findings.

Exchange rates

Most countries have their own currency. In the UK we use pounds sterling which has the symbol £. The United States of America uses the US dollar ($) and many countries in Europe use the euro (€). When you are travelling overseas it is usually a good idea to change your pounds for the currency of the country you plan to visit and this is another situation where you can shop around to get a good deal.

There are many places where you can exchange pounds for foreign currency, such as banks, the Post Office and online agencies. Each will offer their own **exchange rate** and may charge you an amount for exchanging your money; this may be a flat fee or it may be a commission (perhaps 2% or 3% of the value of the amount you are changing). When you return from overseas you can change any remaining foreign money back into pounds or keep it (perhaps for later use if you plan to travel back there again, or as a memento of your trip).

Number, money and measure

Example 13.2

Diana is visiting Florida with her family. She has £200 spending money that she needs to change to US dollars. She visits the bank where the exchange rate is $1·56 to £1 plus a 2% commission. How many dollars will she get?

The commission is 2%, and this is subtracted from the initial amount.

$\frac{2}{100} \times £200 = £4$

£200 − £4 = £196

To change pounds to dollars, multiply by the exchange rate:

£196 × 1·56 = $305·76.

Diana will have $305·76 to spend.

Check: You can check your answer by estimating. For an exchange rate of £1 to $1·56, you should have more $ than you had £; when converting from $ to £, you should have fewer £ than $.

Example 13.3

Ms Tyrell has returned from France with €340. She goes to her local bank to change them into British pounds. Her bank offers her €1 to £0.83 plus a fee of £2.50. How much money will she get?

€340 × 0·83 = £282·20

The bank charges a fee of £2.50, so £282·20 − £2·50 = £279·70.

Ms Tyrell will get £279·70.

Check: Again you can check your answer by estimating. For an exchange rate of €1 to £0·83, you should have fewer £ than you had €.

Exercise 13B

You can research the latest exchange rates online and use these in the following questions.

1. The exchange rate for euros at the local bank is £1 to €1·13. Use this rate to convert the following amounts to euros.

   a £50   b £130   c £312

2. Convert these amounts to British pounds, using these exchange rates:

   €1 = £0·83 $1 = £0·64   ¥1 = £0·0082

   a €80   b €350   c $260
   d $175   e ¥76 500   f ¥123 460

3. Mr Brown decides to take some euros (€) for his holiday.

   The exchange rate at the bank is £1 = €1·13.

   a Mr Brown changes £450 at the bank. How many euros will he receive? Give your answer to the nearest five euros.

   b Mr Brown also has $120 from a previous holiday to change into euros at the bank. The exchange rate at the bank is $1 = €0·75. How many euros will he receive?

   c Mr Brown returns from the holiday with €50. How much is this in pounds?

Money

4 Mr Deckard is travelling to the USA and has £450 to exchange for US dollars. Look at the rates offered by these foreign exchange bureaus and work out which is the best value. Remember to show your calculations.

MoneyX	Change4U	CashCash
£1 : $1·56	£1 : $1·58 2% commission	£1 : $1·57 Flat fee of £3·50

5 Harry went to Spain for a holiday and took €300 spending money. It turned out that the holiday was all-inclusive so he didn't manage to spend anything! Look at the exchange rates offered by these companies and pick the one which offers the best value, explaining why you chose that company.

MoneyX	Change4U	CashCash
€1 : £0·79	€1 : £0·80 3% commission	€1 : £0·81 Flat fee of £5

6 Gillian is planning a trip to Japan to buy some new electronic items. She has budgeted for the following expenses.

Expense	Amount
Flights	£789
Hotel for 7 nights	¥68 000
Food	¥33 500
Spending money	£1600

Use the exchange rates £1 : ¥121·89 and ¥1 = £0·0082 to calculate the estimated total cost of her trip to Japan:

a in pounds (£)   b in yen (¥)

7 Jenny and her friend Dave are planning a trip to the USA to buy cheap jeans. Their flights and accommodation will cost £255 each. A pair of Gevi jeans costs £45 in their local shop. They think the same brand will cost $24·80 in the USA.

a If they buy 10 pairs each, how much will their trip cost in total? Use the exchange rate £1 : $1·55.

b How much money will they save in dollars buying their jeans in the USA?

8 Look at Exercise 13A Question 9. Use the exchange rate of £1 : €1·14 to convert each amount in pounds to euros.

Number, money and measure

9 Mrs Jones runs an import business. She buys flowers in Holland and sells them to florists in Dundee. The exchange rate at the time is £1 : €1·14.

Sales in June

Flower	Import cost per bunch (€)	Selling price per bunch (£)	Number of bunches sold
Roses	6·00	15·00	13
Carnations	3·50	5·00	12
Lilies	5·00	7·50	10
Germinis	2·75	3·00	8
Gypsophila	1·50	2·00	5

a Calculate the total gross profit for Mrs Jones's sales in June.

b Mrs Jones has to pay a 5% import tax. What is her net profit?

> **Challenge**
>
> Plan a holiday abroad, using the internet or brochures from a travel agent. How much will your flights be? How much might you need for accommodation, food and day trips? How much foreign money will you need to take? Research exchange rates and try to find the best deal.

- By working on this topic I have learnt how to keep track of my spending and make the best use of my money.
- I can plan personal spending by budgeting and planning for future expenses. ★ Exercise 13A Q2
- I know how to use exchange rates to convert between different currencies. ★ Exercise 13B Q4

Time

14

Using simple time periods, I can work out how long a journey will take, the speed travelled at or distance covered, using my knowledge of the link between time, speed and distance.

MNU 3-10a

This chapter will show you how to:
- use the relationships between the distance of a journey, the time taken and the speed, to calculate distances, times and average speeds
- use the correct units in speed–time–distance calculations
- use distance–time graphs
- calculate time durations across hours and days.

You should already know:
- the relationships between units of time and how to convert between them
- how to read and use graphs.

Calculating speed

Speed is a measure of how quickly something moves. It is usually measured in metres per second (m/s), kilometres per hour (km/h) or miles per hour (mph). But it can also be measured in any unit of distance per unit of time.

Objects rarely move at a constant speed for their whole journey so the speed we calculate is the **average speed**.

Example 14·1 Katherine swam 10 metres in 5 seconds. This means in 1 second she swam 10 ÷ 5 = 2 metres. Therefore Katherine's average speed was 2 metres per second or 2 m/s.

Example 14·2 A ship sails 60 miles in 3 hours. Calculate its speed in miles per hour.

In 1 hour it sails 60 ÷ 3 = 20 miles. So the ship's average speed is 20 miles per hour. We write this as 20 mph.

Exercise 14A Include the correct unit of speed for each answer (mph, km/h or m/s).

1 Work out the average speed of each of the following.

    a 15 miles in 3 hours   b 40 km in 5 hours

    c 120 miles in 4 hours   d 300 metres in 10 seconds

Number, money and measure

2 A car travels 180 miles in 3 hours. What is the average speed of the car in miles per hour?

3 Dave cycled for 2 hours and covered a distance of 36 km. Calculate his average speed.

4 An aeroplane flies 2400 miles in 5 hours. What was the plane's average speed?

Speed formula

In these calculations we have divided the distance travelled by the time taken. The formula for calculating speed is:

speed = distance ÷ time

$$S = \frac{D}{T}$$

Example 14·3

Calculate the average speed of a car that travelled 135 miles in 3 hours.

$$S = \frac{D}{T}$$

$$S = \frac{135 \text{ miles}}{3 \text{ hours}}$$

$$S = 45 \text{ mph}$$

Exercise 14B

Set your working out like Example 14·3. Remember – answers without units are meaningless.

1 Calculate the average speed of the following journeys.

a 140 km in 4 hours
b 200 metres in 8 seconds
c 72 miles in 3 hours
d 560 miles in 7 hours
e 24 km in 6 hours
f 120 miles in 5 days

2 Colin and Nicola have to catch the ferry at 8:30 am. At 6:30 am the car's sat nav indicates they still have 130 miles to go. What average speed will they have to drive at to catch the ferry?

3 Duncan cycled 300 metres in 30 seconds and Calum sprinted 40 metres in 5 seconds. Who had the faster average speed?

4 A lorry must not exceed 60 mph. If it covers 350 miles in 6 hours how does the lorry's actual average speed compare to its speed limit?

5 The glacier Mer de Glace on Mont Blanc in France was measured to have moved 170 cm in one week. What was its speed in cm per day?

6 A tortoise took a whole day to walk a distance of 600 metres. Find its average speed in metres per hour.

7 Light travels from the Sun to the Earth in approximately 8·5 minutes. If the Earth is about 93 million miles from the Sun find an approximation for the speed of light in miles per minute.

Time

Example 14·4 An athlete exercising on a rowing machine rows 4 km in 20 minutes.

a At this speed how far would she row in 1 hour?

b What is her average speed in km/h?

a One hour is 60 minutes, which is 3 × 20 minutes. If she could maintain this speed the athlete would row 3 times as far in 1 hour as she would in 20 minutes.

3 × 4 km = 12 km

The athlete would row 12 km in 1 hour.

b She would row 12 km in 1 hour so her average speed is 12 km/h.

Exercise 14C

1 Euan ran 5 km in half an hour.

  a If he kept running at the same speed how far would he run in 1 hour?

  b Write down his average speed.

 2 Jessica walked 1 mile in 15 minutes.

  a How far would she walk in 1 hour at this speed?

  b Write down her average speed.

 3 In 10 minutes Audrey ran 2 km on the treadmill.

  a At this speed how far would she run in 1 hour?

  b What is this speed in km/h?

Calculating distance

Example 14·5 How far will a car travel if it is driven for 3 hours at an average speed of 50 mph?

50 mph means 50 miles travelled each hour. A table can illustrate the distance travelled for different times.

Speed	Time	Distance
50 mph	1 hour	50 miles
50 mph	2 hours	100 miles
50 mph	3 hours	150 miles
50 mph	4 hours	200 miles

It can be seen that distance = speed × time:

$$D = S \times T$$

To answer the question set out your working in the following way.

$D = S \times T$

$= 50 \text{ mph} \times 3 \text{ hours}$

$= 150 \text{ miles}$

In 3 hours, the car will travel 150 miles.

Number, money and measure

A note on units: The units of speed and time must be consistent. When using speed in mph or km/h make sure you use time in hours. For m/s use time in seconds.

Exercise 14D

1. Use the formula $D = S \times T$ to calculate the distance travelled in the following journeys. Make sure your answer has the correct units.

   a speed = 20 mph, time = 4 hours
   b speed = 35 km/h, time = 5 hours
   c speed = 8 m/s, time = 6 seconds
   d speed = $4\frac{1}{2}$ mph, time = 2 hours

2. Tony averages a speed of 16 km/h. How far does he travel in 3 hours?

3. An aeroplane flies for 5 hours at an average speed of 450 mph. What distance does it fly?

4. Copy and complete the following table showing distances travelled at a constant speed of 20 mph. You should be able to write in the distances without doing any working.

Speed	Time	Distance
20 mph	1 hour	20 miles
20 mph	2 hours	
20 mph	10 hours	
20 mph	30 minutes	
20 mph	15 minutes	

Time Units

In Question 4 above, the distance travelled in 30 minutes is the distance travelled in half an hour, half the distance travelled in 1 hour.

When using the formula for this type of question we change the time unit so that it is consistent with the speed unit.

Example 14·6

How far does a car travel in 15 minutes if its average speed is 40 mph?

The speed is given in **miles per hour** so we must use time in hours.

Time = 15 minutes
     = $\frac{1}{4}$ of an hour

At 40 mph the car travels 40 miles in 1 hour, so in $\frac{1}{4}$ of an hour it will travel a quarter as far.

Distance = $\frac{1}{4}$ of 40 miles
         = 10 miles

Or using $D = S \times T$
         = 40 mph × $\frac{1}{4}$ h
         = 10 miles

When using a calculator remember to use decimal fractions for common times, that is:

15 minutes = 0·25 hours
30 minutes = 0·5 hours
45 minutes = 0·75 hours
1 hour 30 minutes = 1·5 hours

Time

Exercise 14E The following questions show pupils' working leading to the **wrong** answers. Write out the correct working and answers for each problem.

1 Speed = 40 mph
Time = 30 minutes
D = S × T
= 40 × 30
= 1200 miles

★ 2 Speed = 60 km/h
Time = 15 minutes
D = S × T
= 60 × 15
= 900 km

3 Speed = 20 mph
Time = 45 minutes
D = S × T
= 20 × 0·45
= 9 miles

4 Speed = 10 km/h
Time = 1 h 30 min
D = S × T
= 10 × 1·30
= 13 km

Calculating time

Example 14·7 Ernie drives 120 miles at an average speed of 40 mph. How long does the journey take?

His speed is 40 mph which means he covers 40 miles every hour. The table shows distances travelled at 40 mph for various times.

Speed	Time	Distance
40 mph	1 hour	40 miles
40 mph	2 hours	80 miles
40 mph	3 hours	120 miles
40 mph	4 hours	160 miles

The table shows that it takes 3 hours to travel 120 miles at 40 mph. From the table you can also see that the time taken is the distance divided by the speed.

Example 14·7 shows that time = distance ÷ speed:

$$T = \frac{D}{S}$$

When using this formula you must make sure the units for distance are consistent with the units for speed. For example, do not mix up mph with km.

Example 14·8 How long would it take to drive 240 km at a speed of 60 km/h?

$$T = \frac{D}{S}$$

$$T = \frac{240 \text{ km}}{60 \text{ km/h}}$$

= 4 hours

Number, money and measure

Exercise 14F

1. Calculate the time taken to walk 18 miles at a speed of 3 mph.

2. How long would it take to travel 300 miles at a speed of 50 mph?

3. A cyclist averages 15 km/h. How long will she take to cover a distance of 45 km?

4. Jo runs a 400 metre race at an average speed of 8 m/s. What was her time?

5. Calculate the time taken for each of the following journeys.

Distance	a 150 miles	b 60 km	c 52 miles	d 800 metres	e 720 miles	f 36 km
Speed	30 mph	10 km/h	4 mph	5 m/s	80 mph	4 km/h

Calculating speed, distance and time: Mixed questions

The three formulae for calculating speed, distance and time can be remembered using this triangle:

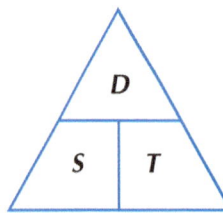

To find the missing variable (speed, distance or time), place your finger on top of the letter (S, D or T) to see the relationship between the other two variables.

$$S = \frac{D}{T} \qquad D = S \times T \qquad T = \frac{D}{S}$$

Exercise 14G

1. Choose the correct formula and work out the missing entries in the following table.

	Distance	Speed	Time
a	40 miles		2 hours
b		30 km/h	5 hours
c	200 metres		40 seconds
d		10 mph	6 hours
e	6 km	3 km/h	
f	300 metres	20 m/s	
g		70 mph	3 hours
h	180 miles		4 hours
i	1600 km	40 km/h	
j	50 miles		$\frac{1}{2}$ an hour

2. Michael walked 12 miles in 3 hours. What was his average speed?

3. How far can a train travel in 5 hours at an average speed of 80 mph?

4. A motorcyclist averages 45 km/h. How long will it take to cover a distance of 135 km?

Distance–time graphs

Example 14·9 The table shows the distance travelled by a car at a constant speed of 10 km/h.

Time (h)	0	1	2	3	4
Distance travelled (km)	0	10	20	30	40

This information can be displayed on a distance–time graph.

Use the graph to work out how far the car travelled in:

a 30 minutes

b $2\frac{1}{2}$ hours

a 30 minutes is $\frac{1}{2}$ an hour; read up from 0·5 h on the horizontal axis and read across to the vertical axis.
Distance travelled is 5 km.

b Distance is 25 km

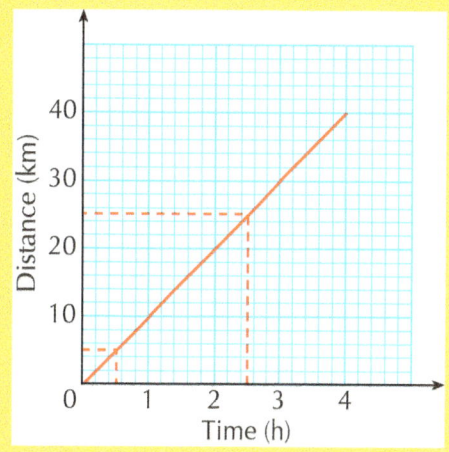

Example 14·10 This time the car stops for an hour.

Time (h)	0	1	2	3	4
Distance travelled (km)	0	10	20	20	40

a Use the graph to work out the average speed of the car:

　i during the first hour

　ii during the fourth hour.

b Write sentences describing the car's journey.

a i The graph shows that the car travelled 10 km during the first hour so its average speed was 10 km/h.

　ii The graph shows that the car travelled 20 km during the fourth hour so its average speed was 20 km/h.

b The car drove at a steady speed of 10 km/h for the first 2 hours. Then it stopped for 1 hour before driving at 20 km/h for 1 hour.

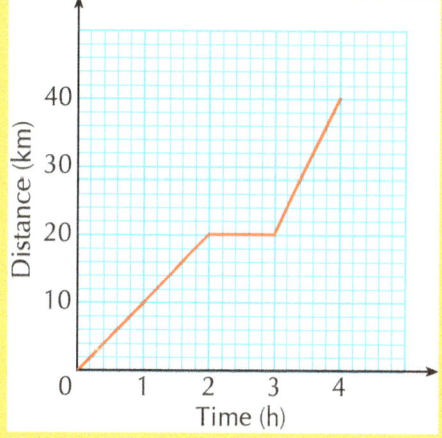

Look at the graph in Example 14·10. Notice that the graph is steepest when the speed is greatest.

Number, money and measure

Exercise 14H

1 Draw a distance–time graph for each of the following journeys.

a

Time (h)	0	1	2	3	4
Distance travelled (km)	0	10	10	20	30

b

Time (h)	0	1	2	3	4
Distance travelled (km)	0	20	30	40	40

c

Time (h)	0	1	2	3	4
Distance travelled (km)	0	10	15	20	40

d

Time (h)	0	1	2	3	4
Distance travelled (km)	0	30	30	40	40

★ 2 Write a description of the journey for each graph in Question 1. Write your descriptions in complete sentences, include the varying speeds of the car in each one and if the car stopped at any time, for how long.

Example 14.11   Calculate the duration between:

a 8:00 am and 11:30 am
b 11:30 am and 7:00 pm
c 9:00 pm and 10:45 am the next day
d 3:30 pm on January 1 and 4:00 pm on January 2

a, b Use a number line to visualise the duration.

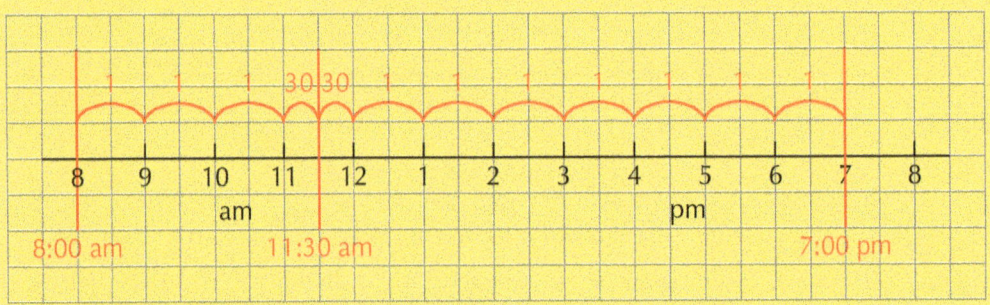

From 8:00 am to 11:30 am is 3 hours 30 minutes

From 11:30 am to 7:00 pm is 7 hours 30 minutes

c From 9:00 pm to midnight is 3 hours, add 10 hours 45 minutes = 13 hours 45 minutes

d From 3:30 pm on January 1 to 3:30 pm on January 2 is 24 hours, add 30 minutes = 24 hours 30 minutes

Time

Exercise 14I

1. Calculate the duration between:
   a. 9:00 am and 11:45 am
   b. 4:30 am and 8:00 am
   c. 1:30 am and 10:30 am
   d. 6:45 am and 11:30 am

2. Calculate the duration between:
   a. 1:00 am and 1:00 pm
   b. 2:30 am and 4:00 pm
   c. 10:00 am and 5:45 pm
   d. 9:30 am and 3:45 pm

3. Calculate the duration between:
   a. 1:00 pm and 3:30 am the next day
   b. 3:00 pm and 11:30 am the next day
   c. 5:30 pm and 10:15 am the next day
   d. 3:45 pm and 2:15 am the next day

4. Calculate the duration between:
   a. 10:00 pm on March 13 and 6:30 am on March 14
   b. 7:45 am on April 24 and 7:45 am on April 26
   c. 6:15 am on June 3 and 11:30 pm on June 4
   d. 2:45 am on December 23 and 6:30 am on December 25.

5. Rowan and Julia ran the West Highland Way race from Milngavie to Fort William. They both started at 1:00 am on Saturday morning. Rowan finished at 4:15 pm on Saturday afternoon. Julia finished at 5:30 am on Sunday morning.

   a. How long did it take for each of them to finish the race?
   b. The West Highland Way is 95 miles long. Calculate their average speeds in miles per hour. Give your answer to 1 d.p.

- By working on this topic I can explain how to use the relationships between speed, distance and time to solve real-life problems.
- I understand that speed is a measure of distance travelled per unit of time. ★ Exercise14A Q2
- I know how to find the average speed of an object given the distance travelled and the time taken. ★ Exercise 14B Q1
- I understand how to find the distance travelled in a whole hour when given the distance travelled in a simple fraction of an hour. ★ Exercise14C Q2
- I know how to find the distance travelled by an object given its average speed and time taken. ★ Exercise 14D Q2
- I understand the importance of using consistent units of time and speed when calculating distance. ★ Exercise14E Q2
- I know how to calculate the time an object will take for a journey given its average speed and the distance travelled. ★ Exercise 14F Q4
- I can solve problems using my knowledge of the relationships between speed, distance and time. ★ Exercise 14G Q1
- I can read and interpret distance–time graphs. ★ Exercise14H Q2
- I can calculate time durations across hours and days. ★ Exercise 14I Q4

Number, money and measure

15

I can solve practical problems by applying my knowledge of measure, choosing the appropriate units and degree of accuracy for the task and using a formula to calculate area or volume when required.

MNU 3-11a

This chapter will show you how to:
- use formulae to calculate the perimeter and area of a range of 2D shapes including triangles and quadrilaterals
- use a formula to calculate the volume of a cuboid
- convert between different units for area and volume.

You should already know:
- how to use standard metric units to measure length, area and volume
- how to find the area of a rectangle by counting squares or by multiplying length × breadth
- how to find the volume of a cuboid by counting cubes or by multiplying length × breadth × height.

Perimeter and area of rectangles

The **perimeter** of a rectangle is the total distance around the shape.

Perimeter = 2 lengths + 2 breadths

This can be written as a formula: $P = 2l + 2b$

The units used to measure perimeter are mm, cm or m.

The **area** of the rectangle is the amount of space inside the shape.

Area = length × breadth

This can be written as a formula: $A = l \times b$ or $A = lb$

The units used to measure area are mm^2, cm^2 or m^2.

When calculating perimeters and areas, you must make sure the units are the same in all parts of your calculation.

Example 15·1 Find the perimeter and area of the rectangle.

$P = 2l + 2b = 2 \times 5 + 2 \times 4 = 10 + 8 = 18$ cm

$A = lb = 5 \times 4 = 20$ cm^2

Measurement

Exercise 15A

1. By measuring the length of the sides of the following rectangles, find:

   i the perimeter   ii the area.

   a 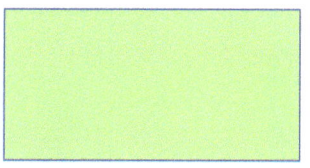   b

   c    d

   e    f

2. For each of the following rectangles, find:

   i the perimeter   ii the area.

   Remember to use the correct units.

   a 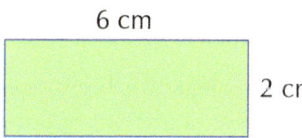   b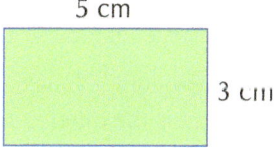

   c 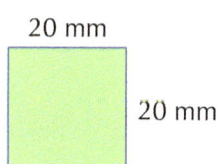   d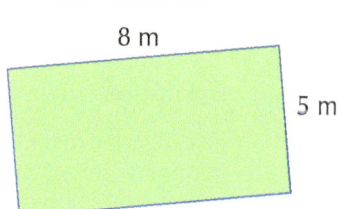

   e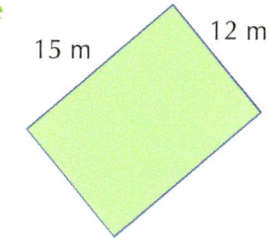

Number, money and measure

3 Find the area of each of the following rectangles.

Remember to use the correct units. If the units are inconsistent, give the area using the units shown in red.

a 5·2 cm 30 mm cm²

b 8 mm 16 mm

c 8·4 m 4·5 m

4 A bungalow has two bedrooms. The first bedroom measures 4·6 m by 3·8 m and the second bedroom is 4·1 m square. Which bedroom has the greater perimeter?

5 Here is a sketch of a plan for a bedroom.

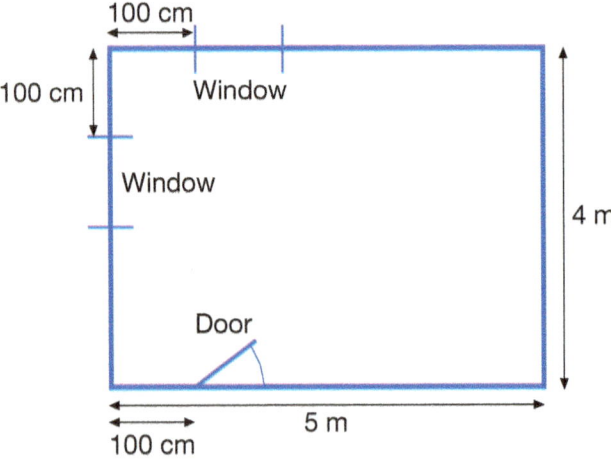

a What is the perimeter of the bedroom, measured in m?
b What is the area of the bedroom, measured in m²?
c How much does it cost to cover the floor with:
   i carpet that costs £30 per square metre
   ii oak laminate that costs £25·99 per square metre?
d What is the difference in price between these two options?

Challenge

1 **Equable rectangles**

Investigate whether a rectangle can have the same numerical value for its perimeter and its area.

2 **Sheep pens**

A farmer has 60 m of fence to make a rectangular sheep pen against a wall.

Find the length and breadth of the pen in order to make its area as large as possible. An example is given.

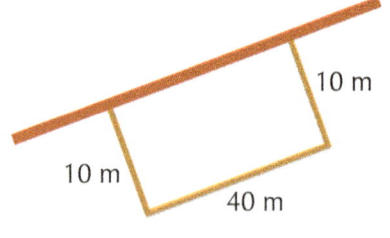

Continued

Measurement

Continued

3 Area

Area is measured in square millimetres (mm²), square centimetres (cm²), square metres (m²) and square kilometres (km²).

This square shows 1 square centimetre reproduced exactly.

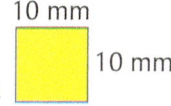

You can fit 100 square millimetres inside this square because a 1 centimetre square is 10 mm by 10 mm.

So 1 cm² = 100 mm².

How many square centimetres are there in 1 square metre?

How many square metres are there in 1 square kilometre?

a What unit would you use to measure the area of each of these?

   i   Football field   ii   Photograph   iii Fingernail
   iv National park   v   Pacific Ocean   vi   Stamp

b Convert

   i   24 cm² to mm²   ii   6 km² to m²
   iii 4000 mm² to cm²   iv   3 456 000 m² to km²

c Look up the areas of some countries on the internet or in an encyclopaedia.

   i   Which are the three biggest countries (in terms of area) in the world?
   ii Which is the biggest country (in terms of area) in Europe?

Area of a triangle

To find the area of a triangle, we need to know the length of its base and its height. The height of the triangle is sometimes known as its **perpendicular height**. The diagram shows that the area of the triangle is half of the area of a rectangle.

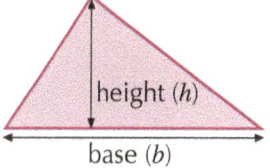

Area 1 = Area 2

and

Area 3 = Area 4

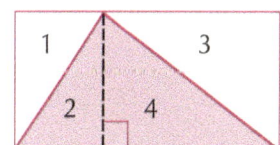

So the area of a triangle is $\frac{1}{2}$ × base × height.

The formula for the area of a triangle is given by:

$$A = \frac{1}{2} \times b \times h = \frac{1}{2} bh = \frac{bh}{2}$$

Example 15·2   Calculate the area of this triangle.

$$A = \frac{b \times h}{2} = \frac{8 \times 3}{2} = \frac{24}{2} = 12 \text{ cm}^2$$

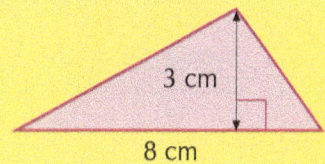

Sometimes the perpendicular height may be shown outside the triangle, as in the next example.

Number, money and measure

Example 15·3 Calculate the area of this triangle.

$$A = \frac{1}{2}bh = \frac{1}{2} \times 6 \times 5 = \frac{1}{2} \times 30 = 15 \text{ cm}^2$$

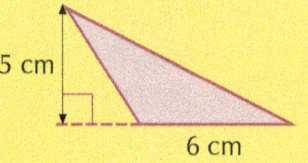

Exercise 15B

1. Calculate the area of each of the following triangles. If the units are inconsistent, give the area using the units shown in red.

   a 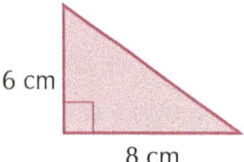   b 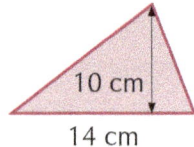   c 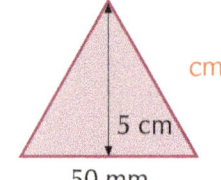 cm²

   d 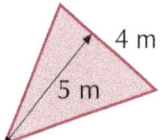   e 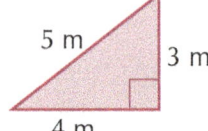   f 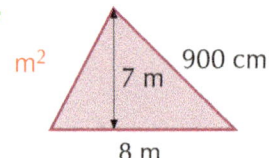 m²

2. Calculate the area of each of the following triangles. If the units are inconsistent, give the area using the units shown in red.

   a 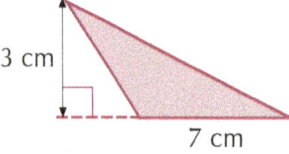   b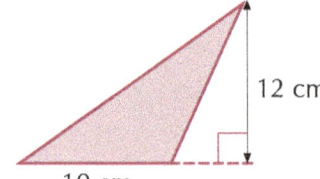

   c 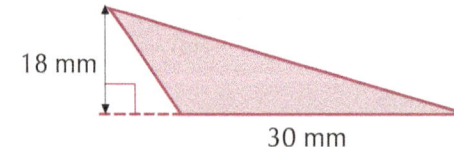   d 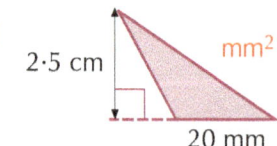 mm²

★ 3. Calculate the area of each of the following triangles.

   a 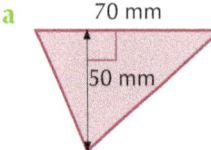   b 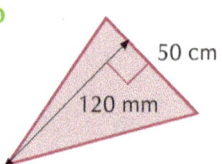   c 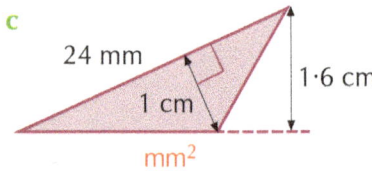 mm²

4. Copy and complete the table for triangles **a** to **e**.

Triangle	Base	Height	Area
a	5 cm	4 cm	
b	7 cm	2 cm	
c	9 m	500 cm	m²
d	12 mm		60 mm²
e		8 m	28 m²

Measurement

Challenge

The diagram shows one side of an **isosceles** triangle.

a Find all the possible positions on this grid for the other vertex.

b Find the area of each triangle you find.

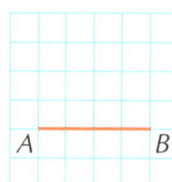

Area of a parallelogram

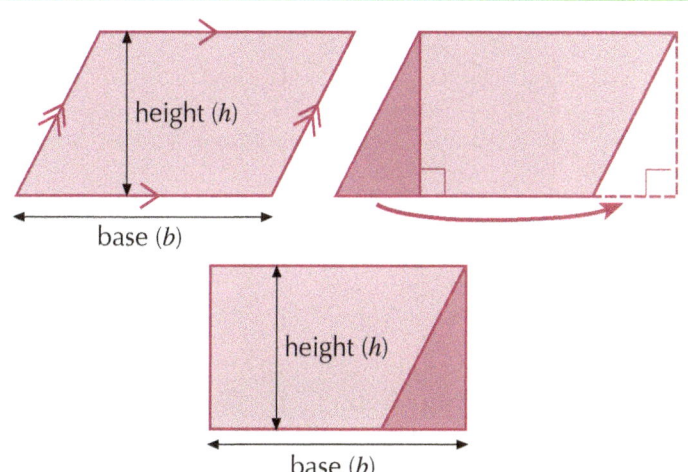

To find the area of a parallelogram, we need to know the length of its base and its height. The height of the parallelogram is sometimes known as its **perpendicular height**. The diagrams show that the parallelogram has the same area as that of a rectangle with the same base and height. So the area of a parallelogram is base × height.

The formula for the area of a parallelogram is given by:

$$A = b \times h = bh$$

Example 15·4 Calculate the area of this parallelogram.

$A = bh = 6 \times 10 = 60$ cm²

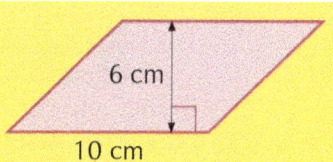

Exercise 15C

1 Calculate the area of each of the following parallelograms. If the units are inconsistent, give the area using the units shown in red.

a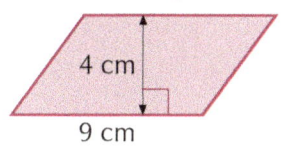

b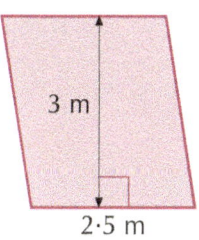

c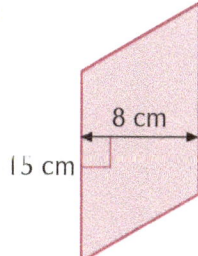

d 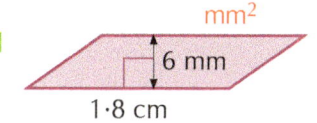 mm²

e 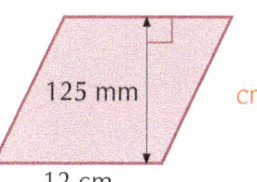 cm²

f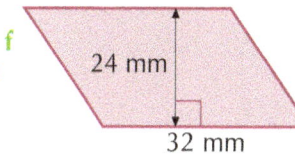

131

Number, money and measure

2 Calculate the area of each of the following parallelograms. If the units are inconsistent, give the area using the units shown in red.

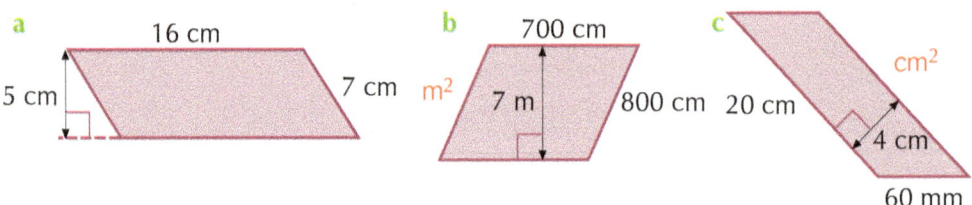

a 16 cm, 5 cm, 7 cm m²

b 700 cm, 7 m, 800 cm

c 20 cm, 4 cm, 60 mm cm²

★ 3 Calculate the area of each of the following parallelograms.

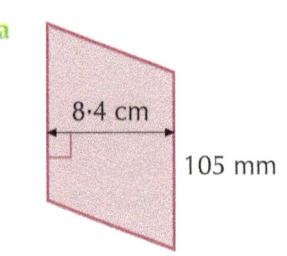

a 8·4 cm, 105 mm

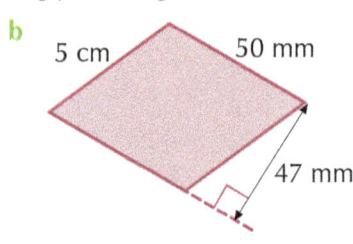

b 5 cm, 50 mm, 47 mm

4 Copy and complete the table below for parallelograms **a** to **e**.

Parallelogram	Base	Height	Area
a	8 cm	4 cm	
b	17 cm	12 cm	
c	8 m	5 m	
d	15 mm		60 mm²
e		8 m	28 m²

5 The area of the parallelogram is 27 cm². Calculate the perpendicular height of the parallelogram.

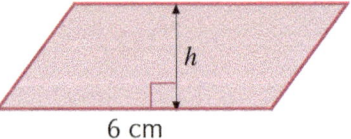

6 cm, h

Measurement

Challenge

1 Calculate the value of h in the given diagram.

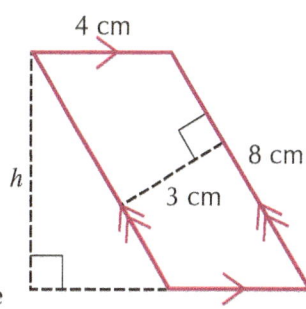

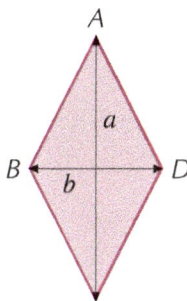

2 The lengths of the two diagonals of the rhombus are $AC = a$ and $BD = b$. The formula for finding the area of the rhombus is:

$$A = diagonal_1 \times diagonal_2 \div 2 = \frac{ab}{2}$$

Use the formula to calculate the area of each of the following rhombuses.

a 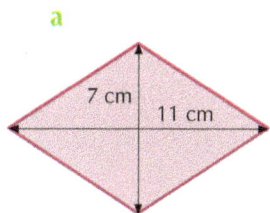   b 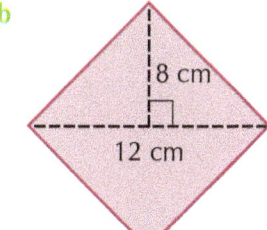   c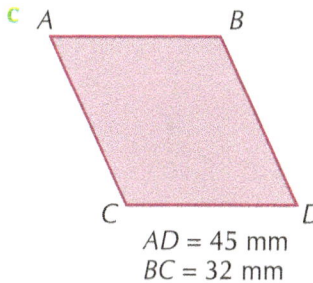
$AD = 45$ mm
$BC = 32$ mm

Area of a trapezium

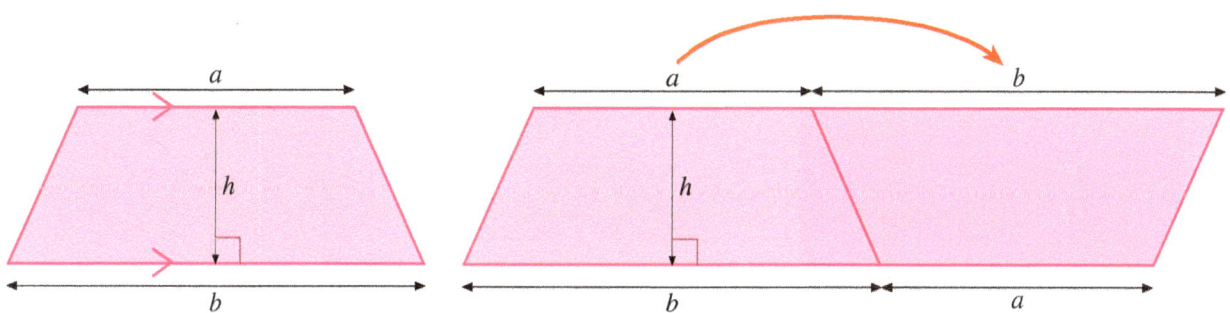

To find the area of a trapezium, we need to know the length of its two parallel sides, a and b, and the perpendicular height, h, between the parallel sides. The diagram shows how two equivalent trapeziums fit together to form a parallelogram. So the area of a trapezium is $\frac{1}{2} \times$ the sum of the lengths of the parallel sides $\times$ the height. The formula for the area of a trapezium is therefore given by:

$$A = \frac{1}{2} \times (a+b) \times h = \frac{1}{2}(a+b)h = \frac{(a+b)h}{2}$$

Number, money and measure

Example 15·5 Calculate the area of this trapezium.

$$A = \frac{(a+b)h}{2} = \frac{1}{2} \times (9+5) \times 4$$

$$= \frac{14 \times 4}{2}$$

$$= 28 \text{ cm}^2$$

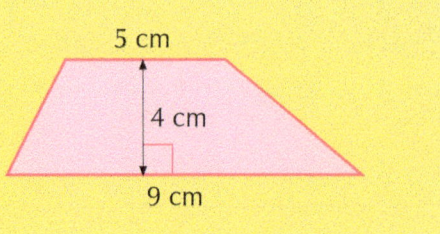

Exercise 15D

1 Calculate the area of each of the following trapeziums. If the units are inconsistent, give the area using the units shown in red.

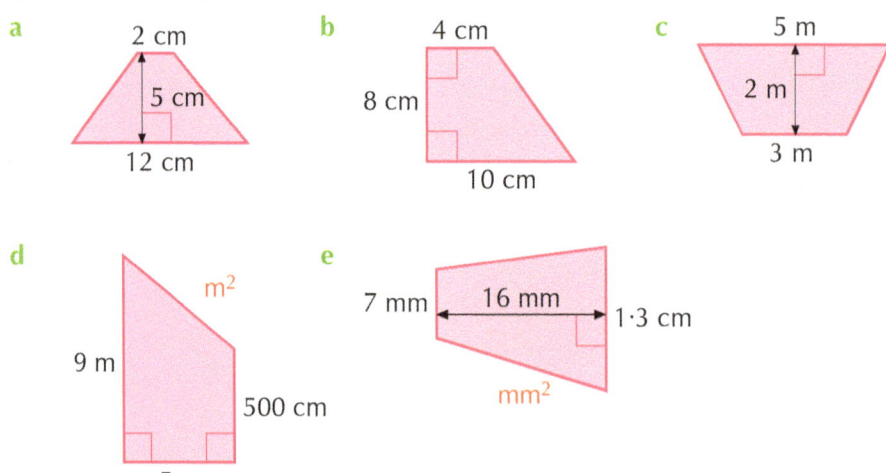

2 Copy and complete the table below for trapeziums **a** to **f**.

Trapezium	Length *a*	Length *b*	Height *h*	Area *A*
a	4 cm	6 cm	30 mm	cm²
b	0.1 m	12 cm	6 cm	cm²
c	9 m	3 m	5 m	
d	5 cm	5 cm		20 cm²
e	8 cm	12 cm		100 cm²
f	600 cm		4 m	32 m²

3 Find the areas, in cm², of the shapes cut out from this mathematical stencil.

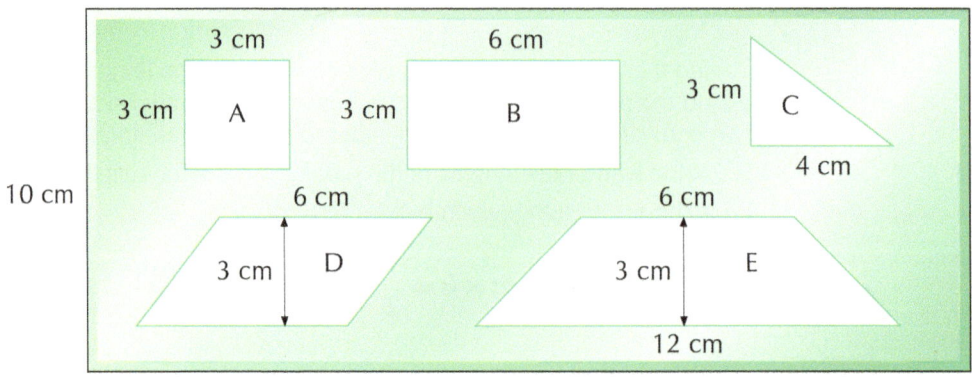

Measurement

4 The side of a swimming pool is a trapezium, as shown in the diagram below. Calculate its area.

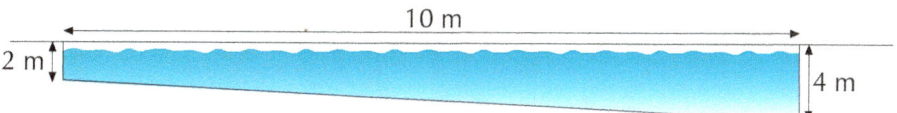

5 The area of this trapezium is 8 cm². Find different values of *a*, *b* and *h*, with *b* > *a*. One possibility is *a* = 1, *b* = 3 and *h* = 4.

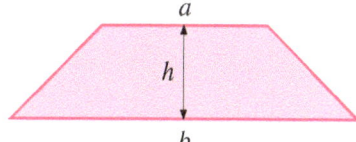

Challenge

1 Kites

The lengths of the two diagonals of the kite are *AC* = *a* and *BD* = *b*. The formula for finding the area of the kite is:

$$A = \frac{ab}{2}$$

Use the formula to calculate the area of each of the following kites.

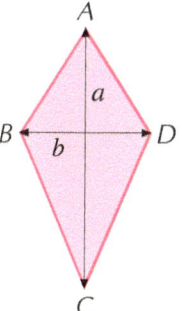

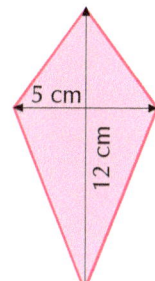 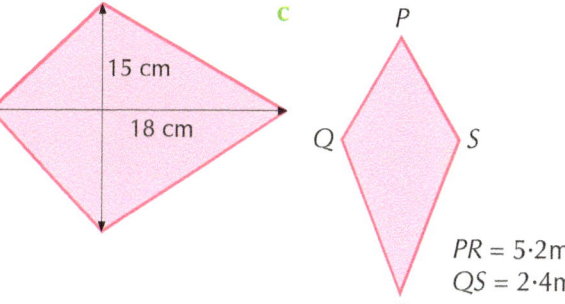

PR = 5·2 m
QS = 2·4 m

Continued

135

Number, money and measure

Continued

2 Pick's theorem

The shapes below are drawn on a 1 cm grid of dots.

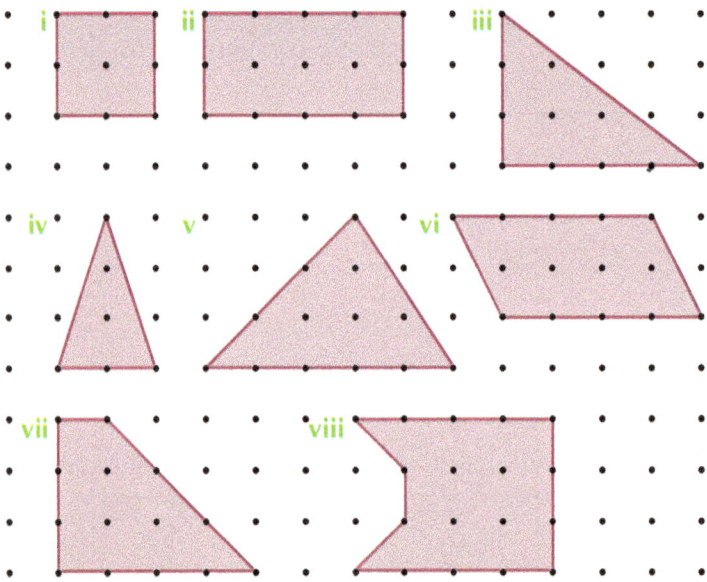

Shape	Number of dots on perimeter of shape	Number of dots inside shape	Area of shape (cm²)
i			
ii			
iii			
iv			
v			
vi			
vii			
viii			

a Copy and complete the table for each shape.

b Find a formula that connects the number of dots on the perimeter P, the number of dots inside I and the area A of each shape.

c Check your formula by drawing different shapes on a 1 cm grid of dots.

Measurement

Task: Design a bedroom

1. Here is a sketch of the plan of a bedroom.

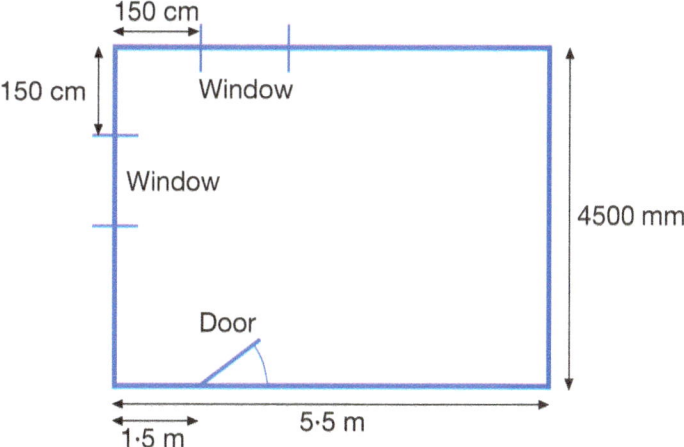

   a. What is the perimeter of the bedroom, measured in m?

   b. What is the area of the bedroom, measured in m²?

   c. How much will it cost to carpet the bedroom if you use carpet that costs £32·50 per m²?

2. Here are sketches of the door and one of the windows.

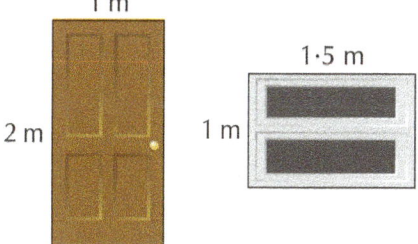

   The height of the bedroom is 2·5 m.

   a. What is the total area of the walls of the bedroom (excluding the door and windows)?

   b. If a 1 litre tin of paint covers 12 m², what is the minimum number of tins needed to paint all of the walls?

   c. How much will the paint cost if a 1 litre tin costs £18·99?

3. Posters cost £6·99 and measure 0.6 m by 45 cm.

   a. How many posters can you buy for £50?

   b. Is there enough wall space to display them all?

Number, money and measure

MTH 3-17b

4 Copy the plan of the bedroom onto cm squared paper. Use a scale of 1 cm to 0·5 m. Decide where you would put the following bedroom furniture. Use cut-outs to help.

Bed — 2 m by 1 m   Bedside table — 0·5 m by 0·5 m   Wardrobe — 1·5 m by 0·5 m   Chest of drawers — 1 m by 0·5 m   Desk — 1 m by 0·5 m

Use catalogues or the internet to find out how much it could cost to buy all the furniture for the bedroom.

5 What is the total cost of your new bedroom?

Volume of a cuboid

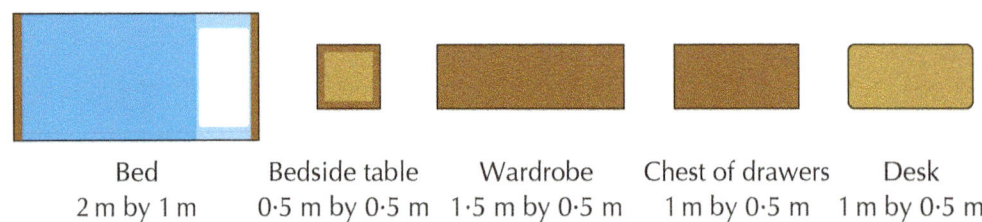

Volume is the amount of space inside a three-dimensional (3D) shape.

The diagram shows a cuboid that measures 4 cm by 3 cm by 2 cm. The cuboid is made up of cubes of edge length 1 cm. The top layer consists of 12 cubes and, since there are two layers, altogether the cuboid has 24 cubes. The volume of the cuboid is therefore found by calculating 4 × 3 × 2 = 24 cubes.

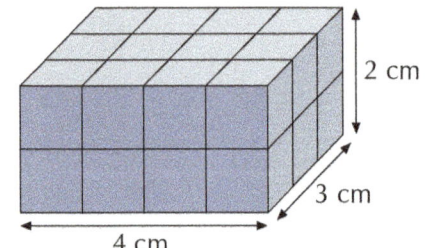

The volume of a cuboid is found by multiplying its length by its breadth by its height:

Volume of a cuboid = length × breadth × height

$$V = l \times b \times h = lbh$$

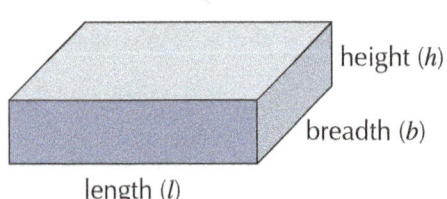

The metric units of volume in common use are:

- cubic millimetre (mm^3)
- cubic centimetre (cm^3)
- cubic metre (m^3).

Measurement

The **capacity** of a 3D shape is the volume of liquid or gas it can hold. The metric unit of capacity is the litre (l) with:

- 100 centilitres (cl) = 1 litre
- 1000 millilitres (ml) = 1 litre.

The following metric conversions between capacity and volume should be learnt:

- 1 l = 1000 cm³
- 1 ml = 1 cm³
- 1000 l = 1 m³.

Example 15·6 Calculate the volume of the following cuboid.

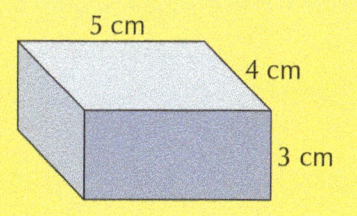

The formula for the volume of a cuboid is:

$V = lbh$

$= 5 \times 4 \times 3$

$= 60$ cm³

Example 15·7 Calculate the volume of the tank shown and then work out the capacity of the tank in litres.

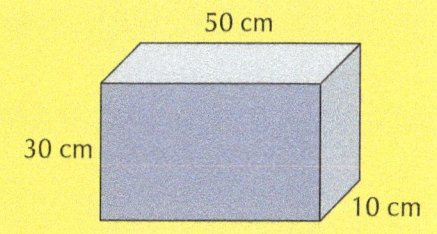

$V = lbh = 50 \times 30 \times 10 = 15\,000$ cm³

Since 1000 cm³ = 1 litre, the capacity of the tank = 15 000 ÷ 1000 = 15 litres.

Exercise 15E

1 Find the volume of each of the following cuboids.

a 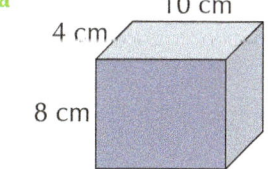   b 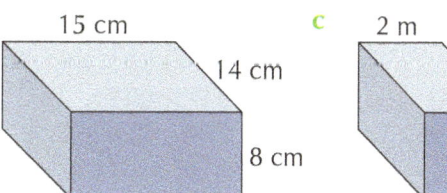   c

2 Find the capacity, in litres, of each of the following cuboid containers.

a 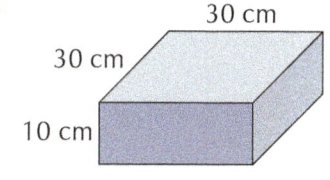   b 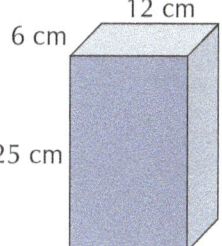   c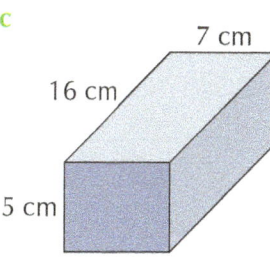

139

Number, money and measure

3 Copy and complete the table of cuboids **a** to **e**.

	Length	Breadth	Height	Volume
a	6 cm	4 cm	1 cm	
b	3·2 m	2·4 m	0·5 m	
c	8 cm	5 cm		120 cm³
d	20 mm	16 mm		960 mm³
e	40 m	5 m		400 m³

4 Calculate the volume of each of the cubes with the following edge lengths.

a 2 cm   b 5 cm   c 12 cm

5 Find the volume of a hall that is 30 m long, 20 m broad and 10 m high.

 6 How many packets of sweets that each measure 8 cm by 5 cm by 2 cm can be packed into a cardboard box that measures 32 cm by 20 cm by 12 cm?

 7 The diagram shows the dimensions of a swimming pool.

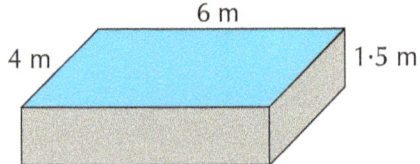

a Calculate the volume of the pool, giving the answer in cubic metres.

b How many litres of water does the pool hold when it is full?

8 Find the volume of this block of wood, giving your answer in cubic centimetres.

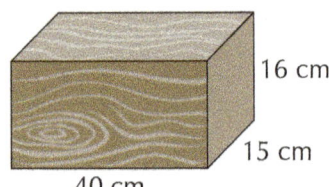

9 The diagram shows the dimensions of a carton of orange juice.

a Calculate the volume of the carton, giving your answer in cubic centimetres.

b How many glasses can be filled with orange juice from four full cartons, if each glass holds 240 ml?

10 A water tank is a cuboid with a square base of side length 2·5 m. The volume of the tank is 38·75 m³. Calculate the height of the tank.

Measurement

Challenge

1. Estimate the volume of various cuboid objects in your classroom. Then copy and complete the table below (some examples have already been filled in).

Object	Estimate of volume	Actual volume
Book		
Storage box		
Cupboard		

2. What does 1 m³ look like? Build a model if you have enough metre sticks.

3. Estimate first, then try to work these out:
   a. How many people could you fit into 1 m³?
   b. How many tins of beans could you fit in a 1 m³ box?
   c. How many people could fit in your classroom if all the furniture was removed?

Metric units for area and volume

The following are the metric units for area, volume and capacity which you need to know. Also given are the conversions between these units.

Area	Volume	Capacity
10 000 m² = 1 hectare (ha)	1 000 000 cm³ = 1 m³	1 m³ = 1000 litres (l)
10 000 cm² = 1 m²	1000 mm³ = 1 cm³	1000 cm³ = 1 litre
1 000 000 mm² = 1 m²		1 cm³ = 1 millilitre (ml)
100 mm² = 1 cm²		10 millilitres = 1 centilitre (cl)
		1000 millilitres = 100 centilitres = 1 litre

The unit symbol for litres is the letter l. To avoid confusion with the digit 1 (one), the full unit name may be used instead of the symbol.

Note:

To change **large** units to **smaller** units, **always multiply** by the conversion factor.

To change **small** units to **larger** units, **always divide** by the conversion factor.

Check: You need fewer larger units to equal more smaller units.

Example 15·8 Convert each of the following as indicated.

  a 72 000 cm² to m²   b 0·3 cm³ to mm³   c 4500 cm³ to litres

  a 72 000 cm² = 72 000 ÷ 10 000 = 7·2 m²   (1 m² = 10 000 cm²)
  b 0·3 cm³ = 0·3 × 1000 = 300 mm³   (1 cm³ = 1000 mm³)
  c 4500 cm³ = 4500 ÷ 1000 = 4·5 litres   (1 litre = 1000 cm³)

Number, money and measure

Exercise 15F

1. Express each of the following in cm².
   - a 4 m²
   - b 7 m²
   - c 20 m²
   - d 3·5 m²
   - e 0·8 m²

2. Express each of the following in mm².
   - a 2 cm²
   - b 5 cm²
   - c 8·5 cm²
   - d 36 cm²
   - e 0·4 cm²

3. Express each of the following in cm².
   - a 800 mm²
   - b 2500 mm²
   - c 7830 mm²
   - d 540 mm²
   - e 60 mm²

4. Express each of the following in m².
   - a 20 000 cm²
   - b 85 000 cm²
   - c 270 000 cm²
   - d 18 600 cm²
   - e 3480 cm²

5. Express each of the following in mm³.
   - a 3 cm³
   - b 10 cm³
   - c 6·8 cm³
   - d 0·3 cm³
   - e 0·48 cm³

6. Express each of the following in m³.
   - a 5 000 000 cm³
   - b 7 500 000 cm³
   - c 12 000 000 cm³
   - d 65 000 cm³
   - e 2000 cm³

7. Express each of the following in litres.
   - a 8000 cm³
   - b 17 000 cm³
   - c 500 cm³
   - d 3 m³
   - e 7·2 m³

★ 8. Express each of the following as indicated.
   - a 85 ml in cl
   - b 1·2 litres in cl
   - c 8·4 cl in ml
   - d 4500 ml in litres
   - e 2·4 litres in ml

9. How many square paving slabs, each of side 50 cm, are needed to cover a rectangular yard measuring 8 m by 5 m?

10. A football pitch measures 120 m by 90 m. Find the area of the pitch in the following units.
    - a m²
    - b Hectares

11. A fish tank is 1·5 m long, 40 cm broad and 25 cm high. How many litres of water will it hold if it is filled to the top?

12. The volume of the cough medicine bottle is 240 cm³. How many days will the cough medicine last?

13. How many lead cubes of side 2 cm can be cast from 4 litres of molten lead?

Measurement

Appropriate units and accuracy

When you are calculating with measurements, you need to think about two things: the units you use and the level of accuracy you need for your answers. Both of these need to be suitable for the size of the object used in your calculations.

It wouldn't make sense to measure the area of your maths book in square metres (m^2), as its measurements are much less than 1 metre. It would be more appropriate to use square centimetres (cm^2) because you would use centimetres to measure the book's length and breadth.

You should also round your answer to an appropriate degree. If your object's measurements are large then your answer could be rounded to the nearest whole number, 10 or 100, but if they are small then it would be more appropriate to round an answer to one or two decimal places.

Example 15·9

The touch screen on a smartphone measures 51 mm × 66 mm. Calculate its area, expressing your answer using an appropriate unit and degree of accuracy.

Area = lb
= 51 × 66
= 3366 mm^2

It would be more appropriate to express this answer in cm^2 and round it to one decimal place. This would be accurate enough to allow you to compare the screen size with other models.

Area = 33·66 cm^2
= 33·7 cm^2 (rounded to 1 dp)

Example 15·10

A shipping container, as used by haulage companies, measures 12·19 m × 2·44 m × 2·59 m. Calculate its volume, expressing your answer using an appropriate unit and degree of accuracy.

The container is a cuboid, so its volume is calculated using the formula:

$V = lbh$
= 12·19 × 2·44 × 2·59
= 77·035 924 m^3

Cubic metres, m^3, is an appropriate unit for this calculation. Rounding the answer either to the nearest whole number or to one decimal place would give a sensible level of accuracy. So, the volume of the container is 77 m^3.

Investigation: Units and accuracy

1 What is the most appropriate unit to measure the following?
   a The area of a postage stamp.
   b The capacity of a swimming pool.
   c The area of the pitch at Hampden Park.

Number, money and measure

d The volume of the biggest room in your school.

e The area of your desk.

f The volume of a box of breakfast cereal.

2 Measure or research the actual size of each object and calculate its area or volume. Choose an appropriate degree of accuracy for your answer.

3 Compare your answer with your estimate. How good was your estimate? Choose some other objects around your class, school or local area and first estimate then calculate their area or volume.

Challenge

1 A farmer has 100 m of fencing to enclose his sheep. He uses the wall for one side of the rectangular sheep-pen. If each sheep requires 5 m² of grass inside the pen, what is the greatest number of sheep that the pen can hold?

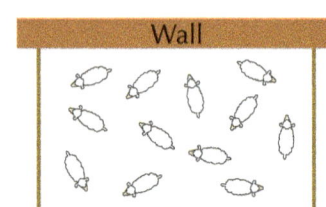

2 What is an acre? Use reference books or the internet to find out.

3 Use the information given to find:

   a the number of square inches in one square yard

   b the number of cubic inches in one cubic yard.

12 inches = 1 foot
3 feet = 1 yard

- By working on this topic, I have learnt how to use a formula to find the area, volume or capacity of a variety of shapes and objects.

- I can use my knowledge of perimeter and area calculations to solve practical problems.

- I can calculate the area of rectangles, triangles, parallelograms and trapeziums presented in different ways. ★ Exercise 15A Q5   ★ Exercise 15B Q3   ★ Exercise 15C Q3
★ Exercise 15D Q4

- I can calculate the volume of cuboids and can use this knowledge to solve practical problems. ★ Exercise 15E Q7

- I can convert between units of area, volume and capacity to express my answers using an appropriate measure. ★ Exercise 15F Q8

- I can give examples of appropriate units to measure real objects with a variety of sizes. *Investigation* pages 143–144

- I can choose an appropriate degree of accuracy to express my answers to problems involving area and volume. *Investigation* pages 143–144

Measurement

16

Having investigated different routes to a solution, I can find the area of compound 2D shapes and the volume of compound 3D objects, applying my knowledge to solve practical problems.

MTH 3-11b

This chapter will show you how to:
- investigate different ways to divide a compound shape into simpler shapes
- calculate the perimeter and area of compound 2D shapes made from squares, rectangles and triangles
- calculate the volume of a compound 3D object made from cuboids
- choose an efficient method to calculate perimeter, area and volume.

You should already know:
- how to use a formula to find the area of a range of 2D shapes
- how to use a formula to find the volume of a cuboid
- how to choose appropriate units of measure for perimeter, area and volume.

Perimeter and area of compound 2D shapes

A **compound shape** is a shape that is made by combining other simple shapes, such as squares, rectangles and triangles.

The examples show you how to find the perimeter and area of a compound shape made from rectangles.

Example 16·1 Find the perimeter and area of the compound shape on the right.

First copy the shape, then find and label the lengths of any sides which are not already shown as shown below.

Now the perimeter and area of the compound shape can be worked out as follows:

$P = 10 + 12 + 4 + 7 + 6 + 5$

$= 44$ cm

Total area = area of A + area of B

$= 6 \times 5 + 12 \times 4$

$= 30 + 48$

$= 78$ cm²

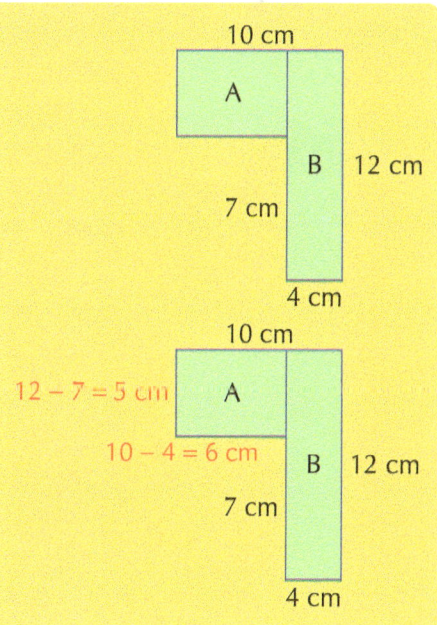

Number, money and measure

Example 16·2 Find the perimeter of this shape. Then calculate its area in two different ways.

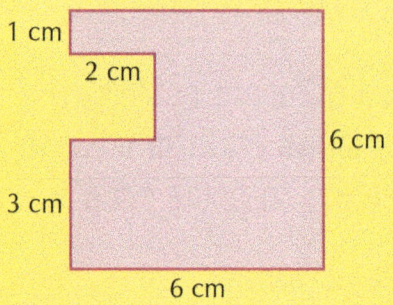

The perimeter is: $P = 6 + 6 + 6 + 3 + 2 + 2 + 2 + 1$

$\qquad\qquad\qquad = 28$ cm

There are several different ways you could divide the shape to find its area. Here are two possibilities:

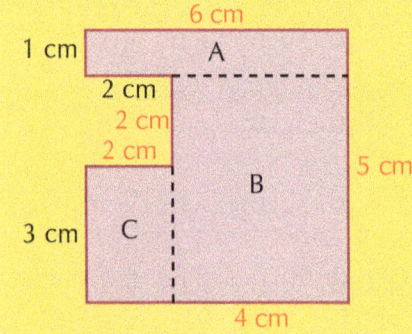 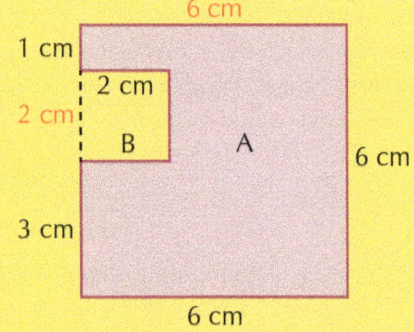

Total area = area of A + area of B + area of C

$= 1 \times 6 + 5 \times 4 + 3 \times 2$

$= 6 + 20 + 6$

$= 32$ cm²

Total area = area A − area B

$= 6 \times 6 - 2 \times 2$

$= 36 - 4$

$= 32$ cm²

Exercise 16A

1. For each of the compound shapes, find: **i** the perimeter **ii** the area.
Start by copying the diagrams and splitting them up into rectangles as shown.

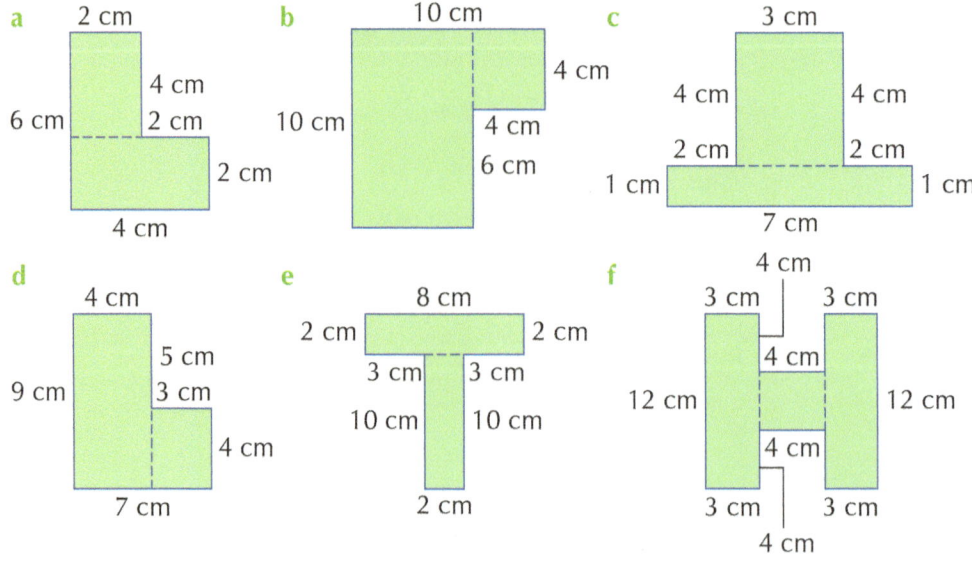

Measurement

2 For each of the following compound shapes, find: **i** the perimeter **ii** the area.

a 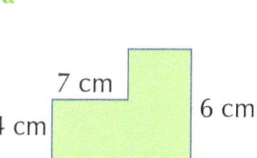   b 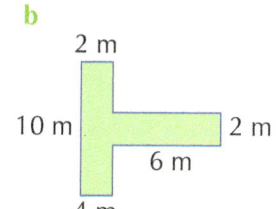   c 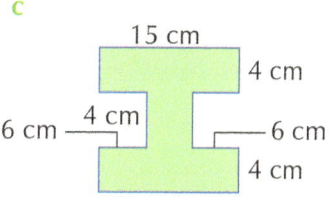   d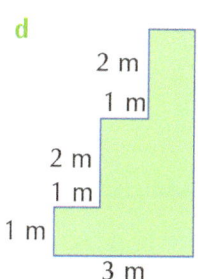

3 Sandra makes a picture frame from a rectangular piece of card for a photograph of her favourite group.

  a Find the area of the photograph.
  b Find the area of the card she uses.
  c Find the area of the border.
     (*Hint:* use subtraction.)

4 A garden is in the shape of a rectangle measuring 16 m by 12 m.

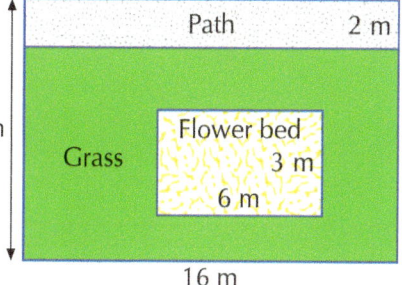

  a Find the area of the garden.
  b Find the area of the path.
  c Find the area of the flower bed.
  d Find the area of the grass in the garden.

5 A room is 6 m long and 4 m wide. A carpet measuring 5 m by 3 m is placed on the floor of the room. Find the area of the floor not covered by carpet.

6 Phil finds the area of this compound shape.

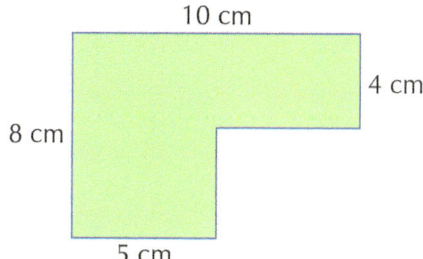

This is his working:

Area = 10 × 4 + 8 × 5

= 40 + 40

= 80 cm²

  a Explain why he is wrong.
  b Calculate the correct answer.

7 How many rectangles with sides measured in whole numbers can you draw with a fixed perimeter of 20 cm but each one having a different area?

Number, money and measure
Area of compound 2D shapes involving triangles

To find the area of a compound shape, made from rectangles and triangles, find the area of each one separately and then add together all the areas to obtain the total area of the shape.

Example 16·3 Calculate the area of this shape.

Divide the shape into a rectangle A and a triangle B:

Area of A = lb = 8 × 4 = 32 cm²

Area of B = $\frac{ab}{2} = \frac{6 \times 8}{2} = \frac{48}{2}$ = 24 cm²

Area of shape = 32 + 24 = 56 cm²

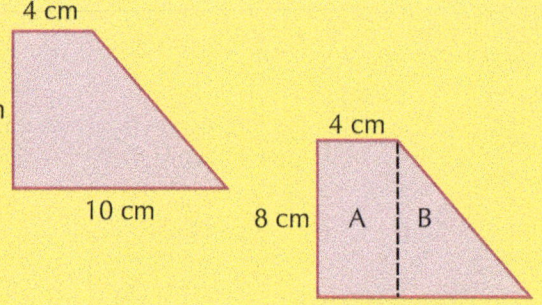

Exercise 16B

1 Calculate the area of each compound shape below.

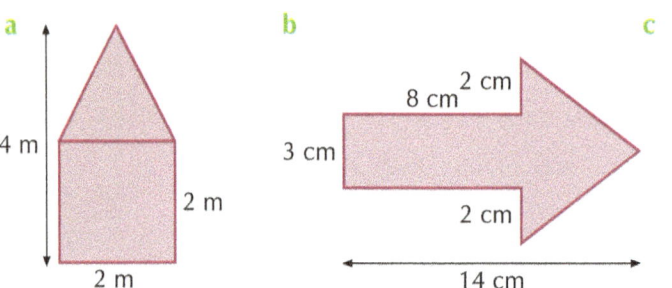 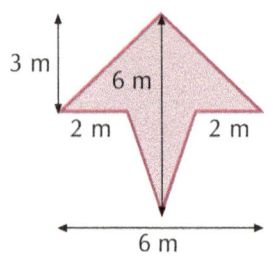

2 Find the area of this computer worktop.

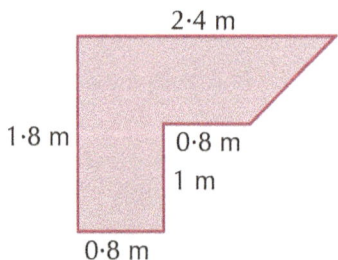

 3 The diagram shows the end wall of a garden shed. The green area is the door.

a Find the area of the door.

b Find the area of the brick wall.

148

Measurement

4 The diagram shows the measurements of a sauce bottle label. Calculate its area.

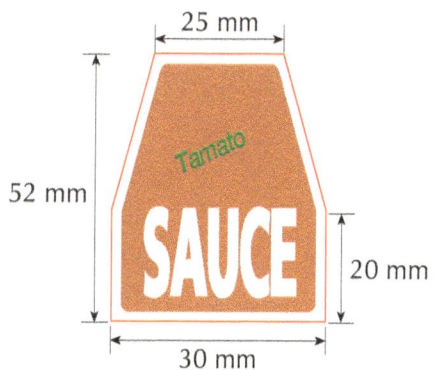

5 Find the solid area of this mathematical stencil, which has the shapes cut out.

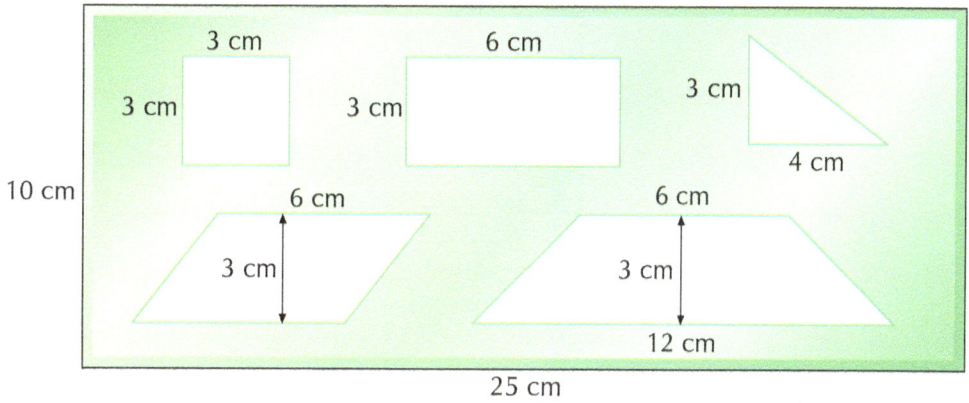

Volume of compound 3D objects

Example 16·4 Calculate the volume of the object shown.

The object is made up of two cuboids A and B with measurements 7 m by 3 m by 2 m (A) and 2 m by 3 m by 6 m (B). So the volume of the object is given by:

$V = (7 \times 3 \times 2) + (2 \times 3 \times 6)$

$= 42 + 36$

$= 78 \text{ m}^3$

Remember to include units in your answer.

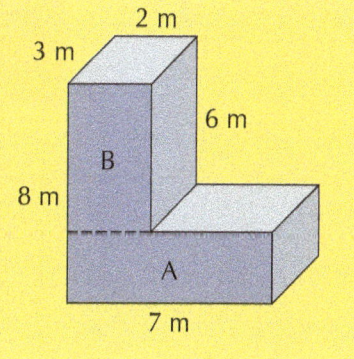

Exercise 16C

1 Find the volume of this 3D object.

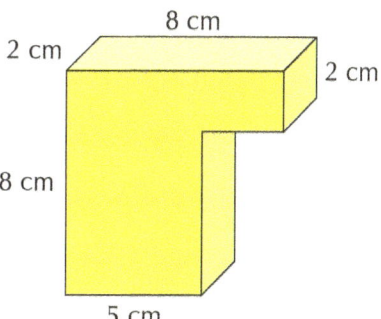

Number, money and measure

2 Calculate the volume of each of the following 3D objects.

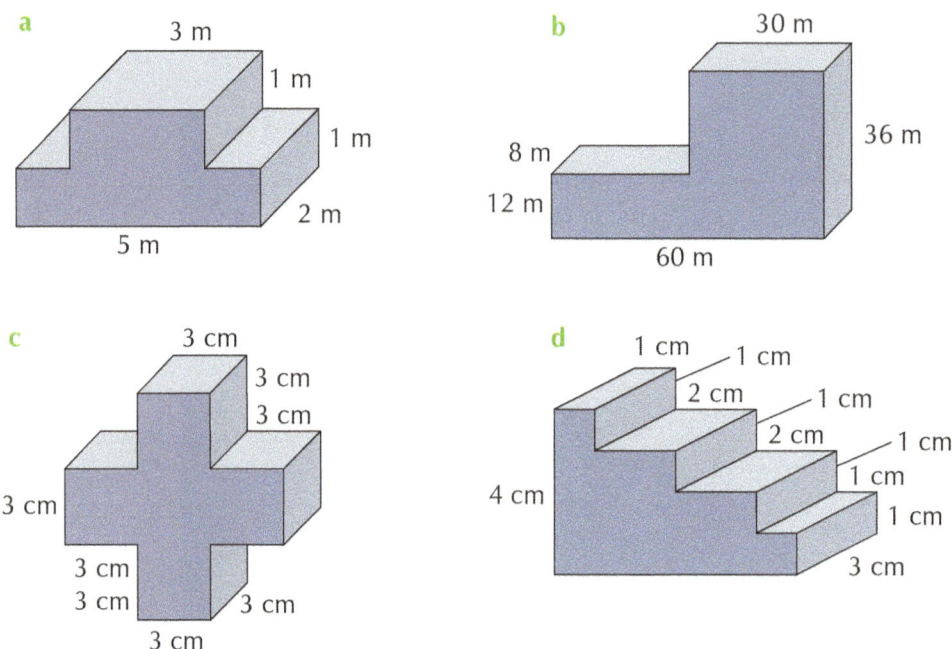

3 The diagram shows the measurements of a magnet. Calculate the volume of material needed to make it.

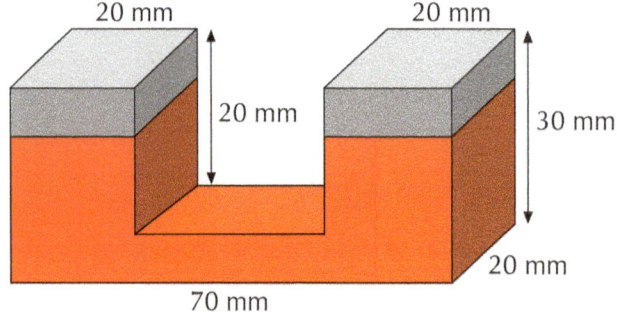

4 A series of plastic toy letters are a maximum size of 8 cm × 5 cm × 2·5 cm.

   a Calculate the volume of plastic needed to make these three letters:

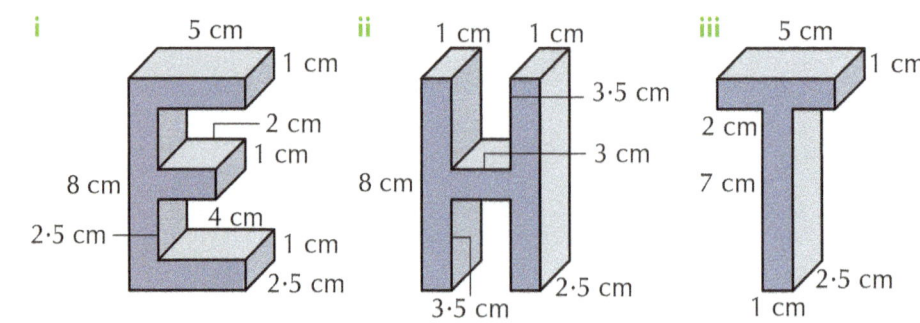

   b Sketch your initials and work out the volume of plastic needed to make each of your letters.

Measurement

Challenge Each step in a concrete staircase for a new building measures 250 mm × 200 mm × 900 mm.

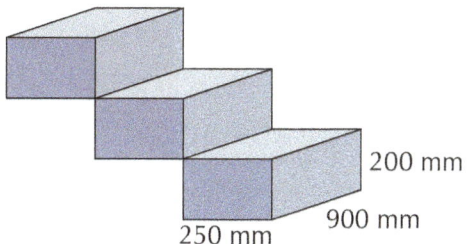

200 mm
250 mm   900 mm

Calculate the volume of concrete required to make a staircase that reaches 2·8 metres between the first two floors.

- By working on this topic, I can explain how to divide compound shapes and objects into simpler shapes and objects to find their area or volume.

- I have learnt how to divide compound 2D shapes into simpler shapes to find their area.
  ★ Exercise 16A Q4

- I have learnt how to divide compound 2D shapes involving triangles into simpler shapes to find their area. ★ Exercise 16B Q3

- I have learnt how to divide compound 3D objects into simpler objects to find their volume.
  ★ Exercise 16C Q4

- I can explain how to divide a compound shape or object into simpler shapes or objects in different ways and then choose an efficient way to calculate its area or volume.

- I can apply the formulae for the area of rectangles and triangles to compound 2D shapes.

- I can apply the formula for the volume of a cuboid to compound 3D objects.

Number, money and measure

17

I have worked with others to research a famous mathematician and the work they are known for, or investigated a mathematical topic, and have prepared and delivered a short presentation.

MTH 3-12a

This chapter will show you how to:
- investigate the work of a famous mathematician or a mathematical topic, using a range of sources and methods
- prepare an effective presentation about your chosen topic
- discuss your topic and your findings with others.

You should already know:
- that maths is used all around us in everyday life
- that new inventions and scientific discoveries depend on maths
- how to work collaboratively with others.

Introducing the task

Since early humans first started looking for patterns in the movement of the stars and the changing of the seasons to help them to decide when to plant their crops or hold their religious festivals, they have turned to mathematicians for help. Mathematics has been at the core of the development of technology and civilisation ever since. From aircraft to mobile phones, the latest medical equipment to the internet and games consoles, they all have one thing in common – their invention has depended upon the mathematics developed by mathematicians.

Your task in this chapter is to work as part of a group to research either a famous mathematician or a mathematical topic and prepare and deliver a short presentation to your class.

Your teacher will give you some guidance about how long to spend on the task.

Famous mathematicians

A huge number of mathematicians have contributed to the mathematics we use today, but some of the more famous ones you could choose from are:

- Pythagoras – best known for the theorem named after him, but he was also interested in music and had some very strange friends.
- Al-Khwarizmi – worked on mathematics and astronomy and wrote the first algebra textbook.
- Fibonacci – introduced decimal numbers into Europe and solved a rabbit problem with a famous sequence.
- Pierre de Fermat – his name is best known for his 'last theorem', which went unproved for over 300 years and led to lots of other mathematical discoveries along the way.
- Blaise Pascal – invented one of the first mechanical calculating machines and, along with Fermat, came up with the ideas of probability theory.

Mathematics – its impact on the world, past, present and future

- John Napier – a Scottish mathematician who invented logarithms to make complex calculations easier to perform. His 'bones' were also an early calculator!
- Isaac Newton – most famous for discovering the existence of gravity, but he also wrote works on calculus, optics, the mechanics of motion and alchemy.
- Florence Nightingale – you might think of her as the 'Lady with the lamp', but she developed a statistical diagram that saved thousands of lives in the Crimean War.
- Charles Babbage – a mathematician and inventor who was obsessed with the capabilities of machines. He designed the first mechanical computer, but never lived to see it built.
- Ada Lovelace – a pioneer for women in the Victorian era, she was among the first to see the potential of Babbage's machines and became an early computer programmer.

There are many others you could choose, so pick one of your own if there is someone you are particularly interested in.

Mathematical topics

You could choose to research a mathematical topic instead of a mathematician. Find a topic that interests you or choose one of these suggestions:

- Art – explore aspects of mathematics used in art, including perspective, symmetry, tessellations and the golden ratio.
- Babylonian mathematics – this ancient civilisation used a different number system from ours and seemed to know about a famous theorem long before the man it is named after.
- Complex numbers – these numbers do something you might have been told is impossible and can be purely imaginary!
- Fractals – infinitely similar shapes that can make amazing patterns and model natural objects to produce realistic images for computer games and animated films.
- Interest rates – find out how understanding percentages can help you to make sure you are making the most of your money.
- Music – investigate the ways in which fractions, the Fibonacci sequence and sound waves come together to create music.
- Navigation – find out how bearings, trigonometry and vectors can help you to find your way in the world.
- π – this mysterious number (called 'pi') exists all 'around' us. First discovered over 4000 years ago, many famous mathematicians have since tried to define it more accurately.
- Prime numbers – these have many modern uses, including internet security, and are the heart of one of the oldest unsolved problems in number theory.
- Probability – investigate how it is used to calculate risk in the insurance industry or how you can use it to help improve your success in games of chance.

Researching your topic

First your group needs to decide if you are going to research a famous mathematician or a mathematical topic. Each person in your group then needs to agree their role in gathering the information you will need.

Lots of information can be found by searching the internet for your mathematician or topic, but you need to make sure you use several different websites to check the information you find is accurate.

Number, money and measure

You should also use at least one source that is not the internet; this could be a book, magazine, film or TV programme. Check the library in your school or local area to see if they have any books on mathematical history or encyclopaedias with mathematical information.

You must ensure that you record the books and websites where you found your information – just because something is available on the internet doesn't mean it can be copied without permission. You will therefore need to include references in your presentation, so the people whose work you've used are given credit. Your teacher will give you more guidance on this.

Presenting your findings

Your presentation could take a variety of forms; you could make a poster, present a short play or a musical item or produce a presentation using ICT. Whatever type of presentation you choose, it must be written in your own words and should include:

- an introduction
- background information about your mathematician's life and interests or the origins of your mathematical topic
- details about the mathematics you researched and how it was discovered
- an example of the mathematics in use
- some examples of where and how we use this mathematics today
- a conclusion about the impact or influence of your mathematician/topic
- references to where you found your information.

Agree a role for everyone in your group so you know exactly what each of you will do to prepare and deliver the presentation. Then rehearse it with your group so you can make any necessary changes.

Be prepared to answer questions from the audience or your teacher. You could get a friend to ask a question you have prepared in advance in case no one else asks a question.

Example 17·1   If you chose Pythagoras as your famous mathematician, then these are some things you should think about for your presentation:

- Where was he born and when did he live?
- Do we know what he looked like?
- What mathematics is he best known for?
- Do we know how or why he developed his ideas?
- What were his ideas about musical notes?
- Who were the Pythagoreans?
- Where is his maths used today outside the classroom?

Mathematics – its impact on the world, past, present and future

Assessment

One way to assess your task is to hold a 'Greatest mathematician' or 'Top topic' competition for your class or even the entire year. Each group gives their presentation in turn and the other groups, or a panel of teachers, mark them on how well they presented it, how much new information or mathematics they learnt and how much they enjoyed it.

Once all the presentations are completed, the whole audience can vote for the overall winner of the 'Greatest mathematician' or 'Top topic'.

- By working on this topic I have learnt that mathematicians have played an important role in developing the society we live in today.
- I can give an example of how mathematics is used in a real-life situation.
- I can use a variety of methods to prepare and deliver an effective presentation.
- I can work collaboratively with others.

Number, money and measure

18

Having explored number sequences, I can establish the set of numbers generated by a given rule and determine a rule for a given sequence, expressing it using appropriate notation.

MTH 3-13a

This chapter will show you how to:
- recognise terms in a sequence of numbers
- calculate terms in a sequence when you know the rule and the first term
- construct the rule for the nth term in a sequence
- calculate the nth term of a sequence.

You should already know:
- how to continue the pattern of simple sequences
- about common sequences like Fibonacci, square numbers and triangular numbers.

Sequences and rules

You can make up many different sequences with **integers** (whole numbers) using simple rules.

Example 18·1   Rule: add 3   Starting at 1 gives the sequence 1, 4, 7, 10, 13, …

Starting at 6 gives the sequence 6, 9, 12, 15, 18, …

Rule: double   Starting at 1 gives the sequence 1, 2, 4, 8, 16, …

Starting at 5 gives the sequence 5, 10, 20, 40, 80, …

With *different* **rules** and *different* **starting points**, there are many possible *different* **sequences**.

The numbers in a sequence are called **terms** and the starting point is called the **1st term**. The rule is often referred to as the **term-to-term rule** or **next term rule**. The terms in a sequence are all generated by the same rule. And the sequence can be continued forever.

Exercise 18A

1 Use each of the following term-to-term rules with the 1st terms: i 1   ii 5.

Create each sequence with five terms in it.

    a add 3   b multiply by 3   c add 5   d multiply by 10

    e add 9   f multiply by 5   g add 7   h multiply by 2

    i add 11   j multiply by 4   k add 8   l add 105

Patterns and relationships

2 Give the next two terms in each of these sequences. Describe the term-to-term rule you have used.

- a 2, 4, 6, ...
- b 3, 6, 9, ...
- c 1, 10, 100, ...
- d 1, 2, 4, ...
- e 2, 10, 50, ...
- f 0, 7, 14, ...
- g 7, 10, 13, ...
- h 4, 9, 14, ...
- i 4, 8, 12, ...
- j 9, 18, 27, ...
- k 12, 24, 36, ...
- l 2, 6, 18, ...

3 Give the next two terms in these sequences. Describe the term-to-term rule you have used.

- a 50, 45, 40, 35, 30, ...
- b 35, 32, 29, 26, 23, ...
- c 64, 32, 16, 8, 4, ...
- d 3125, 625, 125, 25, 5, ...
- e 20, 19·3, 18·6, 17·9, 17·2, ...
- f 1000, 100, 10, 1, 0·1, ...
- g 10, 7, 4, 1, −2, ...
- h 27, 9, 3, 1, $\frac{1}{3}$ ...

4 For each pair of numbers find at least two different sequences, writing the next two terms. Describe the term-to-term rule you have used.

- a 1, 4, ...
- b 3, 9, ...
- c 2, 6, ...
- d 3, 6, ...
- e 4, 8, ...
- f 5, 15, ...

★ 5 Find two terms between each pair of numbers to form a sequence. Describe the term-to-term rule you have used.

- a 4, ..., ..., 10
- b 3, ..., ..., 12
- c 5, ..., ..., 20
- d 1, ..., ..., 8
- e 80, ..., ..., 10
- f 2, ..., ..., 54

Challenge

1 a Make up some of your own sequences and describe them.

b Give your sequences to someone else and see if they can find out what your term-to-term rule is.

2 Use a spreadsheet or graphics calculator for this investigation.

Here is an incomplete rule to find the next term in a sequence:

Add on ☐

Find a 1st term for the sequence and a number to go in the box so that all the terms in the sequence are:

- a odd
- b even
- c multiples of 5
- d numbers ending in 7

Number, money and measure
Finding missing terms

In any sequence, you will have a 1st term, 2nd term, 3rd term, 4th term and so on.

Example 18·2 In the sequence 3, 5, 7, 9, …, what is the 5th term, and what is the 50th term?

You first need to know what the term-to-term rule is. You can see that you add 2 from one term to the next:

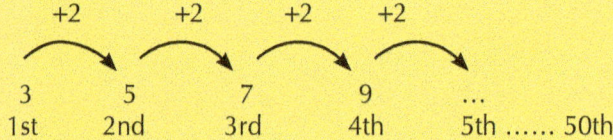

To get to the 5th term, you add 2 to the 4th term, which gives 11.

To get to the 50th term, you will have to add on 2 a total of 49 times (50 − 1) to the first term, 3. This will give $3 + 2 \times 49 = 3 + 98 = 101$.

Exercise 18B

1 In each of the following sequences, find the 5th and the 50th term.

   a 4, 6, 8, 10, …   b 1, 6, 11, 16, …   c 3, 10, 17, 24, …

   d 5, 8, 11, 14, …   e 1, 5, 9, 13, …   f 2, 10, 18, 26, …

   g 20, 30, 40, 50, …   h 10, 19, 28, 37, …   i 3, 9, 15, 21, …

2 In each of the sequences below, find the 1st term, then find the 50th term.

   In each case, you have been given the 4th, 5th and 6th terms.

   a …, …, …, 13, 15, 17, …   b …, …, …, 18, 23, 28, …

   c …, …, …, 19, 23, 27, …   d …, …, …, 32, 41, 50, …

 3 In each of the following sequences, find the missing terms and the 50th term.

Term	1st	2nd	3rd	4th	5th	6th	7th	8th	50th
Sequence A	…	…	…	…	17	19	21	23	…
Sequence B	…	9	…	19	…	29	…	39	…
Sequence C	…	…	16	23	…	37	44	…	…
Sequence D	…	…	25	…	45	…	…	75	…
Sequence E	…	5	…	11	…	…	20	…	…
Sequence F	…	…	12	…	…	18	…	22	…

4 Find the 40th term in the sequence with the term-to-term rule add 5 and a 1st term of 6.

5 Find the 80th term in the sequence with the term-to-term rule add 4 and a 1st term of 9.

6 Find the 100th term in the sequence with the term-to-term rule add 7 and 1st term of 1.

7 Find the 30th term in the sequence with the term-to-term rule add 11 and 1st term of 5.

Patterns and relationships

Challenge

1. You have a simple sequence where the 50th term is 349, the 51st is 354 and the 52nd is 359. Find the 1st term and the 100th term.

2. You are building patterns using black and yellow squares.

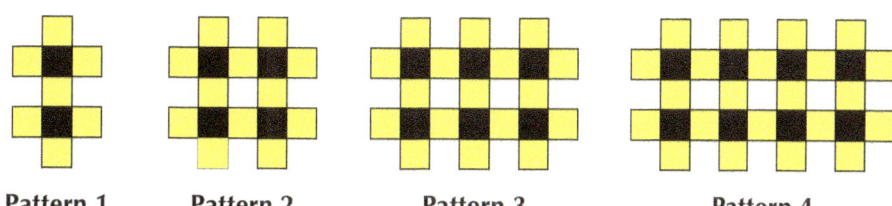

Pattern 1   Pattern 2   Pattern 3   Pattern 4

You have 50 black squares. How many yellow squares will be in the pattern?

The nth term formula

Example 18·3

The first four terms of the sequence 3, 6, 9, 12, … are shown in the table below.

The term-to-term rule is add 3. It is often more useful to have a rule that connects the term of a sequence to its position in the sequence.

This is different from the term-to-term rule and is called the **nth term formula** where the letter n represents the position number. You can describe a sequence by finding the nth term formula, which is the generalisation that will allow you to find any specific term you want.

Position number	1st	2nd	3rd	4th
Terms of sequence	3	6	9	12

a Find the rule that connects the term and the position number.
b Find the nth term formula for the sequence.
c Find the 20th term of the sequence.

a This sequence is like the 3 times table. Each term is 3 times the position number.

1st term $3 \times 1 = 3$
2nd term $3 \times 2 = 6$
3rd term $3 \times 3 = 9$
4th term $3 \times 4 = 12$

b Each term is 3 times the position number so we get **nth term** = $3 \times n$ or **nth term** = $3n$.

c 20th term = 3×20
            = 60

Number, money and measure

Example 18·4

Find the *n*th term formula for the sequence 5, 6, 7, 8, 9, …

A table helps you compare the terms with their position numbers.

Position number	1st	2nd	3rd	4th	5th
Terms of sequence	5	6	7	8	9

Each term is 4 more than the position number.

1st term   1 + 4 = 5

2nd term   2 + 4 = 6

3rd term   3 + 4 = 7

4th term   4 + 4 = 8

5th term   5 + 4 = 9

So the *n*th term formula is *n* + 4.

Exercise 18C

1   For each sequence find:

   i   the missing terms

   ii   the *n*th term formula.

a

Position	1st	2nd	3rd	4th	5th	…	9th	10th
Term	2	4	6	8	10	…		

b

Position	1st	2nd	3rd	4th	5th	…	11th	12th
Term	5	10	15	20	25	…		

c

Position	1st	2nd	3rd	4th	5th	…	20th	21st
Term	4	8	12			…		

d

Position	1st	2nd	3rd	4th	5th	…	9th	10th
Term	7	14		28		…		

e

Position	1st	2nd	3rd	4th	5th	…	50th	51st
Term	10	20	30			…		

f

Position	1st	2nd	3rd	4th	5th	…	100th	101st
Term		12	18		30	…		

 2   Find the *n*th term formula for the following sequences.

   a   8, 16, 24, 32, …

   b   11, 22, 33, 44, …

Patterns and relationships

3   For each sequence find:

   i   the missing terms   ii   the *n*th term formula.

a

Position	1st	2nd	3rd	4th	5th	...	8th	9th
Term	3	4	5	6	7	...		

b

Position	1st	2nd	3rd	4th	5th	...	20th	21th
Term	6	7	8	9	10	...		

c

Position	1st	2nd	3rd	4th	5th	...	15th	16th
Term		12	13		15	...		

d

Position	1st	2nd	3rd	4th	5th	...	100th	101st
Term	21	22		24		...		

4   Find the *n*th term formula for the following sequences.

   a   8, 9, 10, 11, ...   b   15, 16, 17, 18, ...

More difficult sequences

Example 18·5

a   What is the *n*th term formula for the sequence 3, 5, 7, 9, ...?

b   Find the 50th term of the sequence.

a   The term-to-term rule is add 2 which indicates the 2 times table is involved.

This table compares the sequence with the 2 times table:

Position number	1st	2nd	3rd	4th
2 times table	2	4	6	8
Terms of sequence	3	5	7	9

We can see that the terms of the sequence are the same as the sequence of numbers in the 2 times table sequence add 1.

The sequence for the 2 times table has *n*th term formula = $2n$.

Therefore the sequence 3, 5, 7, 9, ... has *n*th term formula = $2n + 1$.

You can check this for all terms in the sequence by substitution:

1st term $2 \times 1 + 1 = 3$
2nd term $2 \times 2 + 1 = 5$
3rd term $2 \times 3 + 1 = 7$
4th term $2 \times 4 + 1 = 9$

b   The *n*th term formula = $2n + 1$

so the 50th term = $2 \times 50 + 1$

            = 101

Number, money and measure

Example 18·6

a Find the *n*th term formula for the sequence 2, 5, 8, 11, ...

b Find the 50th term.

a The term-to-term rule is add 3 which indicates the 3 times table is involved. Create a table to help compare the sequence with the 3 times table.

Position number	1st	2nd	3rd	4th
3 times table	3	6	9	12
Terms of sequence	2	5	8	11

The 3 times table has *n*th term formula = $3n$.

The sequence is 1 less than the 3 times table.

Therefore *n*th term formula = $3n - 1$.

b The 50th term = $3 \times 50 - 1$

$= 149$

Example 18·7

The *n*th term of the sequence 8, 13, 18, 23, 28, ... is given by the expression $5n + 3$.

a Show this is true for the first three terms.

b Use the rule to find the 50th term of the sequence.

a Let $n = 1$, $5 \times 1 + 3 = 5 + 3 = 8$

Let $n = 2$, $5 \times 2 + 3 = 10 + 3 = 13$

Let $n = 3$, $5 \times 3 + 3 = 15 + 3 = 18$

b Let $n = 50$, $5 \times 50 + 3 = 250 + 3 = 253$

so the 50th term is 253.

Example 18·8

Look at the sequence with the following pattern.

Pattern (term) number   1   2   3

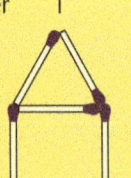 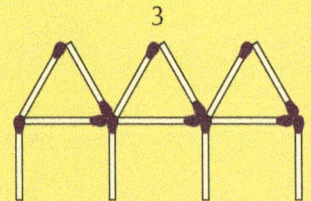

Number of matchsticks   5   9   13

a Find the generalisation (*n*th term) of the pattern.

b Find the 50th term in this sequence.

a What is the term-to-term rule here? It is add 4, so the rule is based on $4n$.

The first term is 5. For the first term $n = 1$ so $4 \times 1 + 1 = 5$, giving:

*n*th term formula = $4n + 1$

b Use the rule to find the 50th term in the pattern.

When $n = 50$, $4n + 1 = 4 \times 50 + 1 = 201$.

Patterns and relationships

Exercise 18D

1 For each of the sequences whose *n*th term is given below, find:

  i the first three terms   ii the 100th term.

  a $2n + 1$ b $4n - 1$ c $5n - 3$
  d $3n + 2$ e $4n + 5$ f $10n + 1$
  g $7n - 1$ h $\frac{1}{2}n + 2$ i $\frac{1}{2}n - \frac{1}{4}$

2 For each of the patterns below, find:

  i the *n*th term for the number of matchsticks

  ii the number of matchsticks in the 50th term.

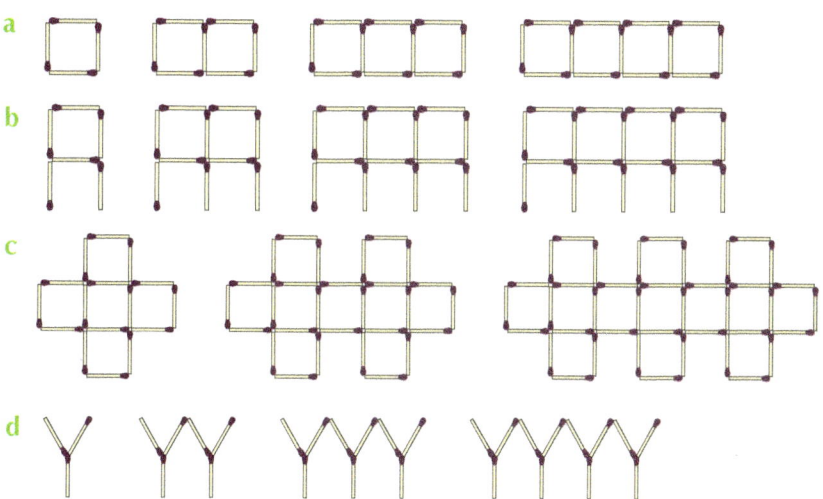

3 Find the *n*th term of each of the following sequences.

  a 3, 9, 15, 21, 27, ...   b 10, 13, 16, 19, 22, ...
  c 7, 13, 19, 25, 31, ...   d 1, 4, 7, 10, 13, ...
  e 4, 11, 18, 25, 32, ...   f 5, 7, 9, 11, 13, ...
  g 9, 13, 17, 21, 25, ...   h 5, 13, 21, 29, 37, ...
  i 11, 21, 31, 41, 51, ...   j 3, 12, 21, 30, 39, ...

Number, money and measure

4 The patterns below contain two different colours of matchsticks. Find the *n*th term for:

　i　the number of red-tipped matchsticks　**ii**　the number of blue-tipped matchsticks

　iii　the total number of matchsticks.

Use your generalisations to describe the 50th term in the patterns by finding:

　iv　the number of red-tipped matchsticks　**v**　the number of blue-tipped matchsticks

　vi　the total number of matchsticks.

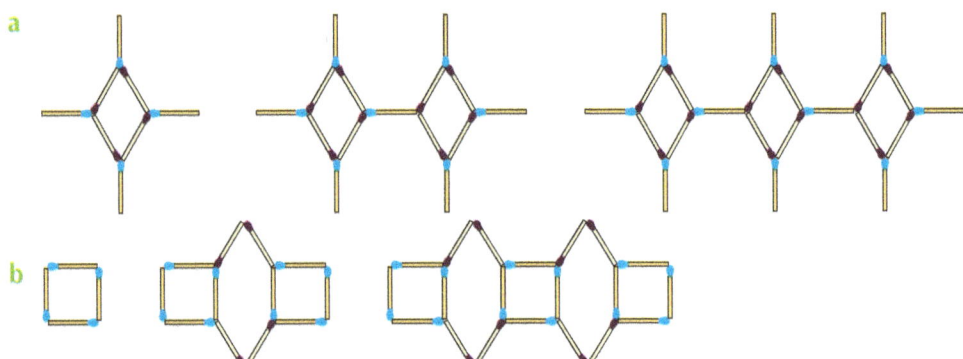

Challenge

1 Describe how the sequences below are generated.

　a 1, 4, 7, 10, 13, 16, …　　**b** 1, 4, 16, 64, 256, 1024, …

　c 1, 4, 8, 13, 19, 26, …　　**d** 1, 4, 9, 16, 25, 36, …

2 Describe how each of the following sequences is generated and write down the next two terms.

　a 40, 41, 43, 46, 50, 55, …　　**b** 90, 89, 87, 84, 80, 75, …

　c 1, 3, 7, 13, 21, 31, …　　**d** 2, 6, 12, 20, 30, 42, …

3 You are given a start number and a multiplier. Write down at least the first six terms of the sequences (for example, start 2 and multiplier 3 gives 2, 6, 18, 54, 162, 486, …).

　a start 1, multiplier 3　　**b** start 2, multiplier 2　　**c** start 1, multiplier 0·1

　d start 1, multiplier 0·5　　**e** start 2, multiplier 0·4　　**f** start 1, multiplier 0·3

- By working on this topic I can explain how to find missing terms in a sequence and find a rule to describe a sequence.
- I understand that a sequence has terms and can explain the term-to-term rule. ★ Exercise 18A Q5
- I can use my understanding of the term-to-term rule to find missing terms of a sequence. ★ Exercise 18B Q3
- I understand the difference between the *n*th term formula and the term-to-term rule and can find the *n*th term formula for simple sequences. ★ Exercise 18C Q2
- I can find the *n*th term formula for more difficult sequences. ★ Exercise 18D Q3

Expressions and equations

19

I can collect like algebraic terms, simplify expressions and evaluate using substitution.

MTH 3-14a

This chapter will show you how to:
- write expressions using letters and other symbols instead of numbers
- collect like terms in algebraic expressions
- add and subtract algebraic terms
- multiply and divide simple algebraic terms
- evaluate expressions by substituting numbers for letters.

You should already know:
- how to apply the rules of BODMAS
- how to use index form
- how to work with negative numbers
- how to simplify a fraction.

Algebraic terms and expressions

In algebra, you will keep meeting three words: **variable, term** and **expression.**

Variable. This is the letter in a term or an expression whose value can vary. Some of the letters most used for variables are x, y, n and t.

In algebra, to make expressions simpler, and to avoid confusion with the variable x, the multiplication sign × is usually left out, and if there is more than one letter they are written in alphabetical order. In the simpler expressions, the numbers go in front of the variables. So:

$3 \times m = 3m \qquad a \times b = ab \qquad w \times 7 = 7w \qquad d \times 4c = 4cd \qquad n \times (d + t) = n(d + t)$

The division sign ÷ cannot be just left out. Instead, it is usually replaced by a short rule in the style of a fraction. So:

$3 \div m = \dfrac{3}{m} \qquad w \div 7 = \dfrac{w}{7} \qquad a \div b = \dfrac{a}{b} \qquad d \div 4c = \dfrac{d}{4c} \qquad n \div (d+t) = \dfrac{n}{d+t}$

Term. This is an algebraic quantity which contains only a letter (or combination of letters) and may contain a number. For example:

- $3n$ means 3 multiplied by the variable n
- $\dfrac{n}{2}$ means n divided by 2
- n^2 means n multiplied by itself (normally said as 'n squared')

Expression. This is a combination of letters (variables) and signs, often with numbers. For example:

- $8 - n$ means subtract n from 8
- $n - 3$ means subtract 3 from n
- $2n + 7$ means n multiplied by 2 then add 7

Number, money and measure

When you give a particular value to the variable in an expression, the expression takes on a particular value.

For example, if the variable n takes the value of 4, then the terms and expressions which include this variable will have particular values, as shown below:

$3n = 3 \times 4 = 12$ $\dfrac{n}{2} = \dfrac{4}{2} = 2$ $n^2 = 4^2 = 16$ $8 - n = 8 - 4 = 4$

$n - 3 = 4 - 3 = 1$ $2n + 7 = 2 \times 4 + 7 = 15$

Rules of algebra

The rules (conventions) of algebra are the same rules that are used in arithmetic. For example:

$3 + 4 = 4 + 3$ $a + b = b + a$

$3 \times 4 = 4 \times 3$ $a \times b = b \times a$ or $ab = ba$

But remember, for example, that:

$7 - 5 \neq 5 - 7$ $a - b \neq b - a$

$6 \div 3 \neq 3 \div 6$ $\dfrac{a}{b} \neq \dfrac{b}{a}$

From one fact, other facts can be stated. For example:

$3 + 4 = 7$ $a + b = 10$

gives $7 - 4 = 3$ and $7 - 3 = 4$ gives $10 - a = b$ and $10 - b = a$

$3 \times 4 = 12$ $ab = 10$

gives $\dfrac{12}{3} = 4$ and $\dfrac{12}{4} = 3$ gives $\dfrac{10}{a} = b$ and $\dfrac{10}{b} = a$

Example 19·1

Show by substitution of suitable numbers that

$a + b - c \neq c + b - a$ (the symbol $\neq$ means 'is not equal to')

Choosing $a = 2$, $b = 5$ and $c = 3$ we get

$a + b - c = 2 + 5 - 3 = 4$

and

$c + b - a = 3 + 5 - 2 = 6$

so $a + b - c \neq c + b - a$

Example 19·2

Work out the value of these expressions when $x = 3$.

a $5x^2$ b $(5x)^2$

a $5x^2 = 5 \times (3 \times 3)$ b $(5x)^2 = (5 \times 3)^2$

$= 5 \times 9$ $= 15^2$

$= 45$ $= 225$

Expressions and equations

Exercise 19A

1. Show by the substitution of suitable numbers that:

   a $m + n = n + m$ b $ab = ba$ c $p - t \neq t - p$ d $\dfrac{m}{n} \neq \dfrac{n}{m}$

2. Show by the substitution of suitable numbers that:

   a If $a + b = 7$, then $7 - a = b$ and $7 - b = a$ b If $ab = 12$, then $a = \dfrac{12}{b}$ and $b = \dfrac{12}{a}$

3. Show by the substitution of suitable numbers that:

   a $a + b + c = c + b + a$ b $acb = abc = cba$

 4. Write down the value of each expression when $t = 5$.

   a $3t^2$ b $(3t)^2$ c $4t^2 + 1$ d $(4t + 1)^2$

5. Show, by substitution, that $9t^2 = (3t)^2$.

6. Only some of these statements are true. List those which are true.

   (*Hint:* Substitute numbers in place of letters and test the statement.)

   a $b + c = d$ is the same as $b = d - c$

   b $a - b = 6$ is the same as $a = 6 + b$

   c $5 = x + 3$ is the same as $5 + x = 3$

   d $5 - 2x = 8$ is the same as $8 = 2x - 5$

   e $ab - ba = c$ is the same as $c = ba - ab$

7. Find the pairs of expressions in each box that are equal to each other and write them down. The first one is done for you.

   | $a + b$ |
   | $b + a$ | $a + b = b + a$
   | ab |

   a
   | $m \times n$ |
   | $m + n$ |
   | mn |

   b
   | $p - q$ |
   | $q - p$ |
   | $-p + q$ |

   c
   | $a \div b$ |
   | $b \div a$ |
   | $\dfrac{a}{b}$ |

   d
   | $6 + x$ |
   | $6x$ |
   | $x + 6$ |

   e
   | $3y$ |
   | $3 + y$ |
   | $3 \times y$ |

Challenge

1. Write down some pairs of values for a and b which make the following statement true.

   $a + b = ab$

   You will find only one pair of integers. There are lots of decimal numbers to find, but each time try to keep one of the variables an integer.

2. Write down some pairs of values for a and b which make the following statement true.

   $a - b = \dfrac{a}{b}$

   You will find only one pair of integers. There are lots of decimal numbers to find, but each time try to keep one of the variables an integer.

   Continued

Number, money and measure

Continued

3. Does $(a + b) \times (a - b) = a^2 - b^2$ work for all values of a and b?

4. Explain why two consecutive integers multiplied together always give an even number.

5. Show that any three consecutive integers multiplied together always give a number in the six times table.

Simplifying expressions

When you have, for example, $5 + 5 + 5$, you can write it simply as 3×5. Likewise in algebra, terms which use the same letter can be added together in the same way. For example:

$m + m + m + m = 4 \times m = 4m$

$p + p + p = 3 \times p = 3p$

Example 19·3

Simplify both of these.

a $d + d + d + d + d$ b $pq + pq + pq + pq + pq + pq$

a There are five ds, which simplify to $5 \times d = 5d$.
b There are six pqs, which simplify to $6 \times pq = 6pq$.

Example 19·4

Simplify these expressions.

a $4a \times b$ b $9p \times 2$ c $3h \times 4i$

Leave out the multiplication sign and write the number to the left of the letters:

a $4a \times b = 4ab$ b $9p \times 2 = 18p$ c $3h \times 4i = 12hi$

When you have a product of two variables, place them in alphabetical order.

Example 19·5

Simplify these expressions.

a $8x \div 2$ b $5m \div 5$ c $9q \div 12$

Set out each expression in a fraction style, and cancel the numerator and the denominator wherever possible. So:

a $\dfrac{8x}{2} = 4x$ b $\dfrac{5m}{5} = m$ c $\dfrac{9q}{12} = \dfrac{3q}{4}$

Exercise 19B

1. Simplify each of the following expressions.

   a $m + m$ b $k + k + k$ c $a + a + a + a$ d $d + d + d$
   e $q + q + q + q$ f $t + t$ g $n + n + n + n$ h $g + g + g$
   i $p + p + p$ j $w + w + w + w$ k $i + i + i + i + i$ l $a + a + a + a$

Expressions and equations

2 Copy and complete each of the following. For example:

$t + t + t + t = 4 \times t = 4t$

a $\quad p + p + p = 3 \times p = \square$

b $\quad m + m + m + m = 4 \times m = \square$

c $\quad k + k + k = \square = \square$

d $\quad h + h + h + h + h = \square = \square$

e $\quad \square = 6 \times m = 6m$

f $\quad \square = 5 \times p = \square$

g $\quad \square = 3 \times g = \square$

h $\quad \square = \square = 7n$

i $\quad \square = \square = 5y$

3 Write each of these expressions using algebraic shorthand.

a $\quad 3 \times n$
b $\quad 5 \times n$
c $\quad 7 \times m$
d $\quad 8 \times t$
e $\quad a \times b$
f $\quad m \times n$
g $\quad p \times 5$
h $\quad q \times 4$
i $\quad m \div 3$
j $\quad 5 \div n$
k $\quad 7 \times w$
l $\quad k \times d$
m $\quad t \times t$
n $\quad 8 \div k$
o $\quad 9 \times m$
p $\quad g \times g$

4 Copy and complete each of these.

a $\quad mp = m \times \square$
b $\quad tv = \square \times v$
c $\quad qr = \square \times \square$
d $\quad \square = k \times g$
e $\quad ab = \square \times b$
f $\quad hp = h \times \square$
g $\quad \square = t \times f$
h $\quad pt = \square \times \square$
i $\quad \square = n \times n$

5 Simplify the following expressions.

a $\quad h \times 4p$
b $\quad 4s \times t$
c $\quad 2m \times 4n$
d $\quad 5w \times 5x$
e $\quad b \div 9c$
f $\quad 3b \times 4c \times 2d$
g $\quad 4g \times f \times 3a$
h $\quad 4m \times 5p \times 3q$

6 Write each of these expressions in as simple a way as possible.

a $\quad 4x \div 2$
b $\quad 12x \div 3$
c $\quad 20m \div 4$
d $\quad 36q \div 3$
e $\quad 4m \div 2$
f $\quad 16p \div 4$
g $\quad 7q \div 7$
h $\quad 5n \div 5$
i $\quad 16m \div 2$
j $\quad 20k \div 10$
k $\quad 36p \div 9$
l $\quad 25t \div 5$
m $\quad 18p \div 12$
n $\quad 16q \div 10$
o $\quad 15m \div 10$
p $\quad 14t \div 6$

★ **7** Write each of these expressions in as simple a way as possible.

a $\quad 3m + 2m$
b $\quad 8q - 5q$
c $\quad 2y + 4y - y$
d $\quad 3 + 5h$
e $\quad 2 + 3r + 2r$
f $\quad 12a + 4$
g $\quad 4t + 2 + 3t$
h $\quad 3b + 4c + 6$

Collecting like terms

Like terms are multiples of the same letter, or of the same combination of letters. Also, they can be the same power of the same letter or the same powers of the same combination of letters. For example:

$x, 4x, \frac{1}{2}x, -3x$ $\quad$ Like terms which are multiples of x

$5ab, 8ab, \frac{1}{2}ab, -2ab$ $\quad$ Like terms which are multiples of ab

$y^2, 3y^2, \frac{1}{4}y^2, -4y^2$ $\quad$ Like terms which are multiples of y^2

Number, money and measure

The multiples are called **coefficients.** So, in the above examples, 1, 4, $\frac{1}{2}$, –3, 5, 8, $\frac{1}{2}$, –2, 1, 3, $\frac{1}{4}$ and –4 are coefficients.

Only like terms can be added or subtracted to simplify an expression. For example:

$3ab + 2ab$ simplifies to $5ab$ $\qquad$ $3y + y$ simplifies to $3y + 1y = 4y$

$8x^2 – 5x^2$ simplifies to $3x^2$ $\qquad$ $7n – n$ simplifies to $7n – 1n = 6n$

Unlike terms cannot be simplified by addition or subtraction. For example:

$9a + 5a + 10b – 4b = 14a + 6b$

The expression $14a + 6b$ cannot be simplified because $14a$ and $6b$ are unlike terms.

Simplifying an expression means making it shorter by combining its terms where possible. This usually involves two steps:

- Collect the like terms into groups of the same sort.
- Combine each group of like terms, and simplify.

Look at these two boxes.

Examples of combining like terms

$3p + 4p = 7p$ $\qquad$ $5t + 3t = 8t$

$9w – 4w = 5w$ $\qquad$ $12q – 5q = 7q$

$a + 3a + 7a = 11a$

$15m – 2m – m = 12m$

Examples of unlike terms

$x + y$ $\qquad$ $2m + 3p$

$7 – 3y$ $\qquad$ $5g + 2k$

$m – 3p$

Like terms can be combined even when they are mixed together with unlike terms.

Examples of different sorts of like terms mixed together

$4t + 5m + 2m + 3t + m = 7t + 8m$

$5k + 4g – 2k – g = 3k + 3g$

Remember: You **never** write the number 1 in front of a variable.

$g = 1g \qquad m = 1m$

Work through Example 19·6.

Example 19·6

a $5p – 2p = 3p$

b $5ab + 3ab = 8ab$

c $3x^2 + 6x^2 = 9x^2$

d $7y – 9y = –2y$ (because $7 – 9 = –2$)

e $–3u – 6u = –9u$ (because $–3 – 6 = –9$)

f $5a + 2a + 3b = 7a + 3b$

g $5p – 2p + 7y – 9y = 3p – 2y$

h $8t + 3i – 6t – i = 8t – 6t + 3i – i$

$\qquad = 2t + 2i$

(put the like terms together before combining)

Expressions and equations

Example 19·7

Simplify $8p + 2q + 3p + 7s + 4q + 9$.

Write out the expression: $8p + 2q \ + \ 3p + 7s \ + \ 4q + 9$

Then collect like terms: $8p + 3p \ + \ 2q + 4q \ + \ 7s + 9$
Next, combine them: $\quad\quad 11p \ \ + \quad 6q \quad + \ 7s + 9$

So, the expression in its simplest form is:

$11p + 6q + 7s + 9$

Example 19·8

Simplify $7x^2 + y^2 + 2x^2 - 3y^2 + 3z - 5$

Write out the expression: $7x^2 + y^2 \ + \ 2x^2 - 3y^2 \ + 3z - 5$

Then collect like terms: $7x^2 + 2x^2 + \ y^2 - 3y^2 \ + \ 3z - 5$
Next, combine them: $\quad\quad 9x^2 \quad - \quad 2y^2 \quad + \ 3z - 5$

So, the expression in its simplest form is:

$9x^2 - 2y^2 + 3z - 5$

Note: Each term has a sign (+ or −). When rearranging an expression, move the sign with the term.

Exercise 19C

1 Simplify each of these.

a $2b + 3b$ b $5x + 2x$ c $6m + m$ d $3m + m + 2m$

e $7d - 3d$ f $8g - 3g$ g $4k - k$ h $3t + 2t - t$

2 Simplify each of these.

a $5g + g - 2g$ b $3x + 5x - 6x$ c $4h + 3h - 5h$ d $4q + 7q - 3q$

e $5h - 2h + 4h$ f $6x - 4x + 3x$ g $3y - y + 4y$ h $5d - 4d + 6d$

i $8x - 2x - 3x$ j $5m - m - 2m$ k $8k - 3k - 2k$ l $6n - 3n - n$

3 From each cloud: **i** group together the like terms   **ii** simplify each cloud.

For example:

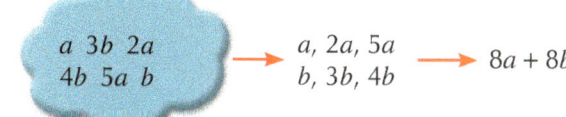

a

    3t g 5t
    8g 9t 7g

b

    m 7p 4m
    9p 10m
    3p

c

    4k 3m −7m
    5w 8m
    7w 7m k

d

    x^2 t $5x^2$
    3t $3x^2$ 4t

e

    y^2 2y 8y
    $7y^2$ $4y^2$
    3y $-4y^2$

f

    7w 7g 3h
    3g 9h 4w
    3w 10g −4w

Number, money and measure

4 Simplify each of these expressions.

- a $3b + 5 + 2b$
- b $2x + 7 + 3x$
- c $m + 2 + 5m$
- d $4k + 3k + 8$
- e $3x + 7 - x$
- f $5k + 4 - 2k$
- g $6p + 3 - 2p$
- h $5d + 1 - 4d$
- i $5m - 3 - 2m$
- j $6t - 4 - 2t$
- k $4w - 8 - 3w$
- l $5g - 1 - g$
- m $t + k + 4t$
- n $3x + 2y + 4x$
- o $2k + 3g + 5k$
- p $3h + 2w + w$
- q $5t - 2p - 3t$
- r $6n - 2t - 5n$
- s $p + 4q - 2q$
- t $3n + 2p - 3n$

5 Simplify each of these expressions.

- a $2t + 3g + 5t + 2g$
- b $4x + y + 2x + 3y$
- c $2m + k + 3m + 2k$
- d $5x + 3y - 2x + y$
- e $6m + 2p - 4m + 3p$
- f $3n + 4t - n + 3t$
- g $6k + 3g - 2k - g$
- h $7d + 4b - 7d - 4b$
- i $4q + 3p - 3q - p$
- j $4g - k + 2g - 3k$
- k $2x - 3y - 5x + 3y$
- l $4d - 3e - 3d - 2e$

6 Simplify each of these expressions.

- a $5x^2 + x^2$
- b $6k^2 + 2k^2$
- c $4m^2 + 3m^2$
- d $6d^2 - 2d^2$
- e $5g^2 - 3g^2 - g^2$
- f $7a^2 - 5a^2$
- g $5f^2 + 2f^2$
- h $3y^2 - y^2$
- i $t^2 + 3t^2$
- j $5h^2 - 2h^2$
- k $6k^2 + k^2 - 7k^2$
- l $7m^2 - 3m^2$

7 Simplify each of these expressions.

- a $8x + 3 + 2x + 5$
- b $5p + 2k + p + 3k$
- c $9t + 3m + 2t + m$
- d $7k + 3t + 2t - 3k$
- e $5m + 4p + p - 3m$
- f $8w + 2d + 3d - 2w$
- g $6x + 2y + 4y + 2x$
- h $8p + 3q + 4q - 3p$
- i $3m + 2t + 3t - m$

8 Simplify the following expressions.

- a $5h + 6h$
- b $4p + p$
- c $9u - 3u$
- d $3b - 8b$
- e $-2j + 7j$
- f $-6r - 6r$
- g $2k + k + 3k$
- h $9y - y$
- i $7d - 2d + 5d$
- j $10i + 3i - 6i$
- k $2b - 5b + 6b$
- l $-2b + 5b - 7b$
- m $3xy + 6xy$
- n $4p^2 + 7p^2$
- o $5ab - 10ab$
- p $5a^2 - 2a^2 - 3a^2$
- q $4fg - 6fg - 8fg$
- r $6x^2 - 3x^2 - 5x^2$

9 Simplify the following expressions.

- a $6h + 2h + 5g$
- b $4g - 2g + 8m$
- c $8f + 7d + 3d$
- d $4x + 5y + 7x$
- e $6q + 3r - r$
- f $4 + 5s - 3s$
- g $c + 2c + 3$
- h $12b + 7 + 2b$
- i $7w - 7 + 7w$
- j $2bf + 4bf + 5g$
- k $7d + 5d^2 - 2d^2$
- l $6st - 2st + 5t$
- m $4s - 7s + 2t$
- n $-5h + 2i + 3h$
- o $4y - 2w - 7w$

★ 10 Simplify the following expressions.

- a $9e + 4e + 7f + 2f$
- b $10u - 4u + 9t - 2t$
- c $b + 3b + 5d - 2d$
- d $4a + 5c + 3a + 2c$
- e $f + 2g + 3g + 5f$
- f $9h + 4i - 9h + 2i$
- g $7p + 8q - 6p - 3q$
- h $14j - 5k + 5j + 9k$
- i $4u - 5t - 6u + 7t$
- j $2s + 5t - 9t + 3s$
- k $5p - 2q - 7p + 3q$
- l $-2d + 5e - 4d - 9e$

Expressions and equations

11 Simplify the following expressions.

a $x^2 + 5x + 2x^2 + 3x$
b $5ab + 3a + 4ab + 7a$
c $4y^2 - 4y + 3y^2 - 5y$
d $8mn - 3n + 3mn + 2n$
e $5t^2 - 8t - 2t^2 - 4t$
f $3q^2 - 5q - 6q^2 + 3q$

12 The expression in each box is made by adding the expressions in the two boxes below it. Copy the diagrams and fill in the missing expressions.

a
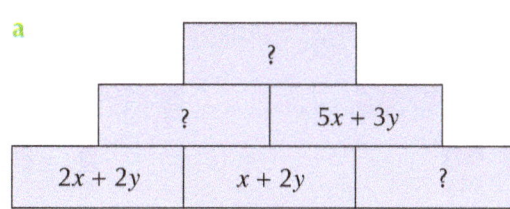

b
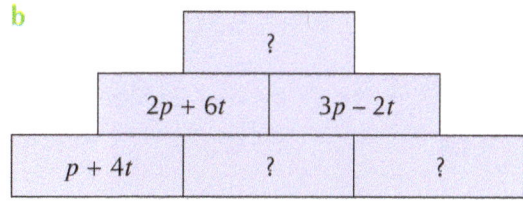

c
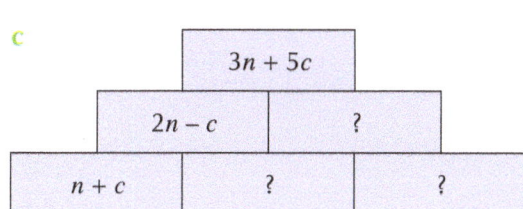

d
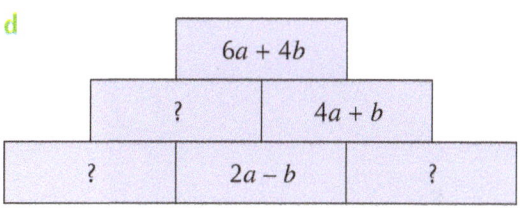

Substituting into expressions

Replacing the letters in an expression by numbers is called **substitution**.

Substituting different numbers will give an expression different values. You need to be able to substitute negative numbers as well as positive numbers into expressions.

Example 19.9

What is the value of $7 + 5x$ when: i $x = 3$ ii $x = -1$

i When $x = 3$, $7 + 5x = 7 + 5 \times 3 = 22$ (Remember the rules of BODMAS.)

ii When $x = -1$, $7 + 5x = 7 + 5 \times (-1) = 7 + -5 = 2$

Exercise 19D

1 Write down the value of each expression for each value of x.

		i	ii	iii
a	$x + 3$	$x = 4$	$x = 5$	$x = -1$
b	$7 + x$	$x = 6$	$x = 2$	$x = -5$
c	$3x$	$x = 7$	$x = 3$	$x = -5$
d	$4x$	$x = 3$	$x = 5$	$x = -1$
e	$3x + 4$	$x = 5$	$x = 4$	$x = -2$
f	$5x - 1$	$x = 3$	$x = 2$	$x = -6$
g	$3x + 5$	$x = 3$	$x = 7$	$x = -1$
h	$4x + 20$	$x = 4$	$x = 5$	$x = -3$
i	$8 + 7x$	$x = 2$	$x = 6$	$x = -2$
j	$93 + 4x$	$x = 10$	$x = 21$	$x = -3$

Number, money and measure

2 If $a = 2$ and $b = 3$, show working to find the value of each of the following.

    a $3a + b$ **b** $a + 3b$ **c** $3a + 5b$ **d** $4b - 3a$

3 If $c = 5$ and $d = 2$, show working to find the value of each of the following.

    a $2c + d$ **b** $6c - 2d$ **c** $3d + 7c$ **d** $3c - 5d$

4 If $e = 4$, $f = 3$ and $g = 2$, show working to find the value of each of the following.

    a $ef + g$ **b** $eg - f$ **c** efg **d** $3e + 2f + 5g$

★ **5** If $h = -4$, $j = 7$ and $k = 5$, show working to find the value of each of the following.

    a hjk **b** $4k + hj$ **c** $5h + 4j + 3k$ **d** $3hj - 2hk$

6 Write down the value of each expression for each value of x.

    a $x^2 - 7$ **i** $x = 6$ **ii** $x = 2$ **iii** $x = -10$

    b $21 + 3x^2$ **i** $x = 7$ **ii** $x = 3$ **iii** $x = -5$

    c $54 - 2x^2$ **i** $x = 3$ **ii** $x = 5$ **iii** $x = -1$

    d $5(3x + 4)$ **i** $x = 5$ **ii** $x = 4$ **iii** $x = -2$

    e $3(5x - 1)$ **i** $x = 3$ **ii** $x = 2$ **iii** $x = -6$

Challenge

1 What values of n can be substituted into n^2 that give n^2 a value less than 1?

2 What values of n can be substituted into $(n - 4)^2$ that give $(n - 4)^2$ a value less than 1?

3 What values of n can be substituted into $\frac{1}{n}$ that give $\frac{1}{n}$ a value less than 1?

4 Find at least five different expressions in x that give the value 10 when $x = 2$ is substituted into them.

Substituting into formulae

Formulae occur in all sorts of situations, often when converting between two sorts of quantity. Some examples are converting between degrees Celsius and degrees Fahrenheit, or between different currencies, such as from pounds (£) to euros (€).

Example 19·10   The formula for converting kilograms (K) to pounds (P) is given by:

$P = 2\cdot2K$

Convert 10 kilograms to pounds.

Substituting $K = 10$ into the formula gives: $P = 2\cdot2 \times 10 = 22$

So, 10 kg is 22 pounds.

Expressions and equations

Example 19·11 The formula for the volume, V, of a box with length l, breadth b and height h, is given by:

$V = lbh$

Calculate the volume of a box whose length is 8 cm, breadth is 3 cm and height is 2 cm.

Substitute the values given into the formula: $V = 8 \times 3 \times 2 = 48$

So, the volume of the box is 48 cm³.

Example 19·12 The formula for converting degrees Celsius (°C) to degrees Fahrenheit (°F) is:

$$F = \frac{9C}{5} + 32$$

Convert 35 °C to °F.

Substituting $C = 35$ into the formula gives:

$$F = \frac{9 \times \cancel{35}^{7}}{\cancel{5}_{1}} + 32 = 63 + 32 = 95$$

So, 35 °C = 95 °F.

Exercise 19E

1. The area, A, of a rectangle is found using the formula $A = LB$, where L is its length and B its breadth. Find A when:
   a. $L = 8$ and $B = 7$
   b. $L = 6$ and $B = 1\cdot5$

2. The surface area, A, of a coil is given by the formula $A = 6rh$. Find A when:
   a. $r = 6$ and $h = 17$
   b. $r = 2\cdot5$ and $h = 12$

3. The sum, S, of the angles in a polygon with n sides is given by the formula $S = 180(n - 2)°$. Find S when:
   a. $n = 7$
   b. $n = 12$

4. If $V = u + ft$, find V when:
   a. $u = 40$, $f = 32$ and $t = 5$
   b. $u = 12$, $f = 13$ and $t = 10$

5. If $D = \frac{M}{V}$ find D when:
   a. $M = 28$ and $V = 4$
   b. $M = 8$ and $V = 5$

6. A magician charges £25 for every show he performs, plus an extra £10 per hour spent on stage. The formula for calculating his charge is $C = 10t + 25$, where C is the charge in pounds and t is the length of the show in hours.

   How much does he charge for a show lasting the following durations?
   a. 1 hour
   b. 3 hours
   c. 2 hours

7. The area (A) of the triangle shown is given by the formula $A = \frac{1}{2}bh$.

   Find the area when:
   a. $h = 12$ cm and $b = 5$ cm
   b. $h = 9$ cm and $b = 8$ cm

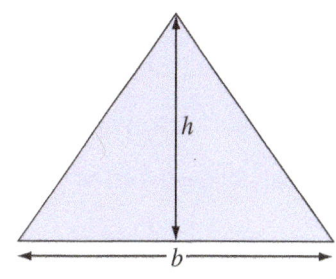

Number, money and measure

8. The area (A) of the trapezium shown is given by the formula $A = \dfrac{h(a+b)}{2}$.

   What is the area when:

   a $h = 12$ cm, $a = 7$ cm and $b = 5$ cm
   b $h = 9$ cm, $a = 1.5$ cm and $b = 8.5$ cm?

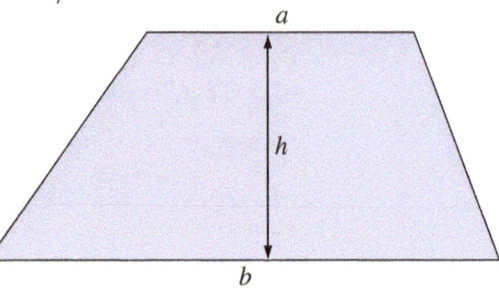

9. The following formula converts temperatures in degrees Celsius (C) to degrees Fahrenheit (F).

   $F = 1.8C + 32$

   Convert each of these temperatures to degrees Fahrenheit.

   a 45 °C   b 40 °C   c 65 °C   d 100 °C

10. If $N = h(A^2 - B^2)$, find N when:

    a $h = 7$, $A = 5$ and $B = 3$ b $h = 15$, $A = 4$ and $B = 2$

11. If $V = hr^2$, find V when:

    a $h = 5$ and $r = 3$ b $h = 8$ and $r = 5$

12. The volume (V) of the cuboid shown is given by the formula:

    $V = abc$

    The surface area (S) of the cuboid is given by the formula:

    $S = 2ab + 2bc + 2ac$

    a Find: i the volume
              ii the surface area

       when $a = 3$ m, $b = 4$ m and $c = 5$ m.

    b Find: i the volume
              ii the surface area

    when $a = 3$ cm, and a, b and c are all the same length. What name is given to this cuboid?

- By working on this topic I can explain how to work with simple algebraic terms.
- I understand that a variable is a letter that can represent a number. ★ Exercise 19A Q4
- I can simplify expressions by adding, subtracting, multiplying and dividing terms. ★ Exercise 19B Q7
- I understand what like terms are and can simplify expressions by rearranging and collecting like terms. ★ Exercise 19C Q10
- I can evaluate expressions by substituting given values of variables. ★ Exercise 19D Q5
- I can follow the rules of algebra to evaluate unfamiliar formulae. ★ Exercise 19E Q11

Expressions and equations

20

Having discussed ways to express problems or statements using mathematical language, I can construct, and use appropriate methods to solve, a range of simple equations.

MTH 3-15a

This chapter will show you how to:
- solve simple equations using a variety of methods
- check your answers by substituting the value back into the equation
- solve equations which have letters on both sides
- construct equations to solve word problems.

You should already know:
- that algebra uses letters to represent numbers
- how to simplify an algebraic expression by collecting like terms
- the order of operations (BODMAS)
- that in order to 'undo' an operation you do the opposite
- how to evaluate an algebraic expression by substitution.

Solving equations

An **equation** is formed when an expression is put equal to a number or another expression.

The equations you will meet contain only one unknown value, which is often represented by x. This is called the **unknown** of the equation. To find the value of the unknown, the equation has to be **solved**. This chapter will show you a number of methods to help solve equations.

Solving equations using the 'cover-up' method

To solve an equation using this method, cover the unknown and work out what value it must take.

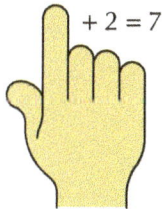
$+ 2 = 7$

Example 20·1   Solve the equation $x + 2 = 7$

What do you add to 2 to get 7?

You know that $5 + 2 = 7$

So, $x = 5$

Number, money and measure

Example 20·2

Solve the equation $2x - 3 = 5$.

From what do you take 3 and end up with 5?

You know that $8 - 3 = 5$

So, $2x = 8$

You know that $2 \times 4 = 8$

So, $x = 4$

You should always check your solution by substituting the value back into the equation and checking that the left-hand side (LHS) of the equation equals the right-hand side (RHS).

To check the answer above substitute $x = 4$ into the equation:

LHS $= 2 \times 4 - 3 = 8 - 3 = 5 =$ RHS

Exercise 20A

1 Solve each of the following equations.
   - a $x + 1 = 11$
   - b $x - 3 = 5$
   - c $x + 4 = 19$
   - d $x - 1 = 13$
   - e $x + 3 = 9$
   - f $x - 3 = 12$
   - g $x + 7 = 12$
   - h $x - 5 = 10$
   - i $x - 12 = 33$
   - j $3 + x = 80$
   - k $73 = x + 8$
   - l $x - 7 = 65$

2 Solve each of the following equations.
   - a $x + 23 = 35$
   - b $x + 13 = 21$
   - c $x + 18 = 30$
   - d $48 + x = 54$
   - e $m + 44 = 57$
   - f $m - 13 = 4$
   - g $k - 12 = 6$
   - h $p - 10 = -7$
   - i $72 + k = 95$
   - j $k - 12 = -5$
   - k $m + 33 = 49$
   - l $x - 13 = -8$
   - m $x + 85 = 112$
   - n $n - 21 = -10$
   - o $m - 15 = -9$
   - p $37 = x + 12$

3 Solve each of the following equations.
   - a $2x + 1 = 11$
   - b $4x - 3 = 5$
   - c $5x + 4 = 19$
   - d $2x - 1 = 19$
   - e $9 = 4x - 3$
   - f $6x - 6 = 12$
   - g $5x + 7 = 12$
   - h $2x - 6 = 10$
   - i $33 = 3x - 12$
   - j $3 + 7x = 80$
   - k $5x + 8 = 73$
   - l $9x - 7 = 65$

4 Solve each of the following equations.
   - a $3x - 2 = 13$
   - b $2m - 5 = 1$
   - c $4x - 1 = 11$
   - d $5t - 3 = 17$
   - e $2x - 3 = 13$
   - f $4m - 5 = 19$
   - g $3m - 2 = 10$
   - h $7x - 3 = 25$
   - i $5m - 2 = 18$
   - j $3k - 4 = 5$
   - k $8x - 5 = 11$
   - l $2t - 3 = 7$
   - m $4x - 2 = 14$
   - n $8y - 3 = 29$
   - o $5x - 4 = 21$
   - p $3m - 1 = 17$

5 Solve each of the following equations.
   - a $2x + 3 = 21$
   - b $3x - 5 = 10$
   - c $5x - 1 = 29$
   - d $4x - 3 = 25$
   - e $3m - 8 = 19$
   - f $5m + 4 = 49$
   - g $7m + 3 = 24$
   - h $4m - 5 = 23$
   - i $6k + 1 = 25$
   - j $5k - 3 = 2$
   - k $3k - 1 = 23$
   - l $2k + 5 = 15$
   - m $7x - 3 = 18$
   - n $4x + 3 = 43$
   - o $5x + 6 = 31$
   - p $9x - 4 = 68$

Expressions and equations

6 The solution to each of the following equations may involve a decimal or a fraction.

a $2x + 7 = 8$ b $5x + 3 = 4$ c $2x + 3 = 8$ d $4x + 7 = 20$

e $5x - 3 = 9$ f $2x - 7 = 10$ g $4x - 5 = 6$ h $10x - 3 = 8$

7 Solve each of the following equations.

a $5 + 2x = 11$ b $7 + 3w = 25$ c $5 + 4g = 17$ d $4 + 5x = 24$

e $7 + 4j = 23$ f $3 + 2x = 13$ g $2 + 3x = 38$ h $8 + 5x = 13$

i $3 + 4m = 11$ j $6 + 2n = 20$ k $4 + 3x = 31$ l $7 + 5x = 52$

Solving equations using inverse mapping

Another way to solve equations is to use **inverse mapping**.

Example 20·3

Solve the equation $5x - 4 = 11$.

The mapping which gives this equation is:

$x \longrightarrow \boxed{\times 5} \longrightarrow \boxed{-4} \longrightarrow 11$

The inverse is:

$? \longleftarrow \boxed{\div 5} \longleftarrow \boxed{+4} \longleftarrow 11$

$3 \longleftarrow 15 \longleftarrow 11$

which gives $x = 3$

Exercise 20B

1 i Write the equation which is given by each of the following mappings.

ii Use inverse mapping to solve each equation.

a $x \longrightarrow \boxed{\times 2} \longrightarrow \boxed{+3} \longrightarrow 11$ b $x \longrightarrow \boxed{\times 3} \longrightarrow \boxed{+1} \longrightarrow 16$

c $t \longrightarrow \boxed{\times 5} \longrightarrow \boxed{+4} \longrightarrow 34$ d $t \longrightarrow \boxed{\times 4} \longrightarrow \boxed{-3} \longrightarrow 13$

e $y \longrightarrow \boxed{\times 2} \longrightarrow \boxed{-1} \longrightarrow 13$ f $y \longrightarrow \boxed{\times 7} \longrightarrow \boxed{+5} \longrightarrow 26$

2 i Write the mapping which is given by each of the following equations.

ii Use inverse mapping to solve each equation.

a $3x + 4 = 19$ b $2x + 4 = 16$ c $4x - 1 = 23$

d $5x - 3 = 27$ e $3x + 1 = 22$ f $6x - 5 = 7$

★ **3** Solve each of the following equations using inverse mapping.

a $2x + 3 = 15$ b $4x + 1 = 29$ c $3x + 4 = 25$

d $5x - 1 = 34$ e $2x - 7 = 21$ f $4x - 3 = 37$

Number, money and measure

4 Solve each of the following equations using inverse mapping.

a $2x + 3 = 17$ b $4x - 1 = 19$ c $5x + 3 = 18$
d $2y - 6 = 18$ e $4z + 5 = 33$ f $6x - 5 = 13$
g $10b + 9 = 29$ h $2r - 3 = 9$ i $3x - 11 = 1$
j $7p + 5 = 82$ k $5x - 7 = 48$ l $9x - 8 = 55$

Challenge

$x \longrightarrow \boxed{\times 2} \longrightarrow \boxed{+1} \longrightarrow \boxed{\times 3} \longrightarrow y$

a Find the value of y when:

  i $x = 5$ ii $x = 4$ iii $x = 3$

b Find the value of x when:

  i $y = 39$ ii $y = 51$ iii $y = 57$

Doing the same thing to both sides

You have already met a few different types of equation, which were solved by using the cover-up method or inverse mapping. Here, you will be shown how to solve them by adding, subtracting, multiplying or dividing **both sides** of an equation by the same number. The aim is to get the variable on its own. This is also called the **balancing method**, because by doing the same thing to both sides, you keep the equation balanced.

Example 20·4   Solve $x + 5 = 9$.

Write down the equation: $x + 5 = 9$
Subtract 5 from both sides: $x = 4$

Example 20·5   Solve $\frac{x}{3} = 5$.

Write down the equation: $\frac{x}{3} = 5$
Multiply both sides by 3: $x = 15$

Example 20·6   Solve the equation $2x = 16$.

Write down the equation: $2x = 16$
Divide both sides by 2: $x = 8$

Example 20·7   Solve the equation $4t + 3 = 23$.

Write down the equation: $4t + 3 = 23$
Subtract 3 from both sides: $4t = 20$
Divide both sides by 4: $t = 5$

Expressions and equations

Example 20·8

Solve the equation $\frac{x}{3} - 4 = 2$.

Write down the equation: $\frac{x}{3} - 4 = 2$

Add 4 to both sides: $\frac{x}{3} = 6$

Multiply both sides by 3: $x = 18$

Exercise 20C

In each of the following questions, show your working. Write each step on a new line.

1 Solve each of the following equations.
- a $2x = 10$
- b $3x = 12$
- c $\frac{x}{5} = 6$
- d $4x = 28$
- e $\frac{t}{5} = 12$
- f $7m = 21$
- g $\frac{k}{3} = 6$
- h $\frac{p}{2} = 18$

2 Solve each of the following equations.
- a $2x + 4 = 12$
- b $2x + 5 = 13$
- c $3x + 4 = 34$
- d $3x + 7 = 19$
- e $4m + 1 = 21$
- f $5k + 6 = 21$
- g $4n + 9 = 17$
- h $2x + 7 = 27$
- i $6h + 5 = 23$
- j $3t + 5 = 26$
- k $8x + 3 = 35$
- l $5y + 3 = 28$
- m $7x + 3 = 10$
- n $4t + 7 = 39$
- o $3x + 8 = 20$
- p $8m + 5 = 21$

3 Solve each of the following equations.
- a $3x - 7 = 14$
- b $2m - 5 = 9$
- c $4x + 6 = 22$
- d $5t - 3 = 42$
- e $2x - 3 = 19$
- f $4m - 8 = 24$
- g $3m - 5 = 16$
- h $7x - 3 = 32$
- i $5m - 2 = 33$
- j $3k - 8 = 22$
- k $8x - 4 = 28$
- l $2t - 3 = 21$
- m $4x - 6 = 30$
- n $8m - 3 = 37$
- o $5x - 4 = 36$
- p $3m - 1 = 29$

4 Solve each of the following equations.
- a $3x + 4 = 10$
- b $5x - 7 = 28$
- c $4x + 5 = 45$
- d $3m - 4 = 11$
- e $5m + 4 = 59$
- f $7m + 3 = 52$
- g $4m - 2 = 26$
- h $6k + 7 = 31$
- i $5y - 3 = 47$
- j $3k - 1 = 26$
- k $2k + 6 = 22$
- l $7x - 3 = 39$
- m $4t + 9 = 25$
- n $5x + 6 = 41$
- o $9x - 6 = 57$
- p $8y - 7 = 25$

5 Solve each of the following equations.
- a $3x + 5 = 11$
- b $\frac{x}{2} + 3 = 11$
- c $4x + 7 = 15$
- d $\frac{x}{5} + 3 = 8$
- e $3x + 4 = 37$
- f $\frac{x}{6} + 1 = 5$
- g $2x + 7 = 15$
- h $\frac{x}{4} + 3 = 13$
- i $3x + 6 = 27$

6 Solve each of the following equations.
- a $5x - 2 = 13$
- b $\frac{x}{3} - 4 = 1$
- c $6x - 1 = 23$
- d $\frac{x}{2} - 3 = 5$
- e $4x - 3 = 33$
- f $\frac{x}{3} - 2 = 2$
- g $4x - 5 = 3$
- h $\frac{x}{6} - 1 = 3$
- i $5x - 4 = 16$

Number, money and measure

7 Solve each of the following equations.

a $4x + 8 = 36$
b $\dfrac{x}{3} + 2 = 4$
c $5x + 2 = 47$

d $\dfrac{x}{4} - 1 = 19$
e $2x - 3 = 17$
f $\dfrac{x}{5} - 4 = 21$

g $3m - 5 = 7$
h $\dfrac{b}{4} + 5 = 7$
i $6q + 1 = 31$

8 Nazia has made a mistake somewhere in her working for each of the equations shown below. Can you spot the line on which the error occurs and work out the correct solution?

a $3x + 8 = 23$
  $3x = 18$
  $x = 6$

b $\dfrac{x}{5} - 3 = 2$
  $\dfrac{x}{5} = 5$
  $5x = 25$
  $x = 5$

c $\dfrac{x}{2} - 8 = 24$
  $\dfrac{x}{2} = 32$
  $x = 16$

d $2x - 5 = 17$
  $2x = 12$
  $x = 6$

Solving equations with letters on both sides

Some equations have letters on both sides and need to be rearranged to have both letters (unknowns) on one side and the numbers on the other side before they can be solved. The best method for solving these is to do the same to both sides.

Example 20·9

Solve $5x + 3 = 3x + 11$

Write down the equation: $5x + 3 = 3x + 11$
Subtract $3x$ from both sides: $2x + 3 = 11$
Subtract 3 from both sides: $2x = 8$
Divide both sides by 2: $x = 4$

Example 20·10

Solve $2x + 7 = 22 - 3x$

Write down the equation: $2x + 7 = 22 - 3x$
Add $3x$ to both sides: $5x + 7 = 22$
Subtract 7 from both sides: $5x = 15$
Divide both sides by 5: $x = 3$

Expressions and equations

Exercise 20D

1 Solve the following equations.

   a $6x + 5 = 2x + 17$
   b $5x + 3 = 3x + 8$
   c $4x - 3 = x + 18$
   d $3x - 7 = x + 2$
   e $5x + 4 = 16 - x$
   f $2x - 3 = 9 + x$
   g $6x - 5 = 2x + 1$
   h $7x + 1 = 2x + 7$
   i $4x - 1 = 2x + 10$
   j $5x - 3 = 7 + 3x$
   k $x + 17 = 4x + 2$
   l $3x - 16 = 26 - 4x$

★ 2 Solve the following equations.

   a $5x + 5 = 3x + 12$
   b $3s - 3 = s + 10$
   c $8x - 1 = 8 + 2x$
   d $4q + 5 = 26 - 2q$
   e $3p - 8 = 37 - 3p$
   f $3 + 7m = 11 + 3m$
   g $6 + 4n = 2n + 20$
   h $7 + x = 52 - 4x$
   i $6b + 9 = 29 - 4b$
   j $9t + 5 = 2t + 68$
   k $15k - 3 = 36 + 2k$
   l $9k + 8 = 2k + 176$
   m $3n + 9 = 65 - n$
   n $8 + 5n = 22 + 3n$
   o $5h - 25 = 89 - h$

Using algebra and diagrams to solve problems

The first step to solve a problem using algebra is to write down an equation. This is called **constructing an equation**. To do this, you must choose a letter to stand for the simplest variable (unknown) in the problem. This might be x or the first letter of a suitable word. For example, n is often used to stand for the number.

Example 20·11

A gardener has a fixed charge of £5 and an hourly rate of £8 per hour. Write down an equation for the total charge £C when the gardener is hired for n hours. State the cost of hiring the gardener for 6 hours.

The formula is:
$C = 5$ (for the fixed charge) plus $8 \times n$ (for the hours worked)
$C = 5 + 8n$

If $n = 6$ then the total charge $C = 5 + (8 \times 6)$
$C = 53$

So the charge is £53.

Example 20·12

I think of a number, add 3 and then double it. The answer is 16. What is the number?

Construct an equation, letting the number be x:

Start: x
Add 3: $x + 3$
Double: $2x + 6$

The answer is 16

So the equation is:
$2x + 6 = 16$

Now solve the equation:

Subtract 6 from both sides: $2x = 10$

Divide both sides by 2: $x = 5$

Number, money and measure

Example 20·13

My son is 25 years younger than I am. Our ages add up to 81. How old are we?

Construct the equation using x as my son's age. (Since this is the lower age.)

So, my age is $x + 25$.

The total of our ages is 81, which gives:

$x + x + 25 = 81$

This simplifies to:

$2x + 25 = 81$

Subtract 25 from both sides, to give:

$2x = 56$

$x = 28$ (Divide through by 2.)

So, my son's age is 28 years, and I am 25 years older, aged 53.

Exercise 20E

1 I think of a number, double it and add 1. The answer is 33.

   a Write down an equation to represent this information.

   b What is the number?

2 I think of a number, double it and add 5. The answer is the same as the number plus 12.

   a Write down an equation to represent this information.

   b What is the number?

3 Tom has 10 more marbles than Jeff. Together they have 56.

   Let the number of marbles that Jeff has be x.

   a Write down an equation which this gives.

   b Solve the equation to find the number of marbles each boy has.

4 Sanjay has 35 more CDs than Surjit. Together they have 89 CDs. Let Surjit have x CDs.

   a Write down an equation which this gives.

   b Solve the equation to find the number of CDs they each have.

5 Gavin has 13 more DVDs than Michelle. Together they have 129 DVDs.

   Let Michelle have x DVDs.

   a Write down an equation which this gives.

   b Solve the equation to find the number of DVDs they each have.

Expressions and equations

6 Joy thinks of a number rule.

'Multiply the number by 3 and add 5.'

    a When Paul gives Joy a number, she replies, '23'. Write down the equation this gives and solve it to find the number which Paul gave to Joy.

    b When Billie gives Joy a number, she replies, '38'. Write down the equation this gives and solve it to find the number which Billie gave to Joy.

7 Paula is three times as old as Angus. Their ages add up to 52.

Let Angus be x years old.

    a Write down an equation which this gives.

    b Solve the equation to find both ages.

8 Moira scored twice as many goals in a season as Heather. Together, they scored 36 goals.

Let Heather score x goals.

    a Write down an equation which this gives.

    b Solve the equation to find how many goals each player scored.

★ **9** Alan spent four times as many minutes on his maths homework as he did on the rest of his homework. He spent two hours on his homework altogether.

Let Alan spend x minutes on the rest of his homework.

    a Write down an equation which this gives.

    b Solve the equation to find out how much time Alan spent on his maths homework.

10 Farmer Giles keeps only sheep and cows on his farm. He has 55 more sheep than cows and has 207 animals altogether.

Let the number of cows be x.

    a Write down an equation which this gives.

    b Solve the equation to find the number of sheep and the number of cows on Farmer Giles's farm.

11 In a school of 845 pupils, there are 29 fewer girls than boys.

Let there be x girls.

    a Write down an equation which this gives.

    b Solve the equation to find the number of girls and the number of boys in the school.

12 On an aircraft carrying 528 passengers, there were 410 more adults than children.

Let there be x children.

    a Write down an equation which this gives.

    b Solve the equation to find the number of children on this aircraft.

Number, money and measure

Challenge

1. Two consecutive numbers add up to 77. What are the two numbers?

   Let the smaller number be n.

2. Two consecutive numbers add up to 135. What is the product of the two numbers?

   Let the smaller number be n.

3. What is the product of three consecutive numbers which add up to 402?

   Let the smallest number be n.

4. **Brick wall problems**

   The numbers in two bricks which are side by side (adjacent) are added together. The answer is written in the brick above. Find the number missing from the brick in the bottom layer.

   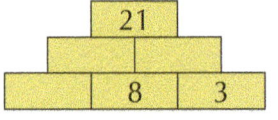

   Let the missing number be x. This gives the second diagram on the right.

   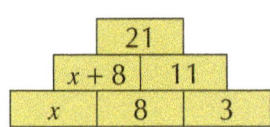

   Adding the terms in all adjacent bricks gives:

   $(x + 8) + 11 = 21$

   $x + 19 = 21$

   $x = 21 - 19$ (take 19 from both sides)

   $x = 2$

   So, the missing number is 2.

   Find the unknown number x in each of these 'brick wall' problems.

   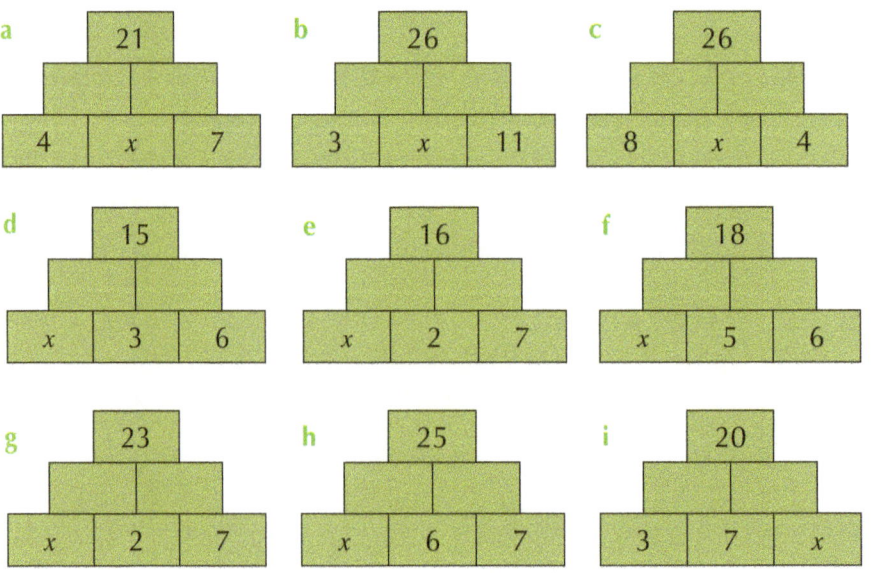

Continued

Expressions and equations

Continued

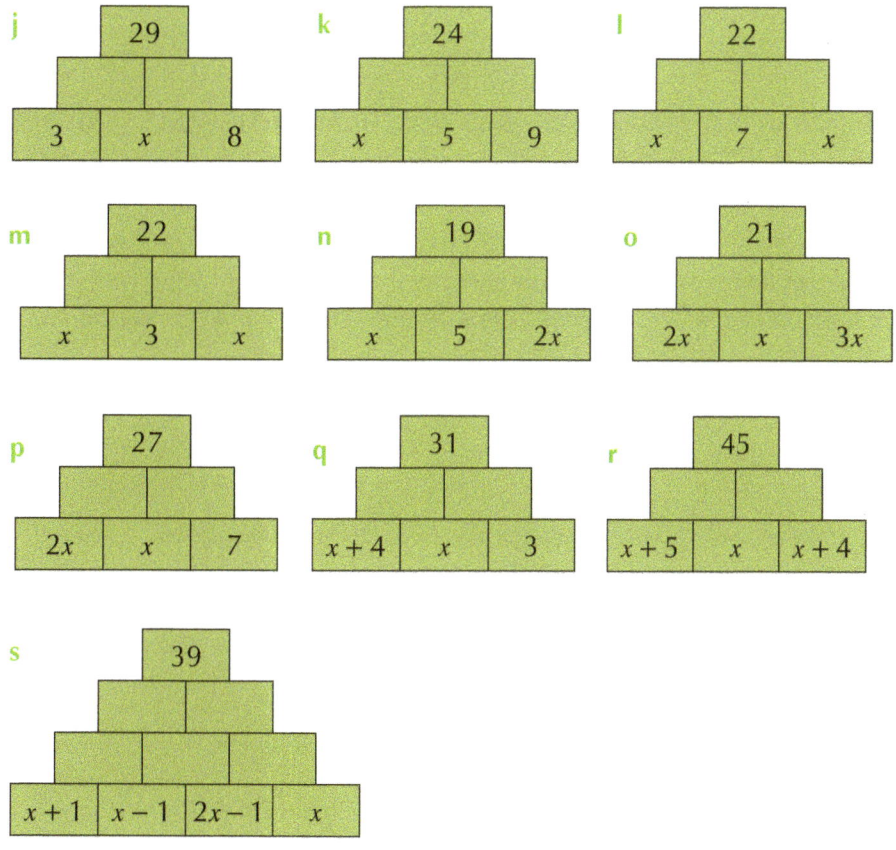

5 Make up some of your own 'brick wall' problems.

- By working on this topic I can explain how to find the value of the unknown in an equation.
- I can solve equations using the 'cover-up' method. ★ Exercise 20A Q3
- By considering what has happened to the unknown I can solve an equation by working backwards using inverse mapping. ★ Exercise 20B Q3
- I can solve an equation by doing the same to both sides until I have the unknown on its own. ★ Exercise 20C Q6
- I can solve equations with letters on both sides by doing the same thing to both sides. ★ Exercise 20D Q2
- Using algebra, I can create my own equation to represent a problem and then solve it. ★ Exercise 20E Q9

Number, money and measure

21
I can create and evaluate a simple formula representing information contained in a diagram, problem or statement.

MTH 3-15b

This chapter will show you how to:
- evaluate a formula in order to solve a problem involving different variables
- create a formula to link two variables using a table
- create a formula to link two variables represented in a problem.

You should already know:
- that algebra uses letters to represent numbers
- how to simplify an algebraic expression by collecting like terms
- the order of operations (BODMAS)
- how to solve an equation
- how to evaluate an algebraic expression by substitution.

Formulae

A **formula** is a rule used to work out a value from one or more values (called **variables** or **inputs**). For example, $A = ab$ is a rule, or formula, used to calculate the area, A, of a rectangle from the lengths, a and b, of two adjacent sides.

A formula also always has a **subject** (an output), which is usually written on the left-hand side of the equals sign. For example, the perimeter P of a rectangle with sides of lengths a and b is given by the formula:

$$P = 2a + 2b$$

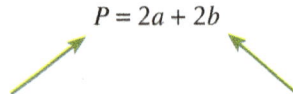

The output is P.
This is also called the subject of the formula.

Inputs are a and b.

When a is 3 cm and b is 5 cm, the perimeter can be caculated:

$P = 2 \times 3 + 2 \times 5$
$ = 6 + 10$
$ = 16$ cm

Expressions and equations

Evaluating a formula

Where you have a **rule** to calculate some quantity, you can write the rule as a formula.

Note: • When evaluating a formula you must use the rules of BODMAS.
• You must also remember to include units in your answer.

Example 21·1

A rule to calculate the cost of hiring a hall for a wedding is £200 plus £6 per person. This rule, written as a formula, is:

$$c = 200 + 6n$$

where c = cost in £
       n = number of people

Use the formula $c = 200 + 6n$ to calculate the cost of a wedding with 70 people.

Cost = 200 + 6 × 70 = 200 + 420 = £620   Remember BODMAS: × before +

Example 21·2

One rule to find the area of a triangle is to take half of the length of its base and multiply it by the vertical height of the triangle. This rule, written as a formula, is:

$$A = \frac{1}{2}bh$$

where A = area
      b = base length
      h = vertical height.

Using this formula to calculate the area of a triangle with a base length of 7 cm and a vertical height of 16 cm gives:

Area $= \frac{1}{2} \times 7 \times 16 = 56 \text{ cm}^2$

Exercise 21A

  1 A cleaner uses the formula:

$$c = 3 + 8h$$

where c = cost in £
       h = number of hours worked

Calculate how much the cleaner charges to work for:

a 5 hours   b 3 hours   c 8 hours

  2 A mechanic uses the formula:

$$c = 8 + 12t$$

where c = cost in £
       t = time, in hours, to complete the work

Calculate how much the mechanic charges to complete the work in:

a 1 hour   b 3 hours   c 7 hours

Number, money and measure

3 A singer uses the formula:

$$c = 25 + 15s$$

where c = cost in £

s = number of songs sung

Calculate how much the singer charges to sing the following:

a 2 songs   b 4 songs   c 8 songs

4 The formula for the average speed of a car is:

$$S = D \div T$$

where S = average speed (miles per hour)

D = distance travelled (miles)

T = time taken (hours)

Use the formula to calculate the average speed for the following journeys.

a 300 miles in 6 hours   b 200 miles in 5 hours

c 120 miles in 3 hours   d 350 miles in 5 hour

5 The formula for a child's dose of medicine is:

$$D = \frac{10C - 10}{C}$$

where D = dose, in millilitres, for a child

C = child's age in years

Use the formula to calculate the dose for each of the following ages.

a 8 years   b 5 years   c $2\frac{1}{2}$ years   d 10 years

6 The formula for the cost of a newspaper advertisement is:

$$C = 5W + 10A$$

where C = charge in £

W = number of words used

A = area of the advertisement in cm²

Use the formula to calculate the charge for the following advertisements.

a 10 words with an area of 20 cm²   b 8 words with an area of 6 cm²

c 12 words with an area of 15 cm²   d 17 words with an area of 15 cm²

Expressions and equations

7 The cost of a badge is given by the formula:

$$C = 60R^2$$

where C = cost in pence

R = radius of badge in cm

Use the formula to calculate the cost of each of these badges.

a Radius 1 cm   b Radius 2 cm   c Radius 2·5 cm

8 The formula $A = 180n - 360$ is used to calculate the sum of the angles in degrees inside a polygon with n sides. Use the formula to calculate the sum of the angles inside each polygon shown below.

a Pentagon, five sides   b Hexagon, six sides

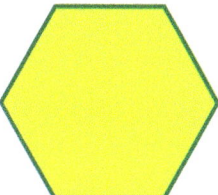

9 The average of three numbers is given by the formula:

$$A = \frac{m + n + p}{3}$$

where A is the average and m, n and p are the numbers.

a Use the formula to find the average of 4, 8 and 15.

b What is the average of 32, 43 and 54?

10 The perimeter of a rectangle is given by the formula:

$P = 2(l + b)$

where P is the perimeter, l is the length and b is the breadth.

a Use the formula to find the perimeter of a rectangle 5 cm by 8 cm.

b Use the formula to find the perimeter of a rectangle 13 m by 18 m.

11 The speed, v m/s, of the train t seconds after passing through a station with an initial speed of u m/s, is given by the formula:

$v = u + 5t$

a What is the speed 4 seconds after leaving a station with an initial speed of 12 m/s?

b What is the speed 10 seconds after leaving a station with an initial speed of 8 m/s?

Number, money and measure
Finding one of the other variables in a formula

Sometimes it is not the subject of the formula which is required but one of the other variables. When this is the case you need to solve an equation.

Example 21·3

When calculating his fee, a plumber uses the following formula:

$$C = 25 + 8t$$

where C = charge in £

t = time, in hours, to complete the job

For one job he charges £73. How many hours did it take to complete?

In this problem, you need to find t when you are given C.

Substituting $C = 73$ into the formula gives: $25 + 8t = 73$

Subtracting 25 from both sides gives: $8t = 48$

Dividing both sides by 8 gives: $t = 6$

So the job took him 6 hours to complete.

Exercise 21B

1. The cost, C, of placing an advertisement in a local newspaper is given by:

   $C = £20 + £2N$

   where N is the number of words used in the advertisement.

   How many words are used in an advertisement which costs:

   a £60   b £38   c £44

2. Lennie, the driving instructor, used the following formula to charge learner drivers:

   $C = £4 + £13H$

   where H is the number of hours in the driving lesson.

   How many hours does a lesson last if it costs:

   a £30   b £69   c £43

3. The speed, S m/s, of a rocket can be found from the formula $S = AT$, where the rocket has acceleration, A m/s^2, for a number of seconds, T.

   Calculate the acceleration of a rocket in each of the following cases.

   a The rocket has a speed of 400 m/s after travelling for 8 seconds.

   b The rocket has a speed of 330 m/s after travelling for 6 seconds.

Expressions and equations

4. The amount of money, M, expected to be collected for a charity was approximated by the following formula:

   $M = £5000T + £20C$

   where T is the number of TV advertisements appearing the day before a charity event was held, and C is the number of collectors.

   a One charity raised £16 000. How many advertisements were placed if there were 50 collectors?

   b Another charity raised £27 000. How many collectors were there if 5 advertisements were placed?

Creating a formula from a table of values

You can create a formula to represent the pattern of a sequence. Doing this allows you to find the value of any term in the sequence by changing the input.

Example 21·4

The following pattern is made from joining sticks of wood together.

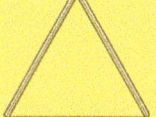 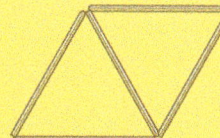 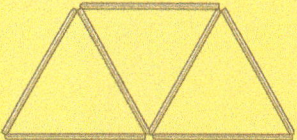

a Create a formula to find the number of sticks when you know the number of triangles.

b How many sticks will be used in 18 triangles?

a First draw up a table which shows the relationship or link between the number of triangles and the number of sticks needed.

Number of triangles, T	1	2	3	4
Number of sticks, S	3	5	7	9

The table shows that the number of sticks increases by 2 each time. This suggests it is linked to the 2 times table. Multiply the number of triangles by 2 and then compare with the number of sticks.

×2 → 2, 4, 6, 8

Number of triangles, T	1	2	3	4
Number of sticks, S	3	5	7	9

+1

You can see that the number of sticks is 1 more in each case.

So, the formula for the number of sticks, S, when you know the number of triangles, T, is:

$S = 2T + 1$

b Using the formula $S = 2T + 1$

Substitute 18 for T: $S = 2 \times 18 + 1$

= 37 sticks

Number, money and measure

Exercise 21C

1. The cost of a taxi journey is given in the table below.

Number of miles, m	1	2	3	4
Cost of journey, C (in £)	8	11	14	17

   Write a formula for the cost, C, of a taxi journey when you know the number of miles, m.

2. David is making a pattern using circles and squares.

   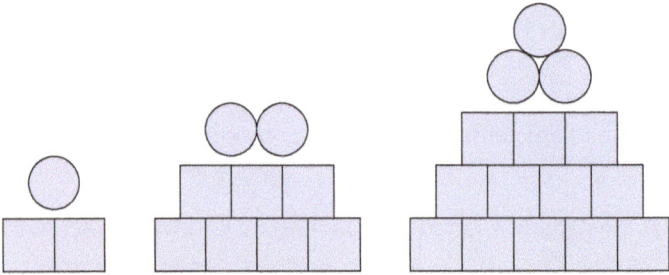

   The table shows the number of circles and squares in each pattern.

Number of circles, C	1	2	3	4	5
Number of squares, S	2	7	12	17	22

   Write a formula for the number of squares, S, when you know the number of circles, C.

3. A joiner charges the following rates based on how long it takes to complete a job.

Length of job in hours, h	1	2	3	4	5
Cost of job, C (in £)	35	55	75	95	115

   Write a formula for the cost of a job, C, when you know the number of hours, h, it will take.

4. Write a formula for y when you know x in each of the following.

   a.
x	1	2	3	4	5
y	7	10	13	16	19

   b.
x	1	2	3	4	5
y	3	7	11	15	19

   c.
x	2	3	4	5	6
y	3	5	7	9	11

   d.
x	3	4	5	6	7
y	11	14	17	20	23

   e.
x	1	2	3	4	5
y	0·3	0·8	1·3	1·8	2·1

Expressions and equations

Creating your own formula to represent a statement or problem

Example 21·5 A mechanic charges £8 per hour plus a call out fee of £15.

Write down a formula to find the cost, £C, of a job lasting h hours.

The cost is calculated by multiplying the number of hours by 8 and adding 15, so the formula is:

$C = 8h + 15$

Example 21·6 Find a formula for the sum, S, of three consecutive whole numbers, n, $n + 1$ and $n + 2$.

$S = n + (n + 1) + (n + 2)$

$S = n + n + 1 + n + 2$

$S = 3n + 3$

Example 21·7 How many months are there in:

a 5 years   b t years

There are 12 months in a year, so

a in 5 years there will be $12 \times 5 = 60$ months

b in t years there will be $12 \times t = 12t$ months

Exercise 21D

  1 Write each of these rules as a formula. Use the first letter of each variable in the formula. (Each letter is printed in **bold**.)

a The **c**ost of hiring a boat is £2 per **h**our.

b The total **d**istance run after each **l**ap around a 300 m track.

c **D**ad's age is always **J**oy's age plus 40.

d The **c**ost of a party is £50 plus £8 per **p**erson.

e The number of bottles of **j**uice needed is 5 plus the number of **p**eople divided by 3.

 2 Using the letters suggested, construct a simple formula in each case.

a The sum, S, of three numbers a, b and c.

b The product, P, of two numbers x and y.

c The difference, D, between the ages of two people. The older person is a years old and the other person is b years old.

d The number of days, D, in W weeks.

e The average age, A, of three boys whose ages are x, y and z years.

Number, money and measure

3 How many days are there in:

    a 1 week   **b** 3 weeks   **c** w weeks

4 A girl is now 13 years old.

    a How old will she be in:

        i 1 year   **ii** 5 years   **iii** t years

    b How old was she:

        i 1 year ago   **ii** 3 years ago   **iii** m years ago

5 A car travels at a speed of 30 mph. How many miles will it travel in the following durations?

    a 1 hour   **b** 2 hours   **c** t hours

6 How many grams are there in the following?

    a 1 kg   **b** 5 kg   **c** x kg

7 How many minutes are there in m hours?

8 Write down the number that is half as big as:

    a 20   **b** 6   **c** b

9 Write down the number that is twice as big as:

    a 4   **b** 7   **c** T

10 If a boy runs at b miles per hour, how many miles does he run in k hours?

11 **a** What is the cost, in pence, of 6 papers at 35 pence each?

    b What is the cost, in pence, of k papers at 35 pence each?

    c What is the cost, in pence, of k papers at q pence each?

12 A boy is b years old and his mother is 6 times as old.

    a Find the mother's age in terms of b.

    b Find the sum of their ages in y years' time.

Expressions and equations

Challenge

Dotty investigations

1 Look at the following two shapes drawn on a dotted square grid.

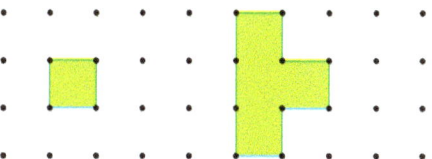

By drawing some of your own shapes (with no dots inside each shape), complete the table below, giving the number of dots on each perimeter and the area of each shape.

Number of dots on perimeter	Area of shape
4	1 cm²
6	
8	
10	4 cm²
12	
14	
16	

2 What is special about the number of dots on the perimeter of all the shapes in the table of Question 1?

3 For a shape with no dots inside, one way to calculate the area of the shape from the number of dots on the perimeter is to:

Divide the number of dots by 2, then subtract 1.

a Check that this rule works for all the shapes drawn in Question 1.

b Write this rule as a formula, where A is the area of a shape and D is the number of dots on its perimeter.

4 Look at the following two shapes drawn on a dotted square grid. They both have one dot inside.

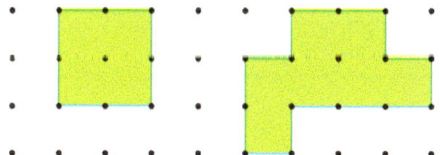

Continued

Number, money and measure

Continued

By drawing some of your own shapes (with only one dot inside each shape), complete the table below.

Number of dots on perimeter	Dots inside	Area of shape
4	1	
6	1	
8	1	4 cm²
10	1	
12	1	
14	1	
16	1	

5 For the shapes in Question 4, find a formula to connect A, the area of each shape with D, the number of dots on its perimeter.

6 a Draw some shapes with an even number of dots on each perimeter and two dots inside each shape.

   b Find the formula connecting A, the area of each shape, with D, the number of dots on its perimeter.

7 a Draw some shapes with an even number of dots on each perimeter and three dots inside each shape.

   b Find the formula connecting A, the area of each shape, with D, the number of dots on its perimeter.

- By working on this topic I have learnt that a formula is a means of generalising a problem where the variables can change.

- By substituting in given values I can confidently find the subject of a formula. ★ Exercise 21A Q6

- When trying to find a variable which is not the subject of a formula, I know that I have to solve an equation. ★ Exercise 21B Q2

- By comparing an increase in one variable with the relevant times table and adjusting accordingly, I can create a formula to link two variables given in a table. ★ Exercise 21C Q1

- I can create a formula to represent a problem or statement. ★ Exercise 21D Q1

Properties of 2D shapes and 3D objects

22

Having investigated a range of methods, I can accurately draw 2D shapes using appropriate mathematical instruments and methods.

MTH 3-16a

This chapter will show you how to:
- investigate different ways to construct triangles based on the information given
- construct a range of 2D shapes using the techniques used to construct triangles
- construct accurate 2D nets of 3D objects.

You should already know:
- how to accurately draw and measure angles using a protractor
- how to use a pair of compasses to construct circles and arcs
- the relationship between 3D objects and their nets.

Constructing triangles

You need to be able to draw a shape exactly from length and angle information given on a diagram, using a ruler, protractor and compasses. This is known as **constructing a shape**.

When constructing shapes you need to be accurate enough to draw the lines to the nearest millimetre and the angles to the nearest degree.

Example 22·1

Here is a sketch of a triangle ABC. It is not drawn accurately.

Construct the triangle ABC.

- Draw line BC 7 cm long.
- Draw an angle of 50° at B.
- Draw line AB 5 cm long.
- Join AC to complete the triangle.

The completed, full-sized triangle is given on the right.

This is an example of constructing a triangle given two sides and the included angle. Moving round the triangle in order, this is **S**ide, **A**ngle, **S**ide – an **SAS** triangle.

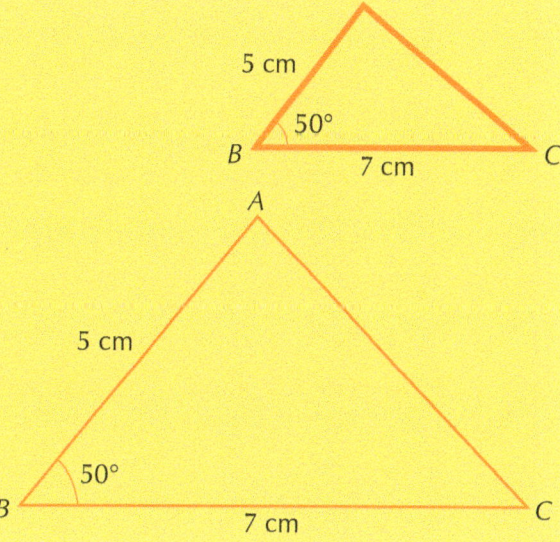

Shape, position and movement

Example 22·2 Construct the triangle XYZ.

- Draw line YZ 8·3 cm long.
- Draw an angle of 42° at Y.
- Draw an angle of 51° at Z.
- Extend both angle lines to intersect at X to complete the triangle.

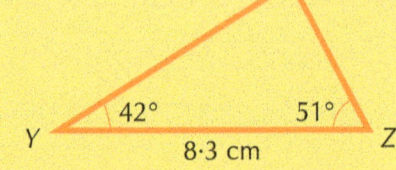

The completed, full-sized triangle is given below.

This is an example of constructing a triangle given two angles and the included side – **ASA**.

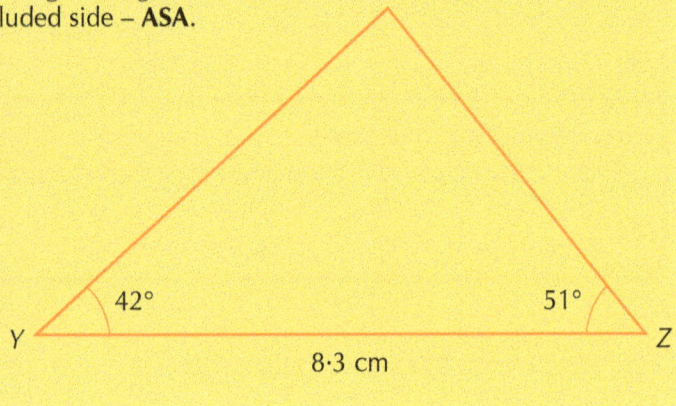

Example 22·3 Construct the triangle PQR.

- Draw line QR 6 cm long.
- Set the compasses to a radius of 4 cm and, with centre at Q, draw a long arc above QR.
- Set the compasses to a radius of 5 cm and, with centre at R, draw a long arc to intersect the first arc.
- The intersection of the arcs is P.
- Join QP and RP to complete the triangle.

This is an example of constructing a triangle given three sides – **SSS**.

The construction lines should be left on the diagram.

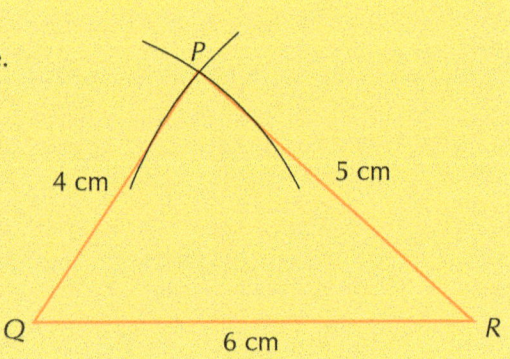

Properties of 2D shapes and 3D objects

Exercise 22A The triangles in this Exercise are not drawn to scale.

1. **a** Construct the triangle PQR.
   b Measure the size of ∠P and ∠R to the nearest degree.
   c Measure the length of the line PR to the nearest millimetre.

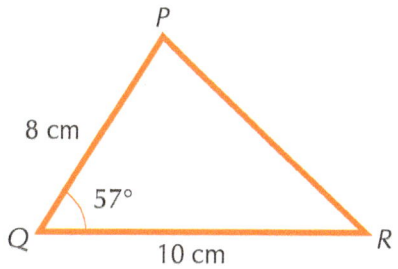

2. Construct each of the following triangles. Remember to label all lines and angles.

   a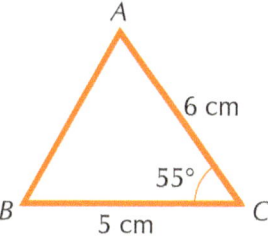

   b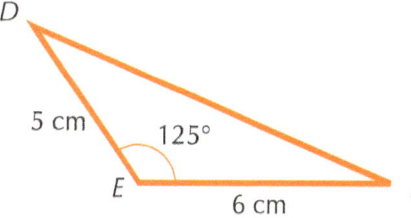

   c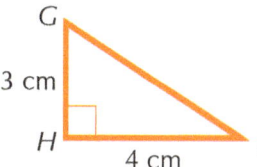

   d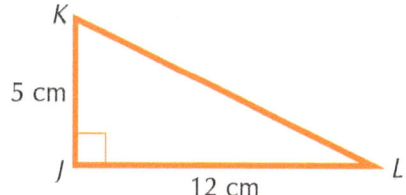

3. Construct each of the following triangles. Remember to label all lines and angles.

   a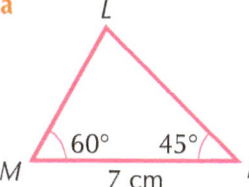

   b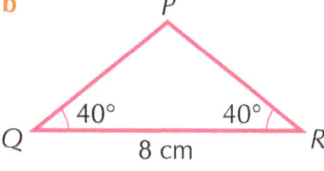

   c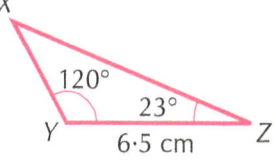

4. Construct each of the following triangles. (Remember: label all the lines and angles.)

   a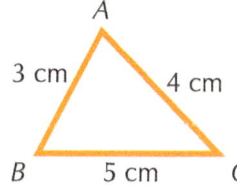

   b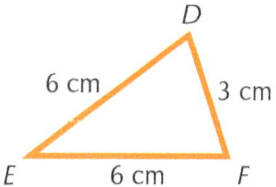

   c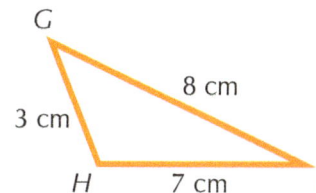

   d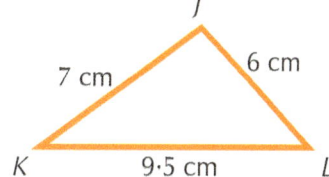

Shape, position and movement

5 Construct the triangle ABC with
∠A = 108°, ∠B = 34° and AB = 7 cm.

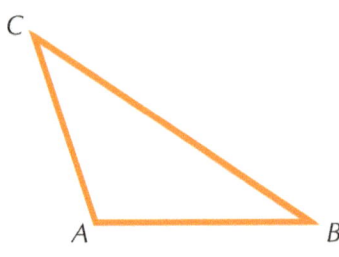

6 Construct each of the following triangles. Remember to label all lines and angles.

a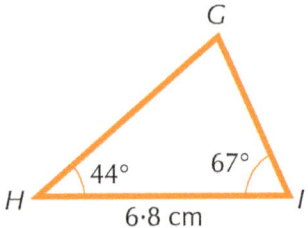

b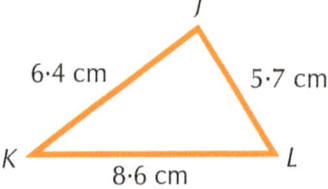

c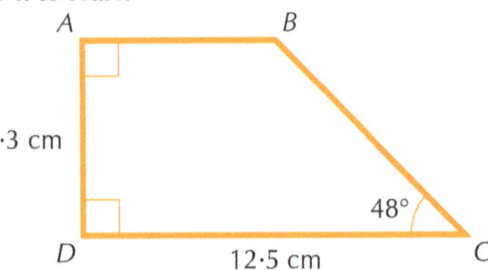

7 Construct the triangle ABC with AB = 5 cm, ∠A = 60° and ∠B = 50°.

8 Construct the triangle XYZ with XY = 7 cm, XZ = 6 cm and ∠X = 45°.

9 Construct the triangle XYZ with XY = 6·5 cm, XZ = 4·3 cm and YZ = 5·8 cm.

10 Construct equilateral triangles with sides of length:

a 8 cm   b 5 cm   c 4·5 cm

11 Paul thinks that he can construct a triangle with sides of length 3 cm, 4 cm and 8 cm, but finds that he cannot draw it.

a Try to construct Paul's triangle.

b Explain why it is not possible to draw Paul's triangle.

Constructing other 2D shapes

You can construct other 2D shapes such as quadrilaterals and regular polygons using the same techniques used to construct triangles. Just look at the information you are given and work out the side lengths or angles you need to find out to complete the construction.

Exercise 22B

The quadrilaterals in this Exercise are not drawn to scale.

1 a Construct the trapezium ABCD.

b Measure the size of ∠B to the nearest degree.

c Measure the length of the lines AB and BC to the nearest millimetre.

Properties of 2D shapes and 3D objects

2 Make an accurate copy of the parallelogram on the right.

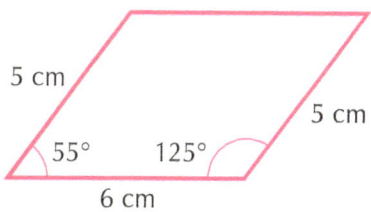

 3 a Construct the quadrilateral PQRS.

b Measure ∠P and ∠Q to the nearest degree.

c Measure the length of the line PQ to the nearest millimetre.

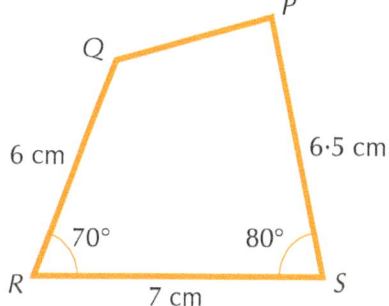

4 Construct the parallelogram ABCD with AB = 7·4 cm, AD = 6·4 cm, ∠A = 50° and ∠B = 130°.

5 Construct the quadrilaterals shown using only a ruler and compasses.

a

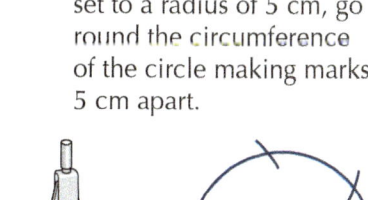

b

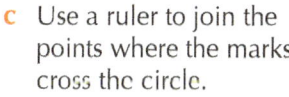

. . . .

Challenge   **1**   How to construct a regular hexagon.

a Draw a circle of radius 5 cm.

b With your compasses still set to a radius of 5 cm, go round the circumference of the circle making marks 5 cm apart.

c Use a ruler to join the points where the marks cross the circle.

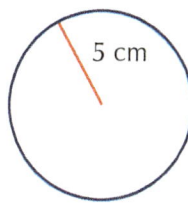

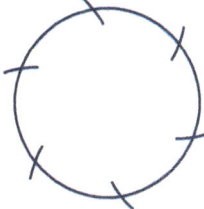
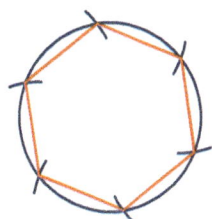

Continued

203

Shape, position and movement

Continued

2 How to construct other regular polygons.

a Draw a circle of radius 5 cm as before and draw in its radius.

b Find the centre angle by dividing 360° by the number of sides; for a pentagon with 5 sides, 360 ÷ 5 = 72°.

Draw an angle of 72° from the radius line and mark where it crosses the circle.

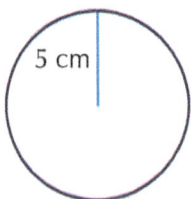
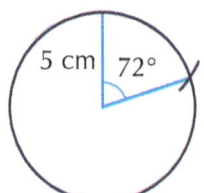

c Set the compasses to the distance between these two marks.

d Go round the circle making marks until you get back to the start.

e Use a ruler to join the points where the marks cross the circle.

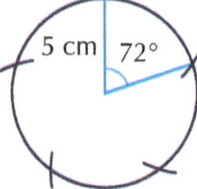
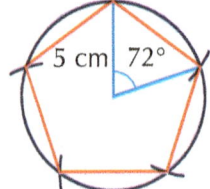

3 Use this method to draw regular polygons with:

i 8 sides **ii** 10 sides **iii** 12 sides

Find out the names of these regular polygons.

4 Use ICT to draw regular polygons.

a These instructions draw a square using LOGO:

fd 50 rt 90
fd 50 rt 90
fd 50 rt 90
fd 50 rt 90

b These instructions draw a regular pentagon using LOGO:

repeat 5 [fd 50 rt 72]

Investigate how to draw different types of triangles, quadrilaterals and other regular polygons using LOGO.

Properties of 2D shapes and 3D objects

Constructing nets of 3D objects

Construct one or more of the 3D objects given in Exercise 22C. For each object, you start by drawing its net accurately on card.

Make sure that you have the following equipment before you start to draw a net: a sharp pencil, a ruler, a protractor, a pair of scissors and a glue-stick or adhesive tape.

The tabs have been included to make it easier if you decide to glue the edges together. The tabs can be left off if you decide to use adhesive tape.

Before folding a net, score the card using the scissors and a ruler along the fold lines. When constructing a shape, keep one face of the net free of tabs and secure this face last.

Exercise 22C Draw each of the following nets accurately on card. Cut out the net and construct the 3D object.

1 Regular tetrahedron

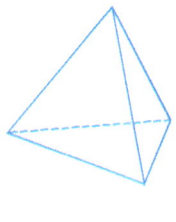
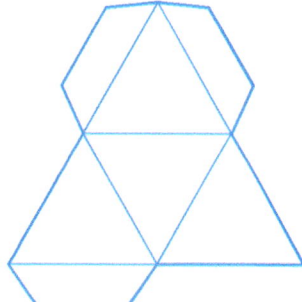

Each equilateral triangle has these measurements:

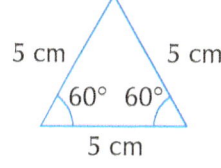

2 Square-based pyramid

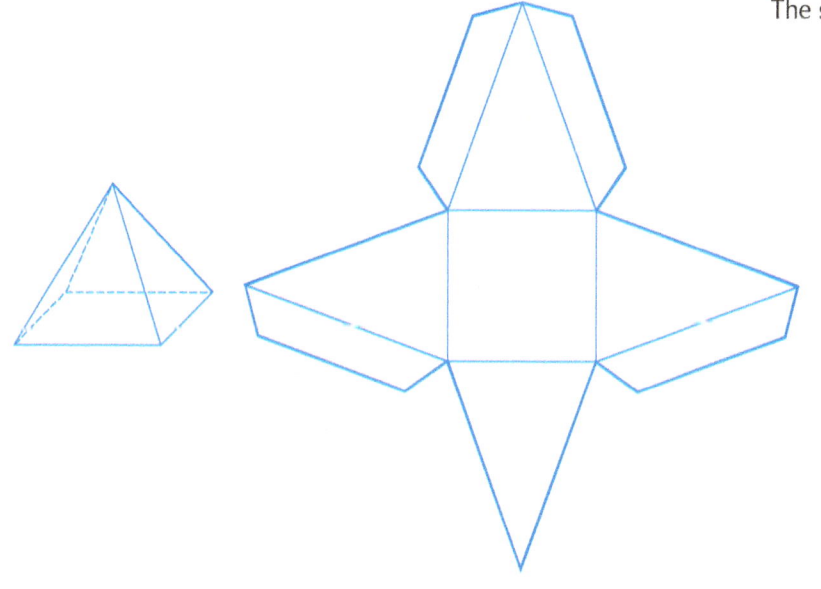

The square has these measurements:

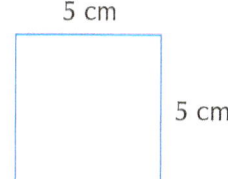

The isosceles triangle has these measurements:

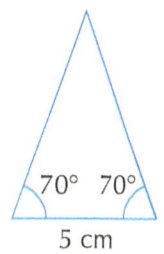

Shape, position and movement

⭐ **3** Triangular prism

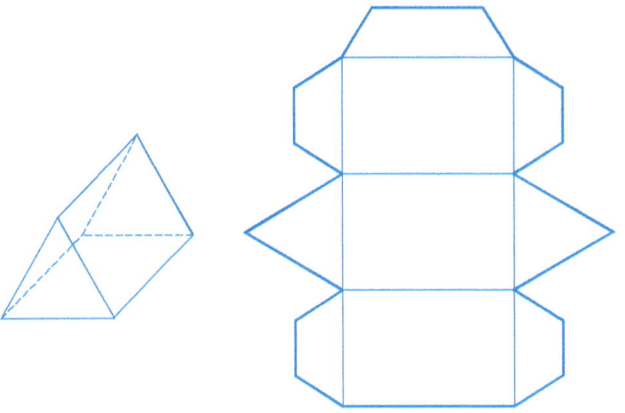

Each rectangle has these measurements:

Each equilateral triangle has these measurements

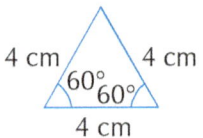

Challenge

The following nets are for more complex 3D objects. Choose suitable measurements and make each object from card.

1 Octahedron   **2** Regular hexagonal prism   **3** Truncated square-based pyramid

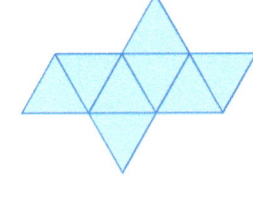 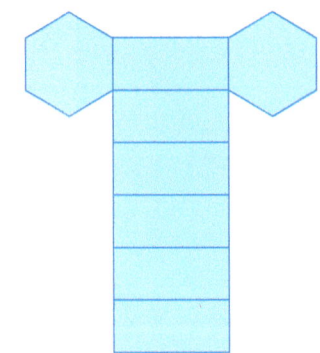 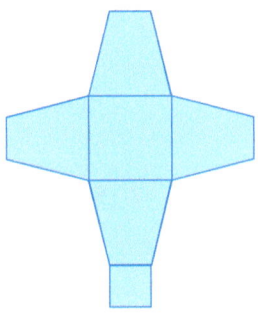

- By working on this topic, I have learnt how to accurately construct a range of 2D shapes and nets of 3D solids using a ruler, protractor and compasses.

- I can explain how to use my knowledge of different types of triangle, using the information provided, to decide the best way to draw them.

- I can use the appropriate equipment to construct accurate drawings of triangles. ⭐ Exercise 22A Q6

- I can apply my knowledge of constructions to accurately draw quadrilaterals and regular polygons. ⭐ Exercise 22B Q3

- I can combine the constructions of 2D shapes to make the net of a 3D object. ⭐ Exercise 22C Q3

Angle, symmetry and transformation

23

I can name angles and find their sizes using my knowledge of the properties of a range of 2D shapes and the angle properties associated with intersecting and parallel lines.

MTH 3-17a

This chapter will show you how to:
- name angles
- calculate the missing angle for complementary and supplementary angles
- calculate the missing angle around a point
- identify and calculate vertically opposite angles
- identify parallel and perpendicular lines
- identify and calculate alternate and corresponding angles
- calculate the size of a missing angle in a triangle
- calculate the missing angles in a quadrilateral
- use your knowledge of angles to investigate angles in other 2D shapes.

You should already know:
- how to describe and classify different types of angles
- how to measure and draw angles accurately.

Naming angles

An **angle** is formed when two lines meet or intersect. The angle can be named using the letters that name the lines.

Example 23·1

Here, lines AB and BC meet, forming angle ABC.

The symbol for angle is ∠, so it can be written as ∠ABC.

It can also be written as ∠CBA.

The middle letter is where the angle is.

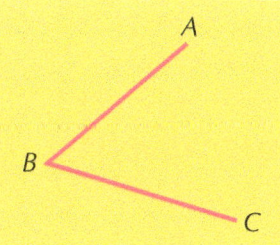

Example 23·2

The name of the angle is ∠SRT.

It can also be written as ∠TRS.

Notice the middle letter is always where the angle is.

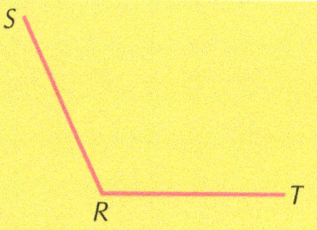

Shape, position and movement

Exercise 23A

1 Name the following angles using three letters.

a 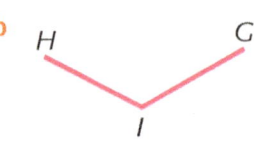   b 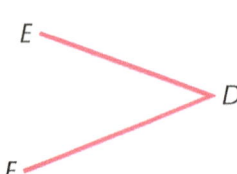   c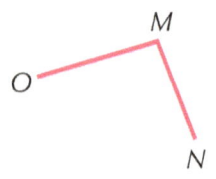

d 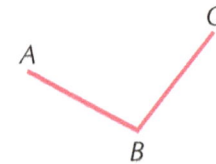   e 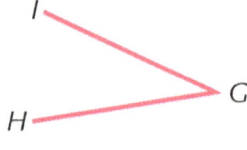   f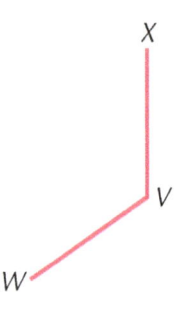

g 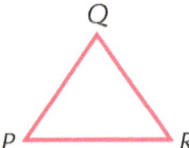   h   i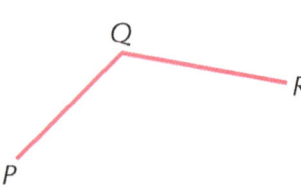

2 Name all three angles in this triangle.

3 Name all the angles in this diagram.

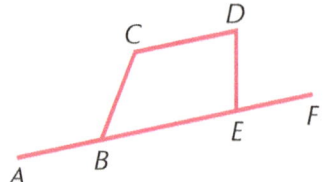

Calculating missing angles

You can calculate the **unknown angles** in a diagram from the information given. Unknown angles are usually denoted by letters, such as *a*, *b*, *c*, … .

Remember: usually the diagrams are not to scale.

Angles making a right angle add to 90°

Angles which add to 90° are called **complementary angles**.

Example 23·3 Calculate the size of the angle *x*.

$x = 90° - 35°$

$x = 55°$

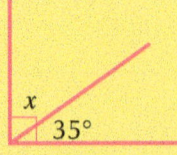

Angle, symmetry and transformation

Angles on a straight line

Angles on a straight line add up to 180°. These are called **supplementary angles**.

Example 23·4   Calculate the size of the angle b.

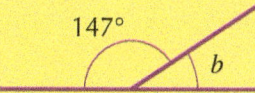

$b = 180° - 147°$

$b = 33°$

Angles around a point

Angles around a point add up to 360°.

Example 23·5   Calculate the size of the angle a.

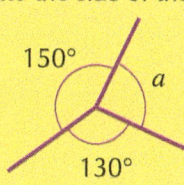

$a = 360° - 150° - 130°$

$a = 80°$

Vertically opposite angles

When two lines intersect, the **opposite angles** are equal.

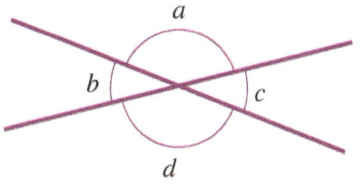

$a = d$ and $b = c$

Example 23·6   Calculate the sizes of angles e and f.

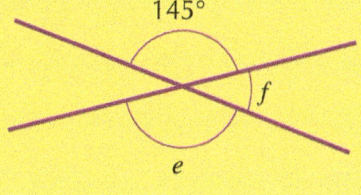

$e = 145°$ (opposite angles)
$f = 35°$ (angles on a straight line)

Exercise 23B

1 Calculate the size of each unknown angle.

a 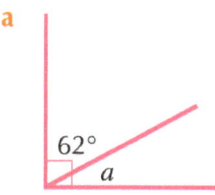   b 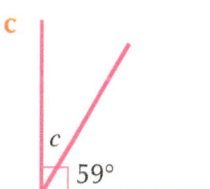   c 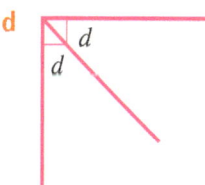   d (see figure)

2 Calculate the size of each unknown angle.

a 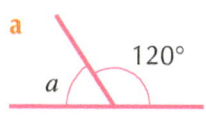   b 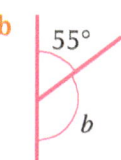   c 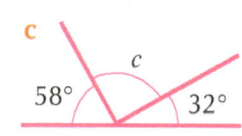   d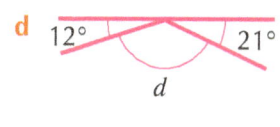

Shape, position and movement

3 Calculate the size of each unknown angle.

a 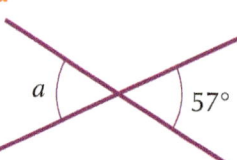 b 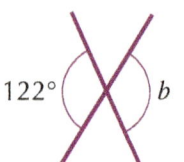 c 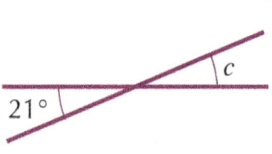 d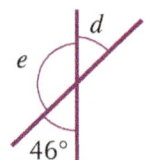

4 Calculate the size of each unknown angle.

a 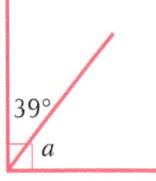 b c d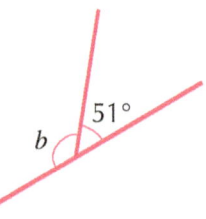

5 Calculate the size of each unknown angle.

a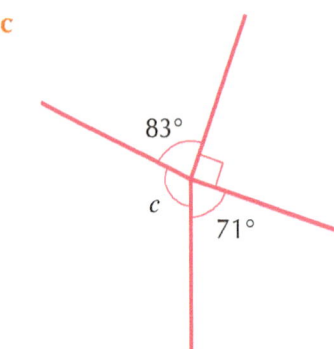

b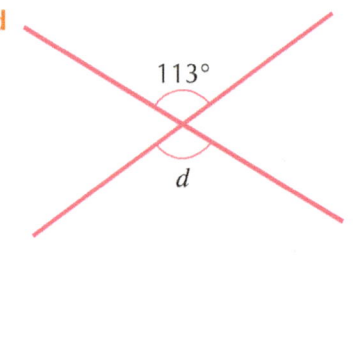

c

d

6 Calculate the size of each unknown angle.

a 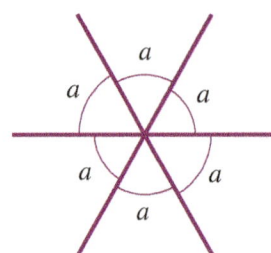 b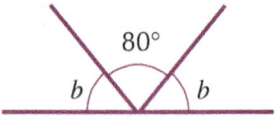

Angle, symmetry and transformation

Parallel and perpendicular lines

Example 23·7

These two lines are **parallel**. If we extended the lines in both directions, they would never meet.

We show that two lines are parallel by drawing arrows on them like this:

Example 23·8

Two lines are **perpendicular** if the angle between them is 90°. This is also called a **right angle**.

90°

We show that two lines are perpendicular by labelling the 90° angle with a square corner.

Exercise 23C

1 Write down which of the following sets of lines are parallel.

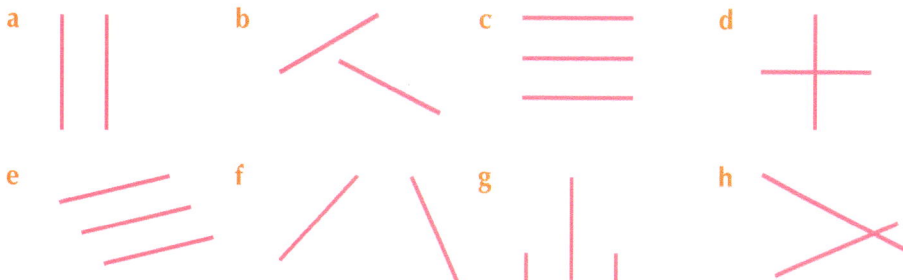

2 Copy each of the following diagrams onto square dotted paper.

On each diagram, use your ruler to draw two more lines that are parallel to the first line. Show that the lines are parallel by adding arrows to them.

a
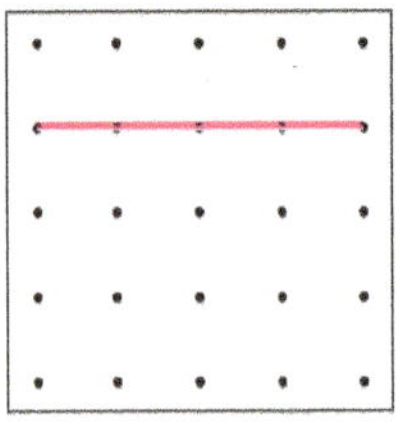
b
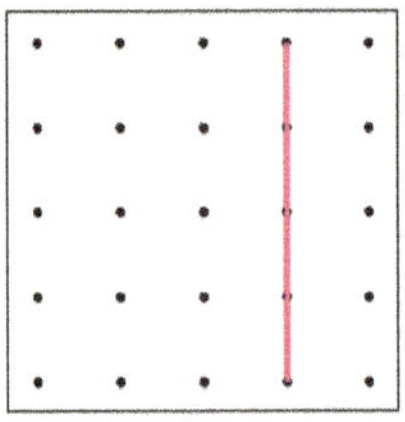
c
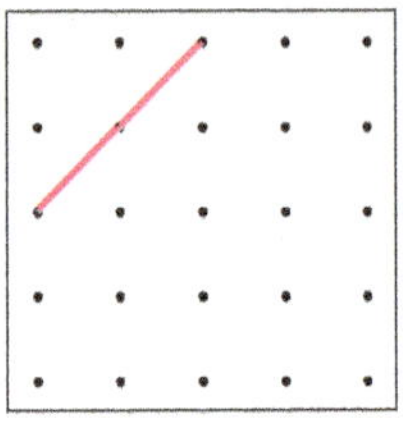

d
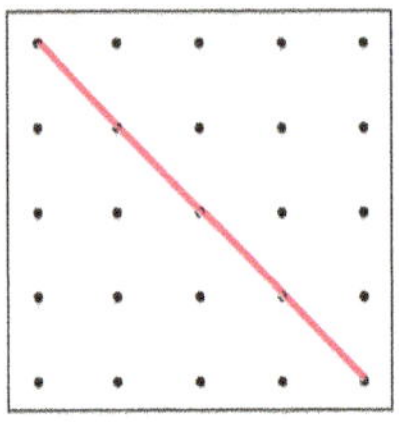
e
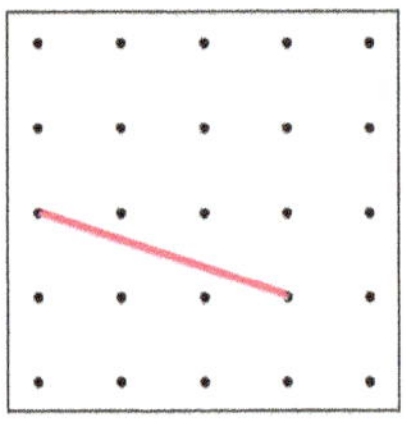
f
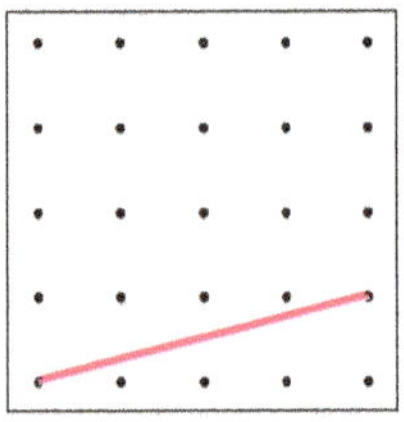

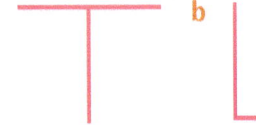

3 Write down which of the following pairs of lines are perpendicular.

a
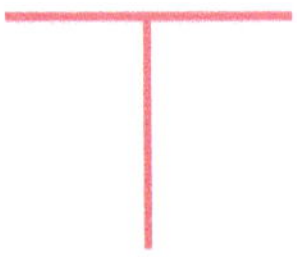
b

c
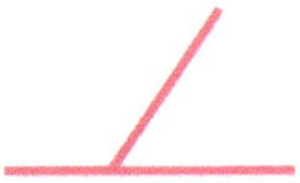
d
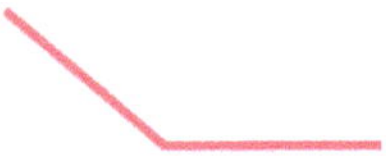

e
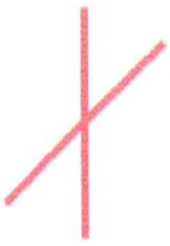
f
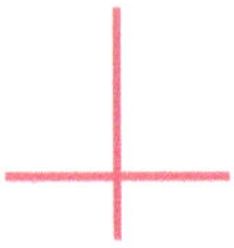
g
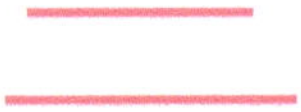
h

4 The line *AB* is perpendicular to the line *CD*.

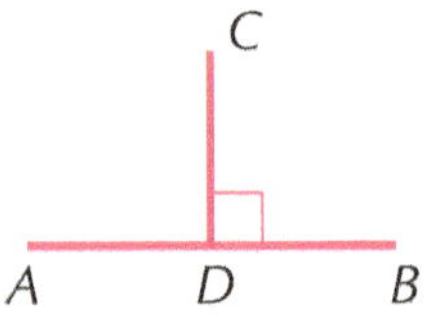

Copy and complete the following.

a
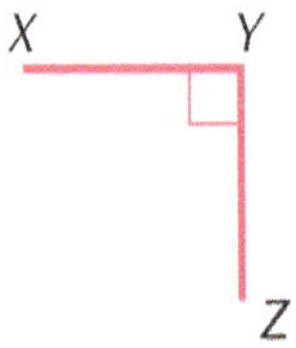

The line *XY* is perpendicular to the line ...

b
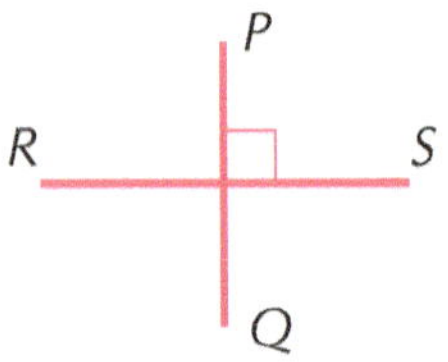

The line ... is perpendicular to the line ...

c
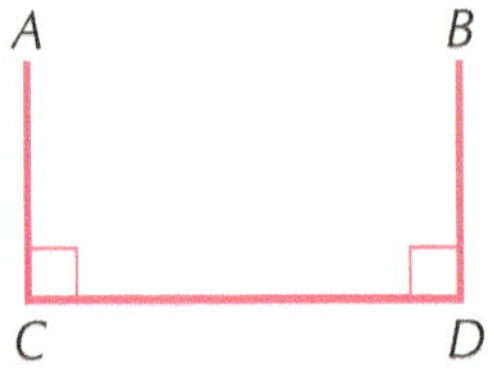

The line *CD* is perpendicular to the lines ... and ...

d
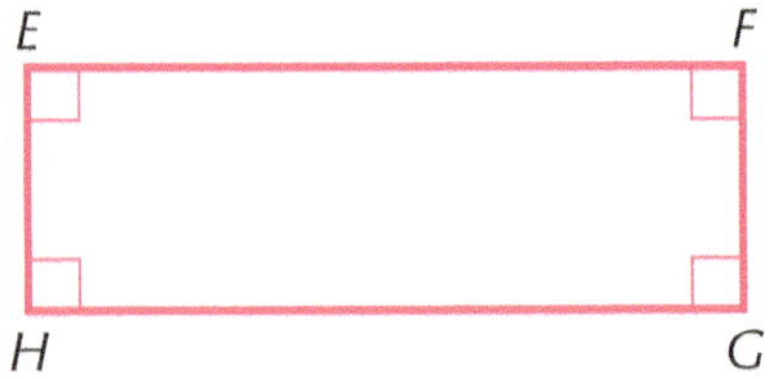

The line *EH* is perpendicular to the lines ... and ...
The line *EF* is perpendicular to the lines ... and ...

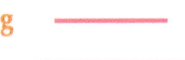

Angle, symmetry and transformation

5 Copy each of the following diagrams onto square dotted paper. Add arrows and square corners to show which lines are perpendicular and which are parallel to each other.

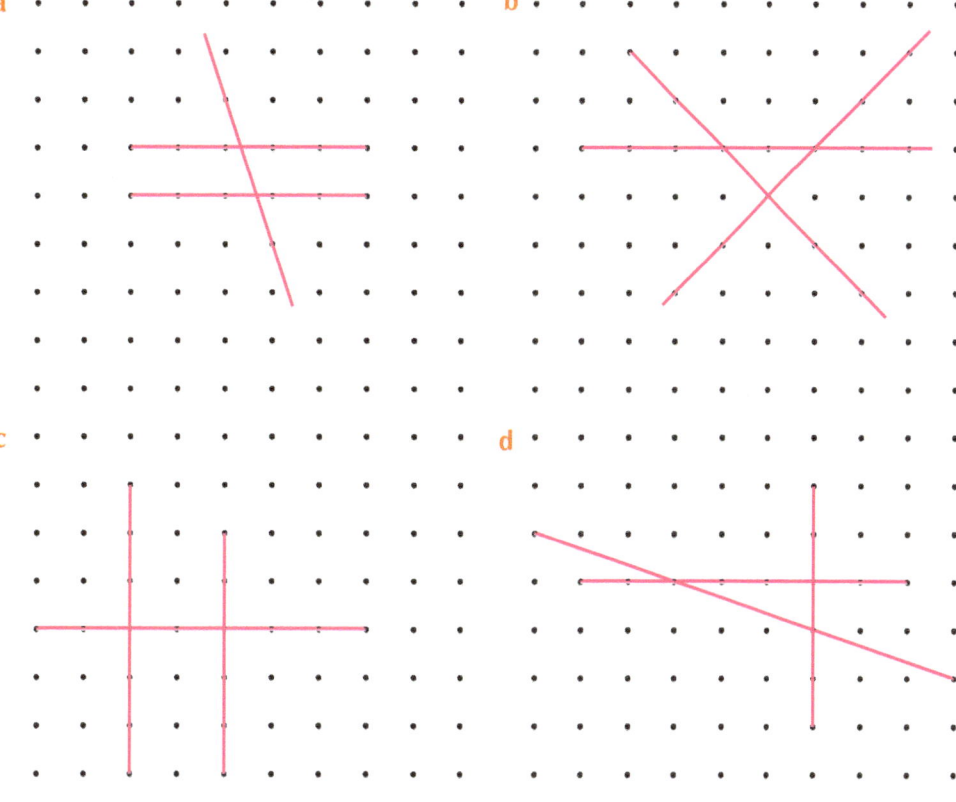

Challenge

1 Draw sketches of at least five objects in your classroom that:

   a contain parallel lines

   b contain perpendicular lines.

   Label any right angles on your sketches with square corners and any parallel lines with arrows.

2 Use your ruler to draw a straight line of any length.

   Now draw two more lines that are both perpendicular to this line.

   Write down what you notice about the two new lines.

Alternate and corresponding angles

The diagram shows two parallel lines with another straight line cutting across them.

The line that cuts across a pair of parallel lines is called a **transversal**.

Notice that the transversal forms eight angles.

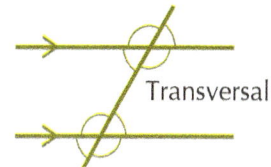

Transversal

Shape, position and movement

Alternate angles

Trace angle *x* on the diagram. Rotate your tracing paper through 180° and place angle *x* over angle *y*.
What do you notice?

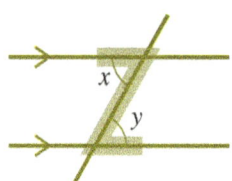

You should find that the two angles are the same size.

The two angles *x* and *y* are called **alternate angles**. (This is because they are on alternate sides of the transversal.) They are sometimes called Z-angles.

This shows that alternate angles are equal.

Example 23·9

Find the size of angle *a* on the diagram.

Alternate angles are equal, so *a* = 120°.

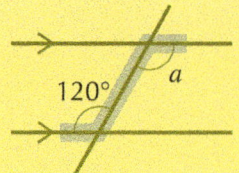

Corresponding angles

Trace angle *x* on the diagram. Slide your tracing paper along the transversal and place angle *x* over angle *y*.
What do you notice?

You should find that the two angles are the same size.

The two angles, *x* and *y*, are called **corresponding angles**. (This is because the position of one angle corresponds to the position of the other.) They are sometimes called *F*-angles.

This shows that corresponding angles are equal.

Example 23·10

Find the size of angle *b* and angle *c* on the diagram.

Corresponding angles are equal, so *b* = 125°.

Angles on a straight line add up to 180°.
Therefore, *c* is 180° − *b*, which gives:

c = 180° − 125° = 55°

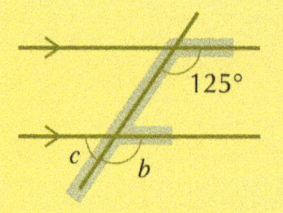

Exercise 23D

1 Which diagrams show a pair of alternate angles?

a

b

c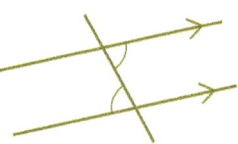

d

e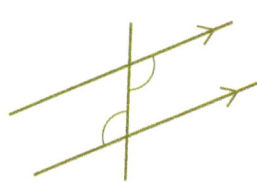

f

Angle, symmetry and transformation

2 Which diagrams show a pair of corresponding angles?

a 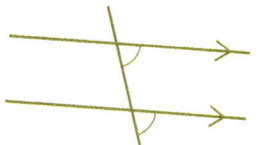   b    c

d 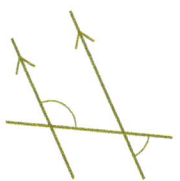   e 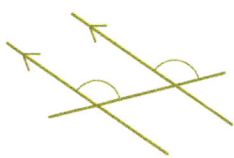   f

3 Work out the size of the lettered angle in each of these diagrams. State whether it is an alternate angle or a corresponding angle.

a 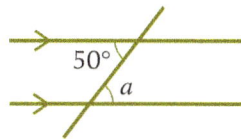   b 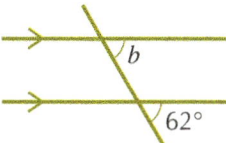   c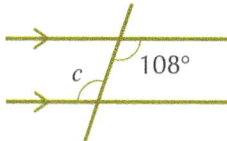

d 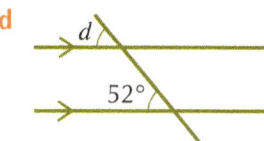   e 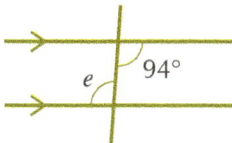   f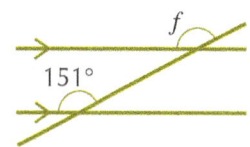

4 Work out the size of the lettered angles in these diagrams.

a 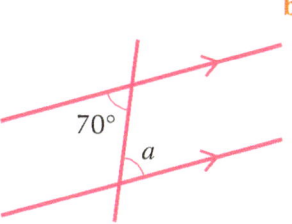   b 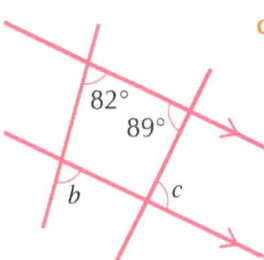   c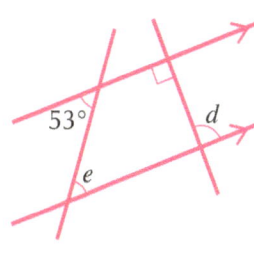

d 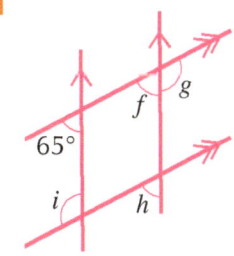   e 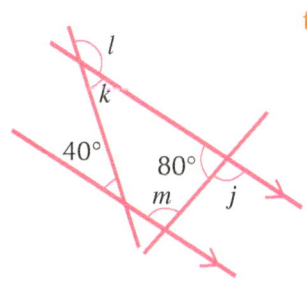   f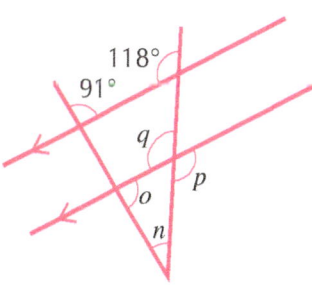

Shape, position and movement

Challenge Calculate the value of x in each diagram.

1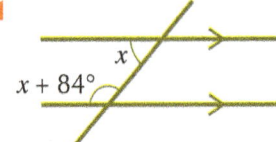
2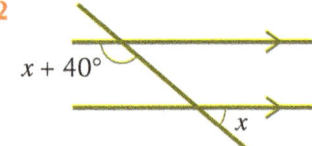
3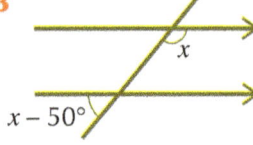

Angles in a triangle

The angles inside a triangle add up to 180°.

$a + b + c = 180°$

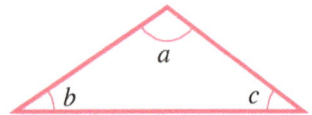

Example 23·11 Calculate the size of the angle c.

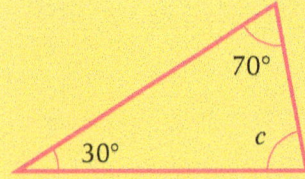

$c = 180° - 70° - 30°$
$c = 80°$

Example 23·12 Calculate the size of the angle d.

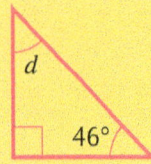

$d = 180° - 90° - 46°$
$d = 44°$

Example 23·13 Work out the size of the angles marked x and y.

Angles in a triangle add up to 180°,

so $x = 180 - 100 - 34$

$= 46°$

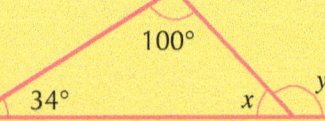

Angles on a straight line add up to 180°, so $y = 180 - 46$

$= 134°$

Angle, symmetry and transformation

Exercise 23E

1 Find the size of the angle marked by a letter in each scalene or right-angled triangle.

a 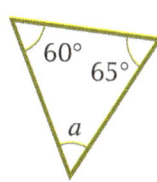 b 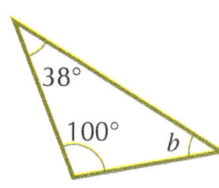 c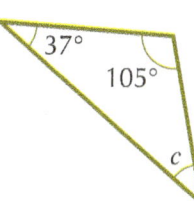

d 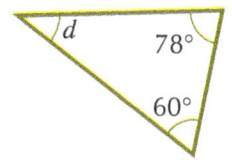 e 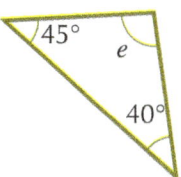 f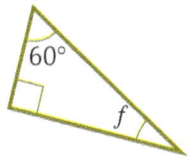

2 Find the size of the unknown angles in each isosceles triangle.

a 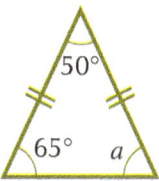 b 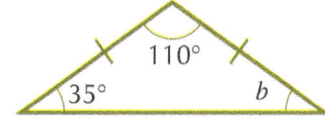 c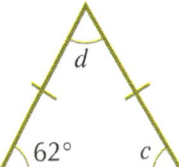

d 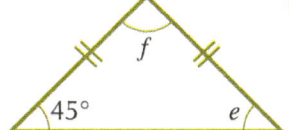 e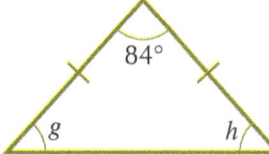

3 Calculate the size of the lettered angles in each of these triangles.

a 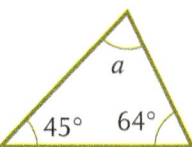 b 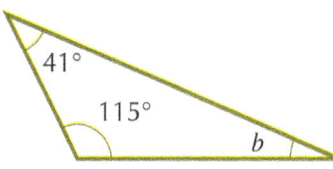 c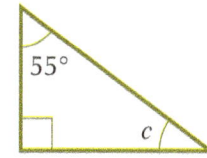

d 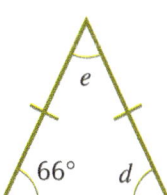 e 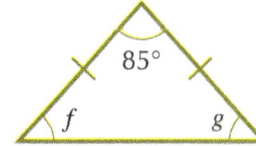 f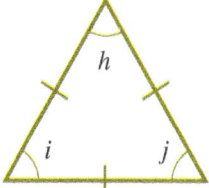

Shape, position and movement

4 Work out the size of the lettered angles in each of these diagrams.

a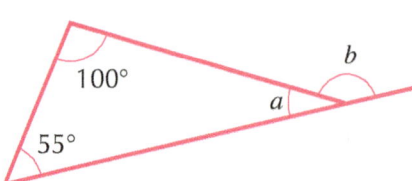

b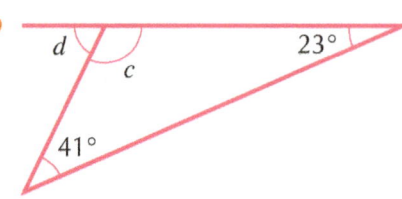

c

d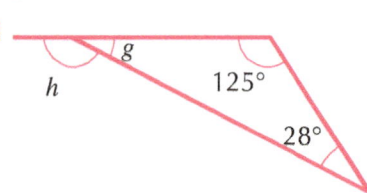

e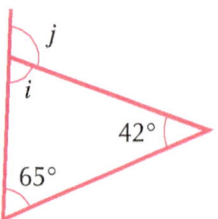

f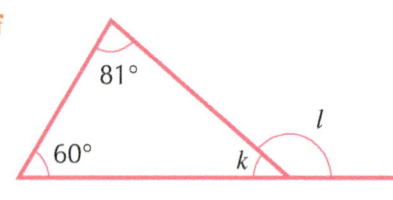

g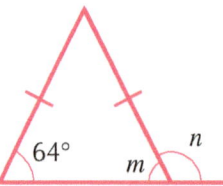

h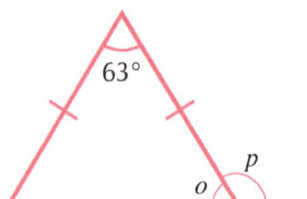

5 Work out the size of the lettered angles in each of these diagrams.

a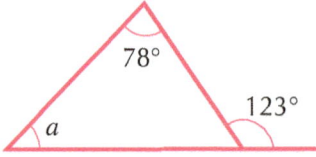

b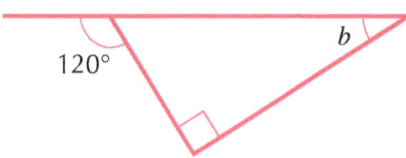

c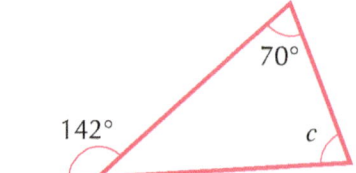

d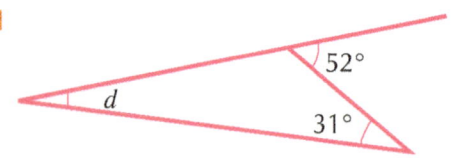

Angle, symmetry and transformation

Challenge

1 Calculate the size of each unknown angle.

a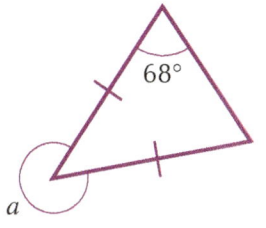

b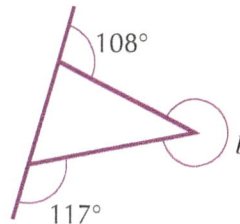

c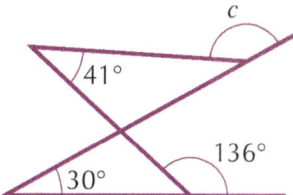

d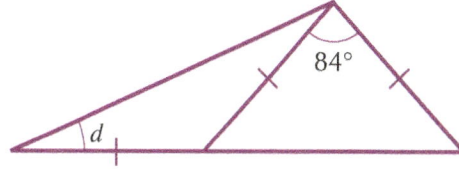

e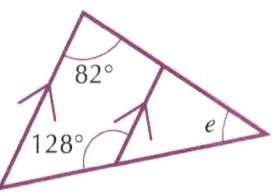

f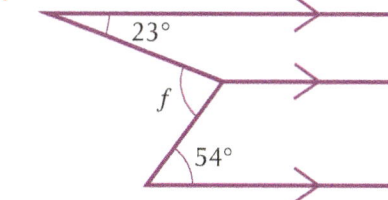

2 One angle in an isosceles triangle is 42°. Calculate the possible sizes of the other two angles.

Angles in a quadrilateral

Draw a large quadrilateral similar to the one on the right.

Measure each interior angle as accurately as you can, using a protractor. Now add up the four angles. What do you notice?

You should find that your answer is close to 360°.

Now draw a different quadrilateral and find the sum of the angles. How close were you to 360°?

For any quadrilateral, the sum of the interior angles is 360°. So, in the diagram:

$a + b + c + d = 360°$

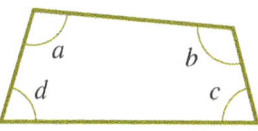

219

Shape, position and movement

Example 23·14 Work out the sizes of the angles marked p and q on the diagram.

The angles in a quadrilateral add up to 360°, which gives:

$p = 360° - 135° - 78° - 83°$
$= 64°$

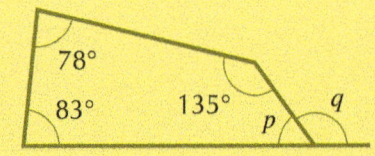

The angles on a straight line add up to 180°, so:

$q = 180° - 64°$
$= 116°$

Exercise 23F

1 Find the size of the angle marked by a letter in each quadrilateral.

a) 82°, 80°, 89°, a

b) 102°, b, 80°, 70°

c) c, 126°, 80°, 86°

d) 69°, 112°, 85°, d

e) e, 106°, 108°, (right angle)

f) 70°, 98°, 47°, f

2 Calculate the size of each unknown angle.

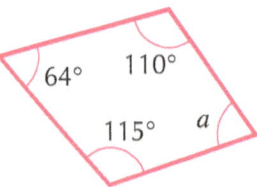
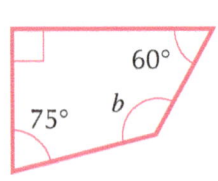
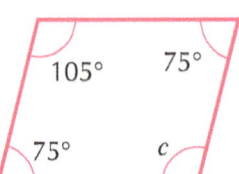
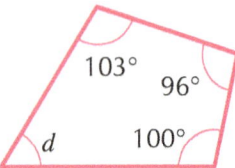

Angle, symmetry and transformation

3 Calculate the size of the lettered angle in each of these quadrilaterals.

a 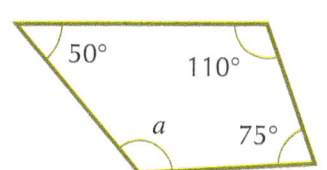　b 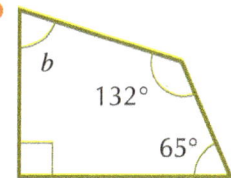　c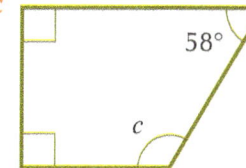

d 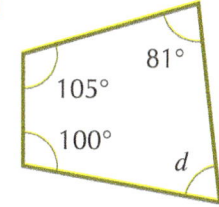　e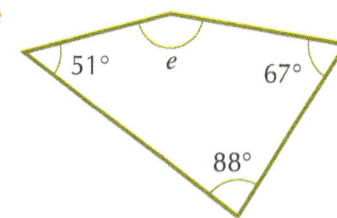

4 Calculate the size of the lettered angles in each of these diagrams. Give an explanation of how you found each angle.

a 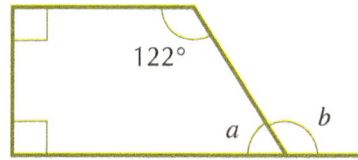　b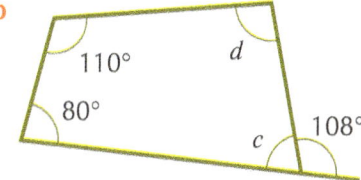

5 Work out the size of the lettered angles in each of these diagrams.

a 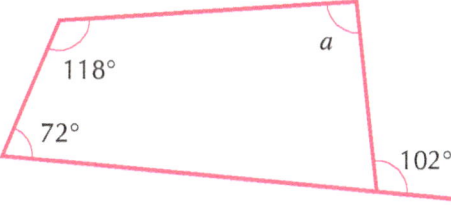　b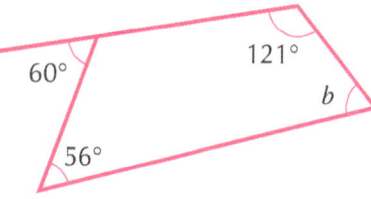

6 Copy these diagrams and fill in all the missing angles.

[**Hint:** The diagonals of a kite intersect at right angles.]

a 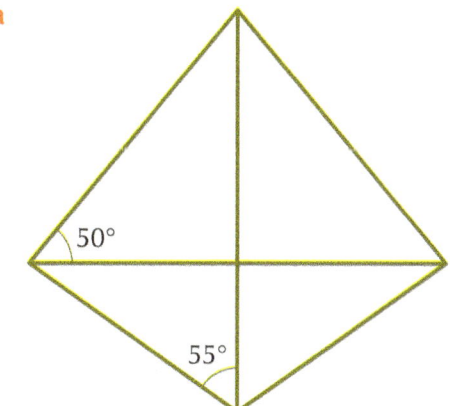　b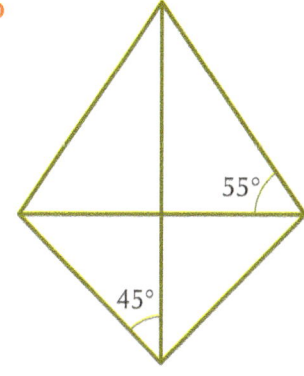

221

Shape, position and movement

Challenge

1. ABCD is a parallelogram with ∠ADC = 63°.

   a What do you know about a parallelogram?

   b Explain how you could find ∠BAD.

   c Write down the size of ∠ABC and ∠BCD.

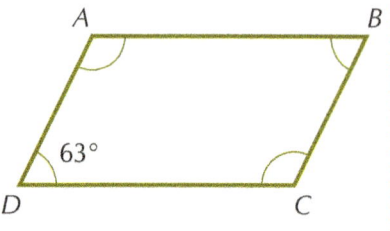

2. ABCD is a kite with ∠DAB = 80° and ∠BCD = 60°.

   a Make a sketch of the kite and draw its line of symmetry.

   b What do you know about angles p and q?

   c Use this information to work out angles p and q.

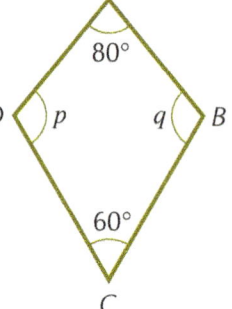

3. PQRS is a trapezium.

   a Work out the size of the angle marked p.

   b Write down anything you notice about the angles in the trapezium.

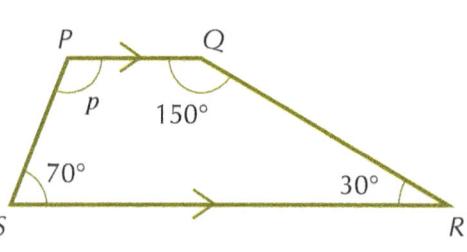

4. A quadrilateral can be split into two triangles, as shown in the diagram. Explain how you can use this to show that the sum of the angles in a quadrilateral is 360°.

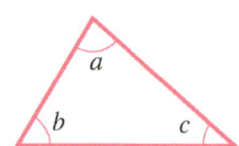

Interior and exterior angles of polygons

Interior angles

The angles inside a **polygon** are known as **interior angles**.

For a triangle, the sum of the interior angles is 180°:

$a + b + c = 180°$

Angle, symmetry and transformation

Example 23·15   Find the sum of the interior angles of a pentagon.

The diagram shows how a pentagon can be split into three triangles from one of its vertices. The sum of the interior angles for each triangle is 180°.

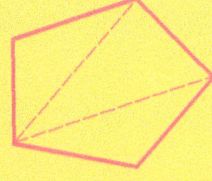

So the sum of the interior angles of a pentagon

= 3 × 180° = 540°

From this, we can deduce that each interior angle of a regular pentagon is:

$$\frac{\text{sum of interior angles}}{\text{number of sides}}$$

540° ÷ 5 = 108°

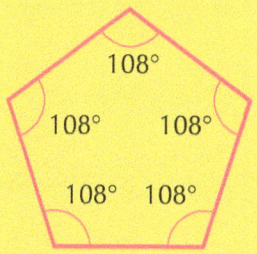

Remember: A regular polygon has equal sides and equal angles.

Exercise 23G

1 a Copy each polygon below.

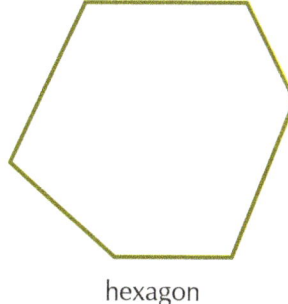   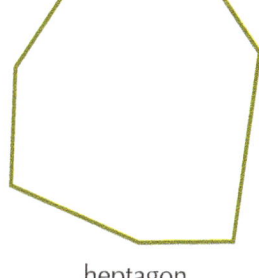   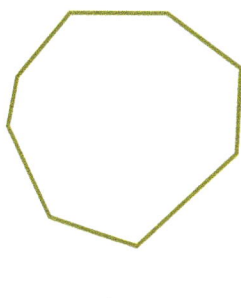

hexagon   heptagon   octagon

b Split each polygon into triangles and use this to find the sum of the interior angles of:

   i a hexagon   **ii** a heptagon   **iii** an octagon.

c Copy and complete the table below. The pentagon has been done for you.

Name of polygon	Number of sides	Number of triangles inside polygon	Sum of interior angles
triangle			
quadrilateral			
pentagon	5	3	540°
hexagon			
heptagon			
octagon			
n-sided polygon			

223

Shape, position and movement

2 For each of the following polygons calculate:

   i the number of sides   ii the sum of the interior angles
   iii the unknown angle.

a   b 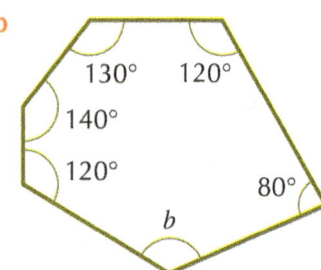   c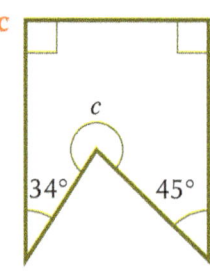

3 Find angle *x* in the pentagon on the right.

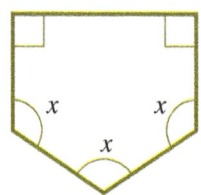

4 Calculate the size of the interior angle for each of the following.

   a A regular hexagon   b A regular octagon

5 A regular dodecagon is a polygon with 12 equal angles.

   a Calculate the sum of the interior angles.
   b Work out the size of each interior angle.
   c Work out the size of each exterior angle.

6 Calculate the value of *x* in each of the following polygons.

a 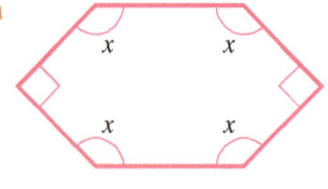   b 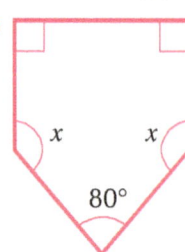   c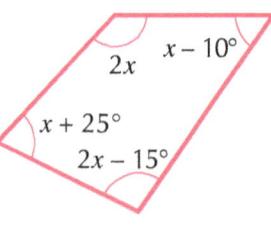

Angle, symmetry and transformation

Challenge

Exterior angles

If we extend a side of a polygon, the angle formed outside the polygon is known as an **exterior angle**.

In the diagram, a is an exterior angle of the quadrilateral.

At any vertex of a polygon, the interior angle plus the exterior angle = 180° (angles on a straight line).

In the diagram, $a + b = 180°$.

This is shown for the quadrilateral, but remember it is true for *any* polygon.

On the diagram, all the sides of the pentagon have been extended to show all the exterior angles.

If you imagine standing on a vertex and turning through all the exterior angles on the pentagon, you will have turned through 360°.

This is true for all polygons. The sum of the exterior angles for any polygon is 360°.

For a regular pentagon, each exterior angle is: 360° ÷ 5 = 72°

1 Calculate the size of the lettered angle in each of the following polygons.

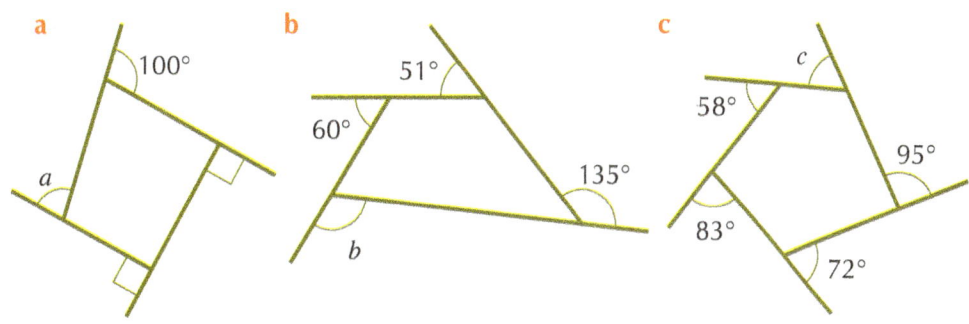

- By working on this topic I can use my knowledge of angles and the angle properties of lines to solve problems.

- I can name angles. ★ Exercise 23A Q2

- I can find angles in a corner, on a line and around a point. ★ Exercise 23B Q5

- I can recognise parallel and perpendicular lines. ★ Exercise 23C Q5

- I can identify alternate and corresponding angles and use my knowledge of their properties to calculate them. ★ Exercise 23D Q3

- I can find the angles in a triangle. ★ Exercise 23E Q3

- I can find the angles in a quadrilateral. ★ Exercise 23F Q3

- I can find the angles in a range of 2D shapes. ★ Exercise 23G Q2

225

Shape, position and movement

24

Having investigated navigation in the world, I can apply my understanding of bearings and scale to interpret maps and plans and create accurate plans, and scale drawings of routes and journeys.

MTH 3-17b

This chapter will show you how to:
- measure accurately on a map or plan, and then use a given scale to calculate the actual distance
- choose an appropriate scale to draw an accurate plan or scale drawing
- use bearings and relate them to compass directions
- locate a point or mark a route on a map when given bearings and distances
- use maps, plans and scale drawings to solve related problems.

You should already know:
- how to give directions for movement based on moves forward, back, turns and angles
- how to use an 8-point compass rose and angles to give directions for movement
- how to use bearings from North to describe positions and directions
- how to use simple models, maps and plans to calculate real sizes of objects and buildings and distances between places on maps
- how to make scale drawings of an object.

Scale drawings

A **scale drawing** is a smaller drawing of an actual object. A scale must always be clearly given by the side of the scale drawing.

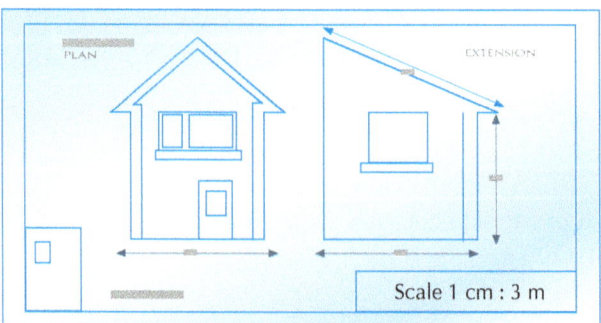

Angle, symmetry and transformation

Example 24·1 Shown is a scale drawing of Rebecca's room.

- On the scale drawing, the length of the room is 5 cm, so the actual length of the room is 5 m.
- On the scale drawing, the breadth of the room is 3·5 cm, so the actual breadth of the room is 3·5 m.
- On the scale drawing, the breadth of the window is 2 cm, so the actual breadth of the window is 2 m.

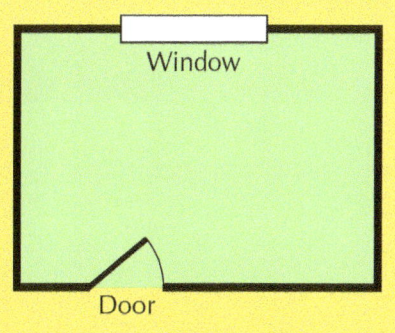

Scale: 1 cm to 1 m

Exercise 24A

1 The lines shown are drawn using a scale of 1 cm to 10 m. Write down the length each line represents.

   a
   b
   c
   d
   e

2 The diagram shows a scale drawing for a school hall.
   a Find the actual length of the hall.
   b Find the actual breadth of the hall.
   c Find the actual distance between the opposite corners of the hall.

Scale: 1 cm to 5 m

3 The diagram shown is Ryan's scale drawing for his mathematics classroom. Nathan notices that Ryan has not put a scale on the drawing, but he knows that the length of the classroom is 8 m.
   a What scale has Ryan used?
   b What is the actual breadth of the classroom?
   c What is the actual area of the classroom?

Shape, position and movement

4 Copy and complete the table below for a scale drawing in which the scale is 4 cm to 1 m.

	Actual length	Length on scale drawing
a	4 m	
b	1·5 m	
c	50 cm	
d		12 cm
e		10 cm
f		4·8 cm

**5** The plan shown is for a bungalow.

   a Find the actual dimensions of each of the following rooms.

   i the kitchen

   ii the bathroom

   iii bedroom 1

   iv bedroom 2

   b Calculate the actual area of the living room.

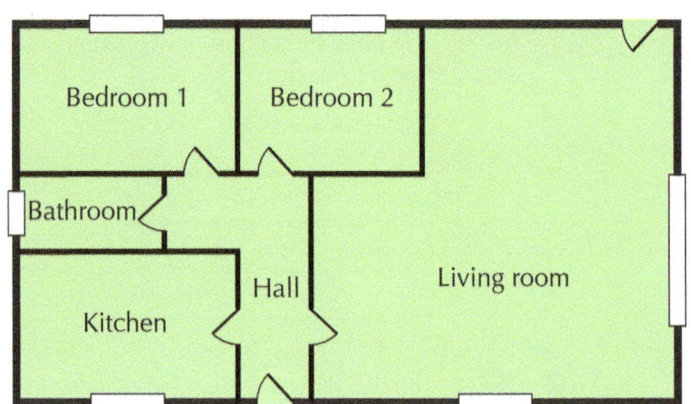

Scale 1 cm to 2 m

★ **6** The diagram shows the plan of a football pitch. It is not drawn to scale. Use the measurements on the diagram to make a scale drawing of the pitch (choose your own scale).

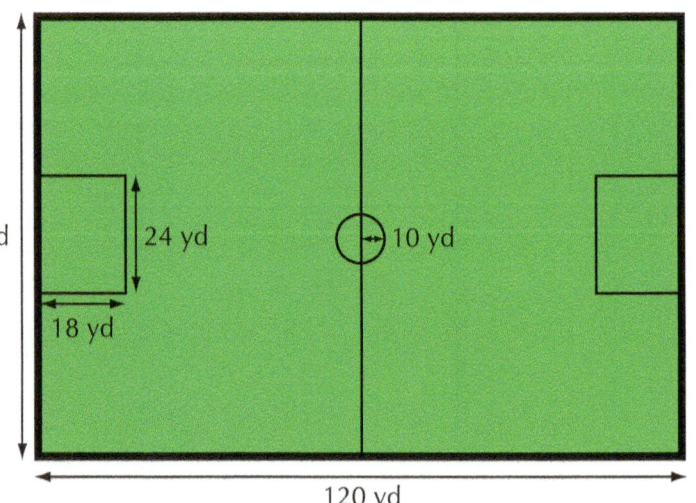

Angle, symmetry and transformation

Challenge

1. On centimetre-squared paper, design a layout for a bedroom. Make cut-outs for any furniture you wish to have in the room (use a scale of 2 cm to 1 m).

2. (You will need a metre rule or a tape measure for this activity.) Draw a plan of your classroom, including the desks and any other furniture in the room (choose your own scale).

3. *Scales as ratios*. A plan has a scale that is given as 1 cm to 2 m. When the scale is changed to the same units it becomes 1 cm to 200 cm. This can be written as the ratio 1 : 200. So a scale of 1 cm to 2 m can also be written as a scale of 1 : 200. Scales on maps are sometimes written in this way. Write each of the scales below as a ratio.

   a 1 cm to 1 m   b 1 cm to 4 m   c 4 cm to 1 m
   d 1 cm to 1 km   e 2 cm to 1 km

Map scales

Maps are **scale drawings** used to represent areas of land.

Distance on a map can be shown in two different ways, as in the examples below.

The first way is to give a **map scale**. This gives a distance on the map and an equivalent distance on the ground. This is shown in Example 24·2.

The second way is to give a map ratio. A map ratio has no units and will look something like 1 : 10 000. Example 24·3 explains how to use a map ratio.

A direct distance is 'as the crow flies'. String can be used to estimate map distance along paths and roads.

Example 24·2

The map shows part of south-east England. Find the actual direct distance between Maidstone and Dover.

The direct distance between Maidstone and Dover on the map is 5·5 cm. The scale is 1 cm to 10 km. So 5·5 cm represents 5·5 × 10 km = 55 km.

The actual distance between Maidstone and Dover is 55 km.

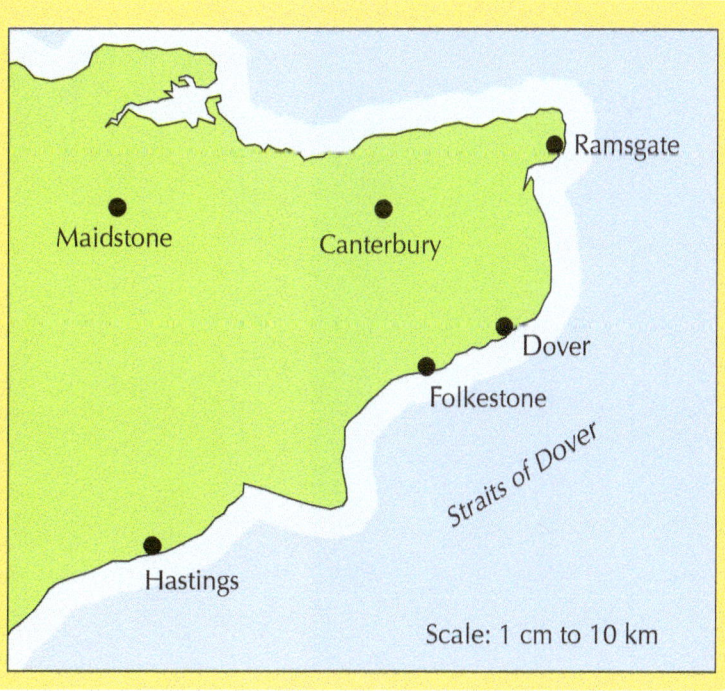

Scale: 1 cm to 10 km

Shape, position and movement

Example 24·3

The scale of the map in Example 24·2 is 1 cm to 10 km.

There are 100 cm in a metre, and 1000 metres in a kilometre.

So 10 km = 10 × 100 × 1000 = 1 000 000 cm.

Therefore 1 cm to 10 km can also be written as 1 cm to 1 000 000 cm.

So the map ratio will be given as 1 : 1 000 000.

When the ratio is used to find actual distances, each centimetre on the map will represent 1 000 000 cm or 10 km on the ground. Similarly, each inch on the map will represent 1 000 000 inches or 15·78 miles on the ground.

Exercise 24B

1 Write each of the following map scales as a map ratio.

   a 1 cm to 2 m   **b** 1 cm to 5 m   **c** 5 cm to 1 m

   d 2 cm to 1 km   **e** 1 cm to 2·5 km

2 Using the map in Example 24·2, find the actual direct distance between:

   a Maidstone and Hastings   **b** Canterbury and Ramsgate

   c Folkestone and Dover

3 The map ratio on a map is 1 : 50 000. The direct distance between two towns on the map is 4 cm. What is the actual direct distance between the two towns?

4 The map ratio on a map is 1 : 100 000. Peter has just been for a walk. Using a piece of string, he measures the map distance of his walk and finds it to be 12·5 cm. How far did he walk?

5 The map ratio on a map of Europe is 1 : 20 000 000. The actual direct distance between Paris and Rome is 1100 kilometres. What is the direct distance on the map?

6 The map shows Edinburgh city centre. Find the actual distance (as the crow flies) between the railway station (use the symbol) and:

   a Stills Gallery   **b** Sheriff Court.

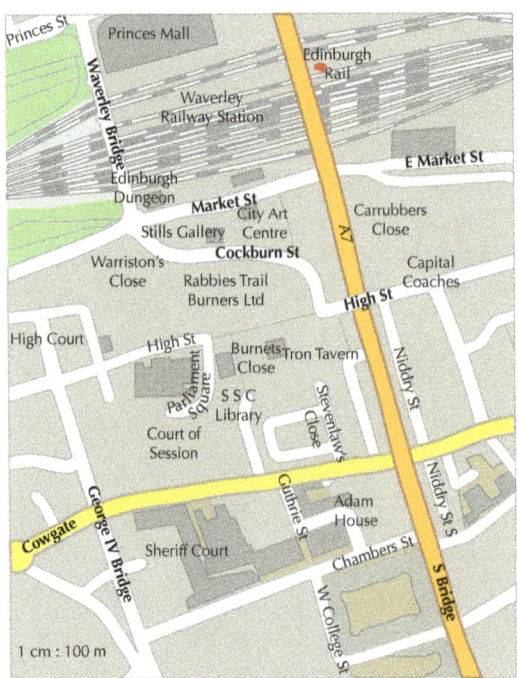

Angle, symmetry and transformation

7 The map shows the route taken by a group of hillwalkers in the Cairngorms. They start at the car park at the Linn of Dee and walk to Cairn a'Mhaim.

Use a piece of string to find the actual distance of the walk between Linn of Dee and Cairn a'Mhaim.

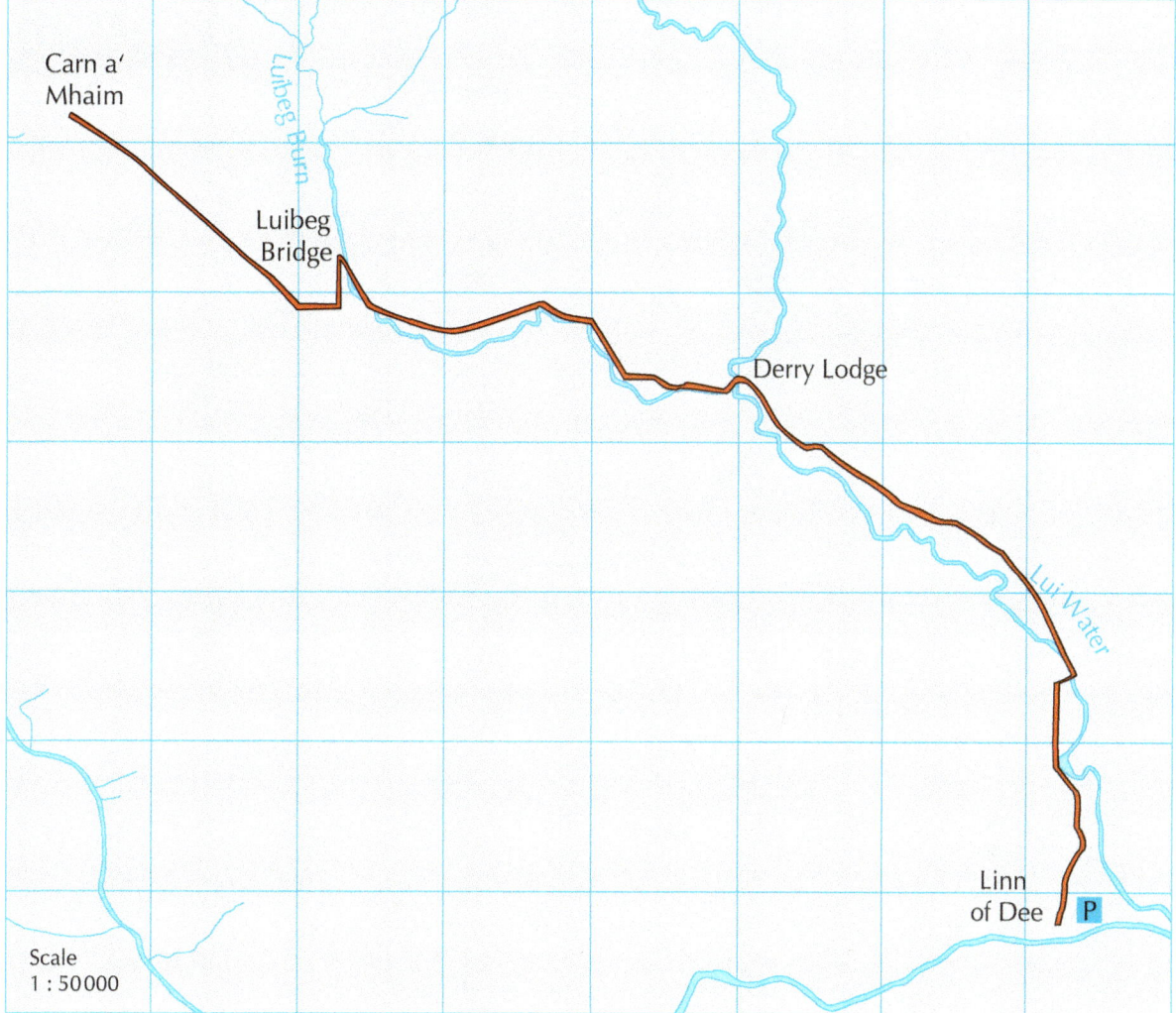

Shape, position and movement

 8 The map shows Uluru (a large rock in the Australian desert). A group of tourists follow the base walk, making one complete circuit of the rock.

  a Use a piece of string to find the distance walked by the tourists.

  b Another group of tourists drive around the rock on a circuit drive. How much further does the second group travel in one complete circuit?

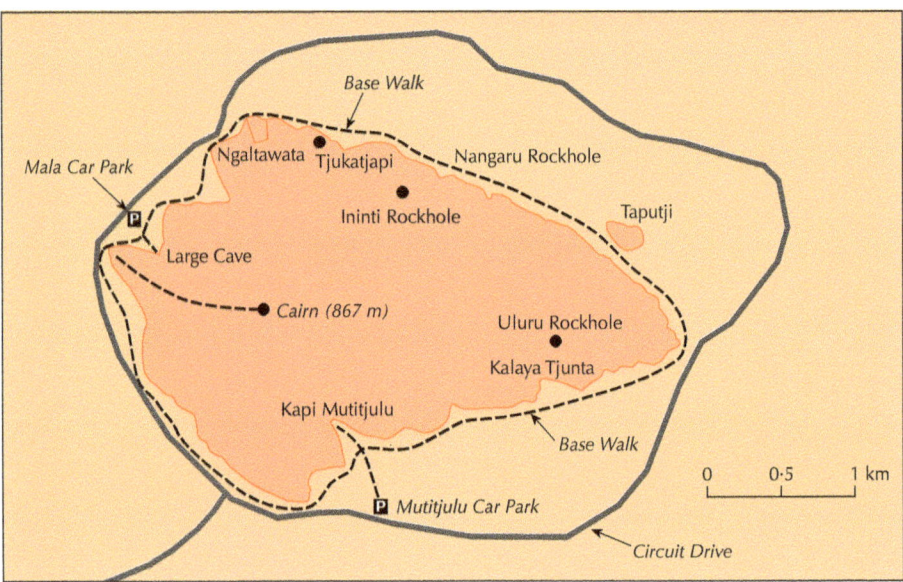

Challenge

1 Write each of the following imperial map scales as a map ratio.

  a 1 inch to 1 mile
  b 1 inch to 1 yard
  c 3 inches to $\frac{1}{2}$ mile
  d 1 inch to 50 yards

2 Ask your teacher for a map of Great Britain. Using the scale given on the map, find the direct distances between various towns and cities.

3 Use the internet to look for maps of the area where you live and find the scale that is used. Print out copies of the maps and find the distances between local landmarks.

Bearings

There are four main directions on a compass – north (N), south (S), east (E) and west (W). These directions are examples of **compass bearings**. A **bearing** is a specified direction in relation to a fixed line. The line that is usually taken is due north. The symbol for due north is: N

You have probably seen this symbol on maps in Geography.

Angle, symmetry and transformation

Bearings are mainly used for navigation purposes at sea, in the air and in sports such as orienteering. A bearing is measured in degrees (°) and the angle is always measured **clockwise** from the **north line**. A bearing is always given using three digits and is referred to as a **three-figure bearing**. For example, the bearing for the direction east is 090°.

Example 24·4 On the diagram, the three-figure bearing of B from A is 035° and the three-figure bearing of A from B is 215°.

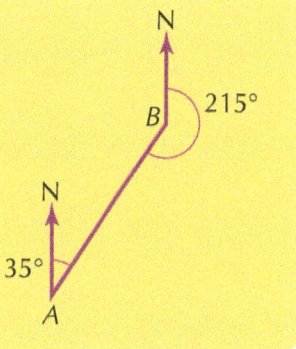

Example 24·5 The diagram shows the positions of Glasgow and Stirling on a map.

The bearing of Stirling from Glasgow is 050° and the bearing of Glasgow from Stirling is 230°. To find the bearing of Glasgow from Stirling, use the dotted line to find the alternate angle of 50° and then add 180°. Notice that the two bearings have a difference of 180°. Such bearings are often referred to as 'back bearings'.

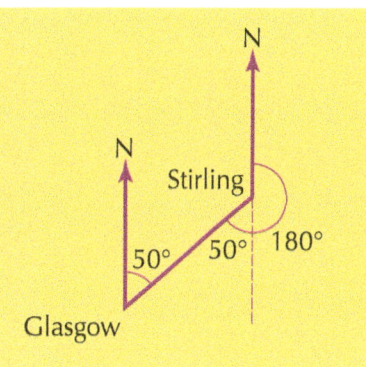

Exercise 24C

 1 Write down each of the following compass bearings as three-figure bearings.

    a South   **b** West   **c** North-east   **d** South-west

2 Write down the three-figure bearing of B from A for each of the following.

    a 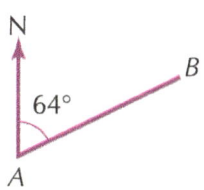   **b**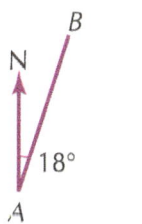

    c 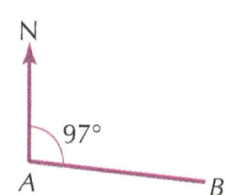   **d**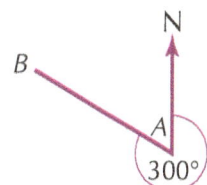

Shape, position and movement

3 Find the three-figure bearing of X from Y for each of the following.

a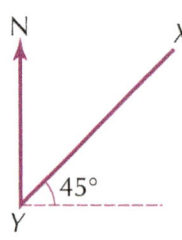

b

c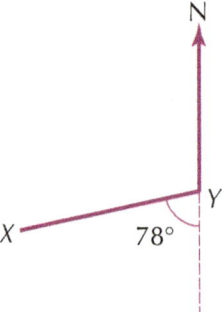

d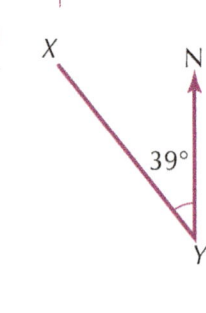

4 Draw a rough sketch to show each of the bearings below (mark the angle on each sketch).

  a From a ship A, the bearing of a lighthouse B is 030°.
  b From a town C, the bearing of town D is 138°.
  c From a gate E, the bearing of a trigonometric point F is 220°.
  d From a control tower G, the bearing of an aircraft H is 333°.

5 The two diagrams show the positions of towns and cities in Scotland. Find the bearing of each of the following:

  a i Perth from Dundee
     ii Dundee from Perth.

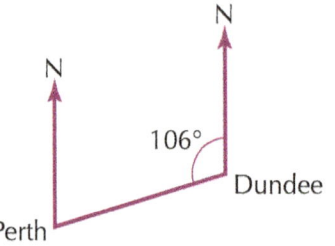

  b i Fort William from Inverness
     ii Inverness from Fort William.

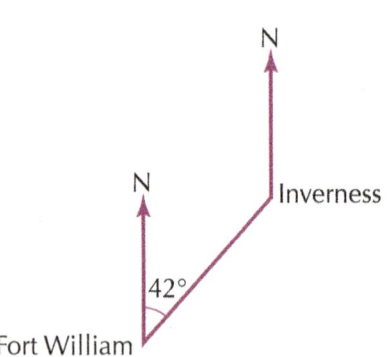

Angle, symmetry and transformation

6 Terry and Barbara are planning a walk on Ilkley Moor in Yorkshire. The scale drawing below shows the route they will take, starting from Black Pots.

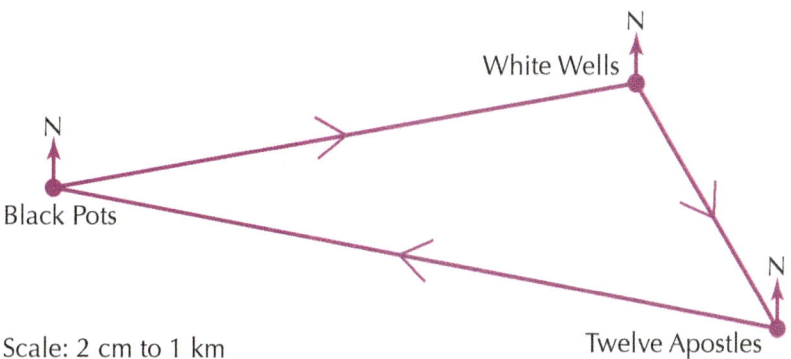

Scale: 2 cm to 1 km

a What is the total distance of their walk if they keep to a direct route between the landmarks?

b They have to take three-figure bearings between each landmark because of poor visibility. Use a protractor to find the bearing of:

  i White Wells from Black Pots

  ii Twelve Apostles from White Wells

  iii Black Pots from the Twelve Apostles.

7 From Glasgow airport, a plane heading for London flies 150 km on a bearing of 120° and then 450 km on bearing of 170°. Make a scale drawing of the plane's journey.

8 Which town is 5 km away from Inverness on a bearing of 077°?

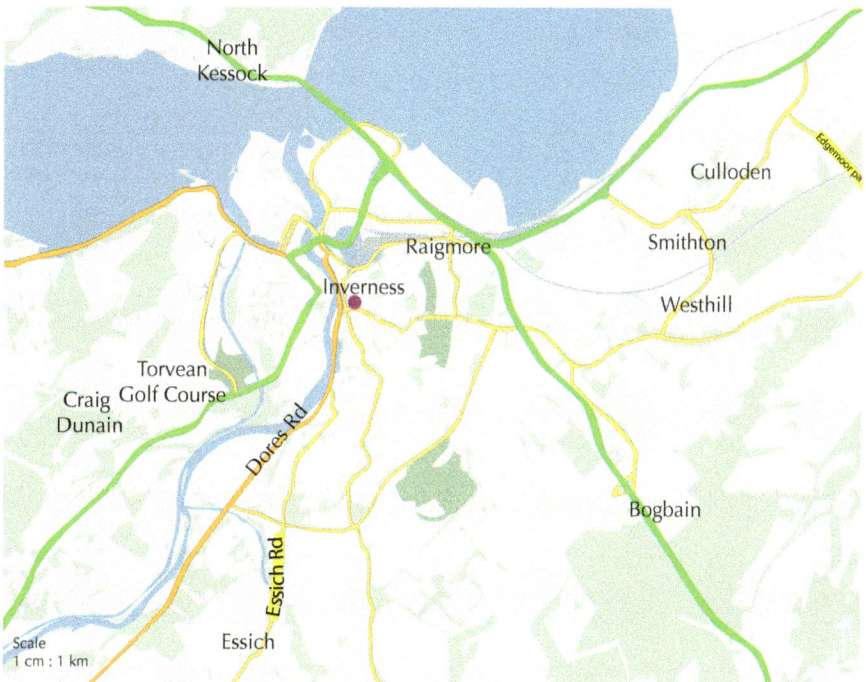

235

Shape, position and movement

Challenge

1. A liner travels from a port X on a bearing of 140° for 120 nautical miles to a port Y. It then travels from port Y on a bearing of 250° for a further 160 nautical miles to a port Z.

   a Make a scale drawing to show the journey of the liner (use a scale of 1 cm to 20 nautical miles).

   b Use your scale drawing to find:

      i the direct distance the liner travels from port Z to return to port X.

      ii the bearing of port X from port Z.

2. The bearing of Stansted airport from Heathrow airport is 040°.

   The bearing of Gatwick airport from Stansted airport is 190°.

   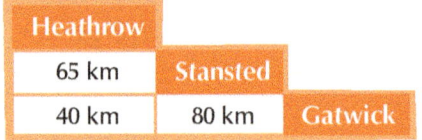

   The distances between the airports is given in the table.

   a Use this information to make a scale drawing to show the positions of the airports (use a scale of 1 cm to 10 km).

   b Use your scale drawing to find the bearing of Gatwick airport from Heathrow airport.

3. A newspaper printed this information about London and Madrid:

   From London to Madrid, the angle from north is **195° clockwise**.

   Madrid is **1300 km** from London.

   Show this information on a scale drawing.

   The position of London is shown for you.

- By working on this topic I can explain how to use scale and bearings to interpret scale drawings, plans and maps.
- I can use a given scale to calculate the actual distance from a map or plan. ★ Exercise 24A Q2
- I can draw a plan or scale drawing without being given a scale. ★ Exercise 24A Q6
- I can solve problems using maps, plans and scale drawings. ★ Exercise 24B Q8
- I understand that compass directions can be written using three figures. ★ Exercise 24C Q1
- I can show a point or route on a map when given bearings and distances. ★ Exercise 24C Q7

Angle, symmetry and transformation

25

I can apply my understanding of scale when enlarging or reducing pictures and shapes, using different methods, including technology.

MTH 3-17c

This chapter will show you how to:
- calculate the scale factor when given the dimensions of two similar objects
- understand that the scale factor is greater than 1 when an object is enlarged
- understand that the scale factor is less than 1 when an object is reduced
- calculate dimensions of an enlargement or reduction, given dimensions of an original and a scale factor
- use appropriate technology to measure or represent an enlargement or reduction.

You should already know:
- what scale is and its purpose
- what is meant by the terms 'real length' and 'real breadth'
- how to calculate the real length or height of an object
- how to interpret models, maps and plans.

Scale factor

Rectangles B, C and D are all **enlargements** of rectangle A.

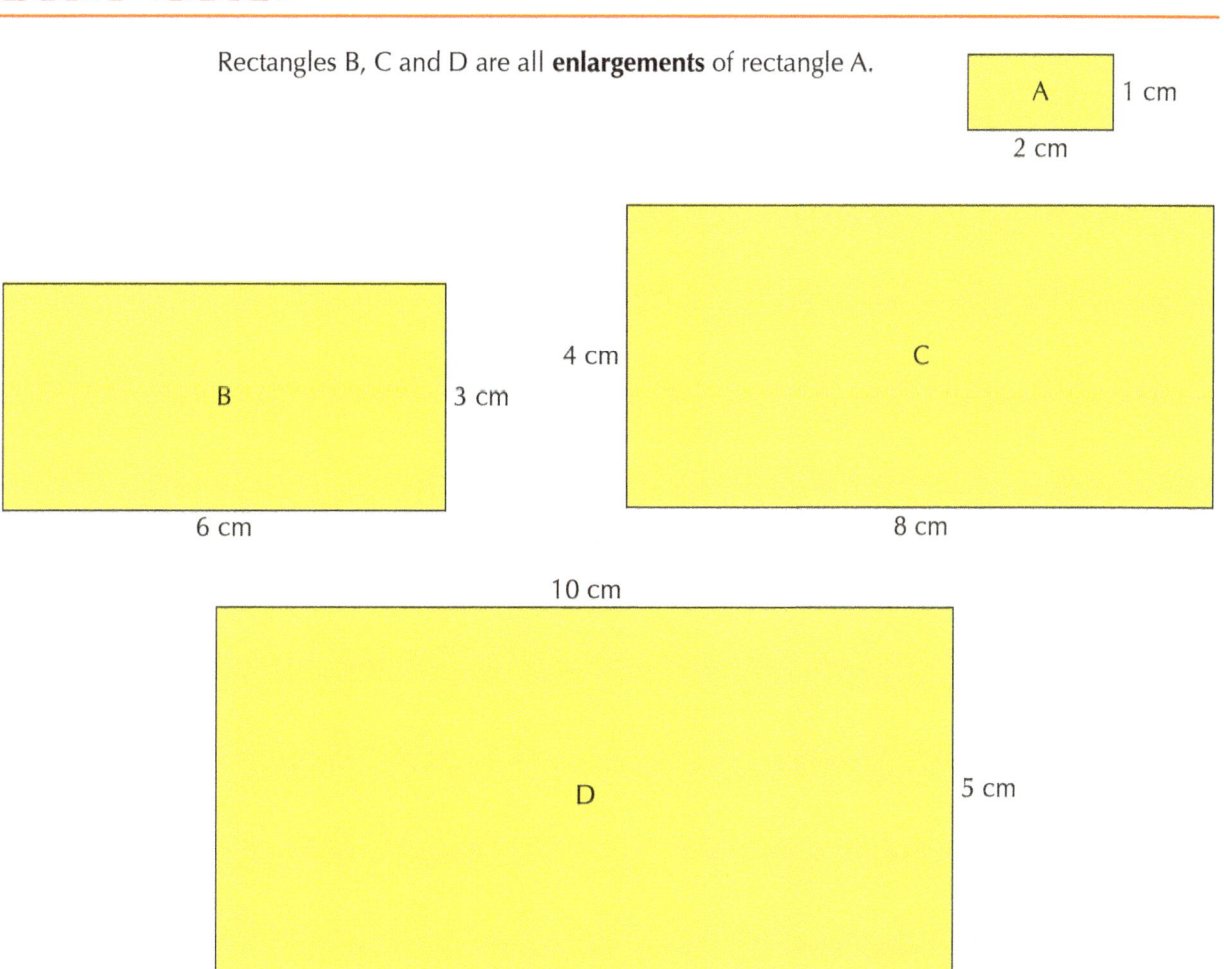

237

Shape, position and movement

Scale factor is the term used to describe the size of the enlargement.

Rectangle B has been enlarged by a scale factor of 3. This means that the lengths of the sides of rectangle B are three times longer than those of rectangle A. Rectangle C has a scale factor of 4 and rectangle D has a scale factor of 5.

Now look at rectangle E. Rectangle E has a scale factor of $\frac{1}{2}$. This means that rectangle E is half the size of rectangle A and so is a **reduction**.

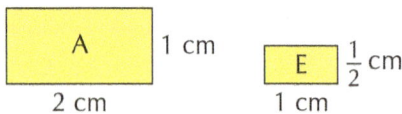

So when enlarging a shape, the scale factor is greater than 1 and when reducing a shape, the scale factor is less than 1. You can use this knowledge to check the answers of any calculations.

Example 25·1

The original rectangle has dimensions 3 cm by 8 cm. The rectangle is enlarged and the new dimensions are 9 cm by 24 cm. Calculate the scale factor.

To calculate the scale factor of this enlargement use the formula:

$$\text{Scale factor} = \frac{\text{enlarged length}}{\text{original length}}$$

You must use the lengths of corresponding sides. This means you can use two different combinations: 3 cm and 9 cm or 8 cm and 24 cm.

Using the combination 3 cm and 9 cm:

$$\text{Scale factor} = \frac{9}{3} = 3$$

Check: this is an enlargement so the scale factor must be greater than 1. ✓

It is very important to get the two lengths the correct way around. Using the rectangles in Example 25·1, if the lengths are put the wrong way around you would get: scale factor $= \frac{3}{9} = \frac{1}{3}$. This doesn't work because the rectangle is being enlarged so the scale factor must be greater than 1.

Angle, symmetry and transformation

Example 25·2 The original rectangle has dimensions 9 cm by 27 cm. The rectangle is reduced and the new dimensions are 6 cm by 18 cm. Calculate the scale factor.

To find the scale factor of this reduction use the formula:

$$\text{Scale factor} = \frac{\text{reduced length}}{\text{original length}}$$

Using the combination 18 cm and 27 cm:

$$\text{Scale factor} = \frac{18}{27}$$

18 ÷ 27 = 0·667 but it is more accurate to leave this as a fraction and simplify.

So the scale factor $= \frac{18}{27} \div \frac{9}{9} = \frac{18 \div 9}{27 \div 9} = \frac{2}{3}$

Check: this is a reduction so the scale factor must be less than 1. ✓

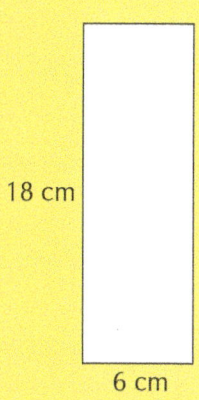

Example 25·3 The manager of the national football team investigates set formations on a board with a scaled down version of a football pitch. The standard size of a football pitch is 105 m long and 68 m wide. The scaled down version is 1·05 m long by 0·68 m.

Calculate the scale factor used when reducing the size of the football pitch for the board.

Use the length of both pitches.

$$\text{Scale factor} = \frac{\text{reduced length}}{\text{original length}} = \frac{1·05}{105} = \frac{1·05}{105} \div \frac{1·05}{1·05} = \frac{1·05 \div 1·05}{105 \div 1·05} = \frac{1}{100}$$

So the scale factor is $\frac{1}{100}$ or 0·01.

239

Shape, position and movement

Exercise 25A

1 Calculate the scale factor in each example.

a

b

c

d

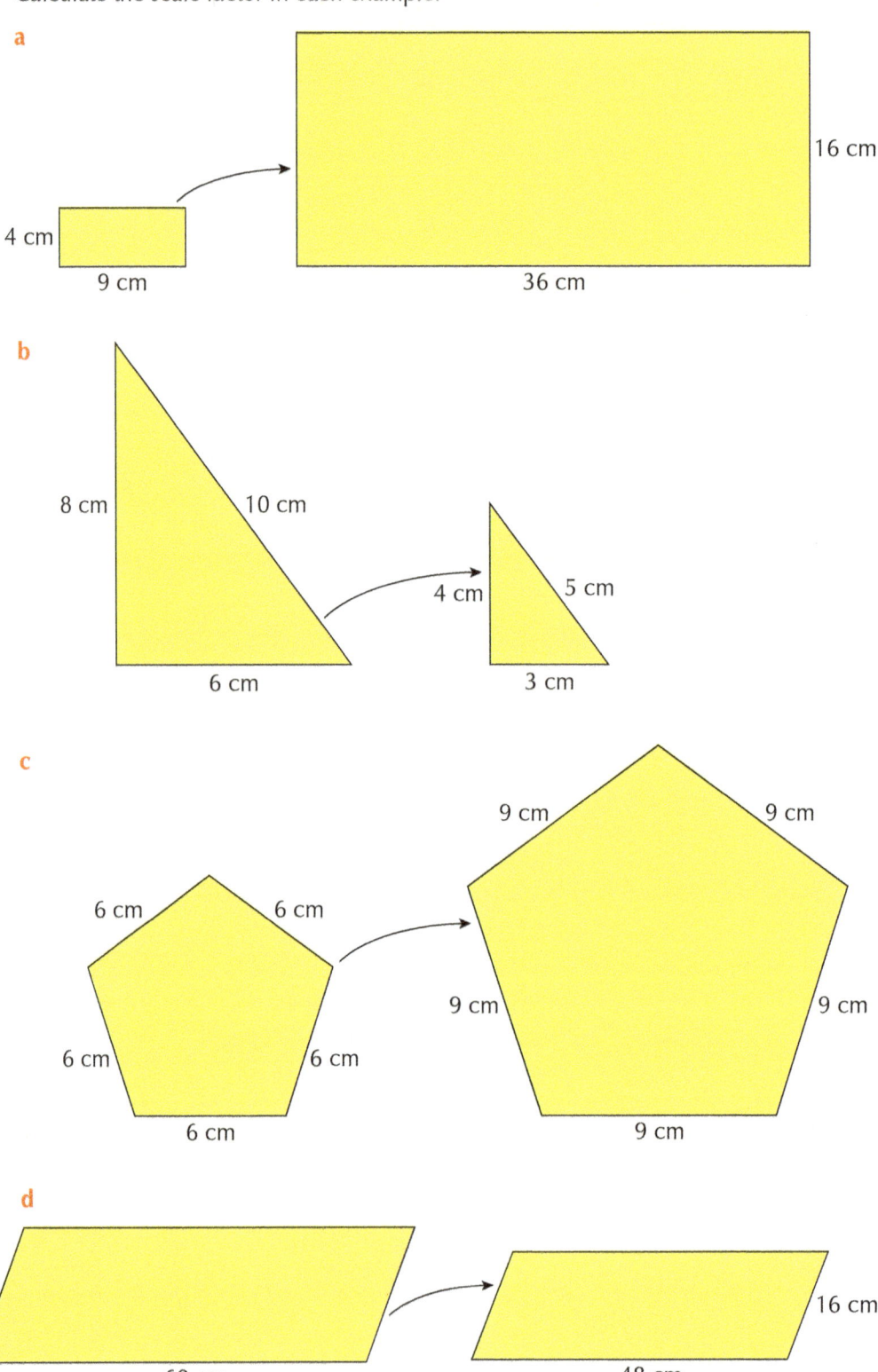

Angle, symmetry and transformation

2 Shape A is the original shape.

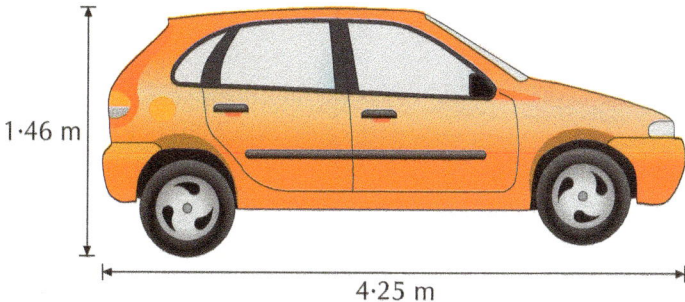

  a Which shape is an enlargement of A with a scale factor of 3?

  b Which one is not an enlargement of rectangle A? Explain your answer.

3 An architect makes a plan of a millionaire's mansion. There is a tennis court in the garden. The tennis court is 7 m wide and 21 m long. On the scale plan, the tennis court is 3·5 cm by 10·5 cm. What scale factor did the architect use for this scale plan?

4 The dimensions of a real car are shown.

A toy manufacturer makes a model of this car. The model has a length of 42·5 cm and a height of 14·6 cm. What scale factor did the toy manufacturer use?

Challenge

The second triangle is an enlargement of the first triangle. Calculate the value of x.

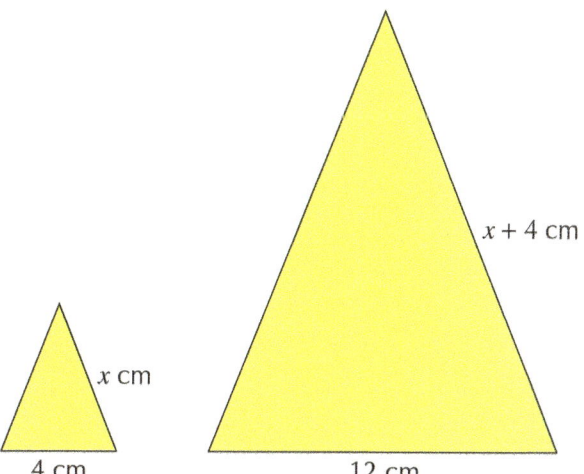

Shape, position and movement
Enlargements and reductions

Enlargements and reductions are types of **transformation**. The lengths of the sides are multiplied by the scale factor.

Example 25·4 Enlarge the small shape on the right by a scale factor of 3.

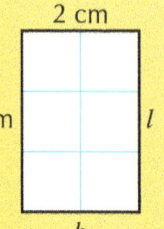

The scale factor is 3 so this means that the length of each side is multiplied by 3. The angles remain the same so you need to draw the enlargement with four right angles.

Original length = 3 cm   Enlargement length = 3 cm × 3 = 9 cm
Original breadth = 2 cm   Enlargement breadth = 2 cm × 3 = 6 cm

So the enlargement looks like this:

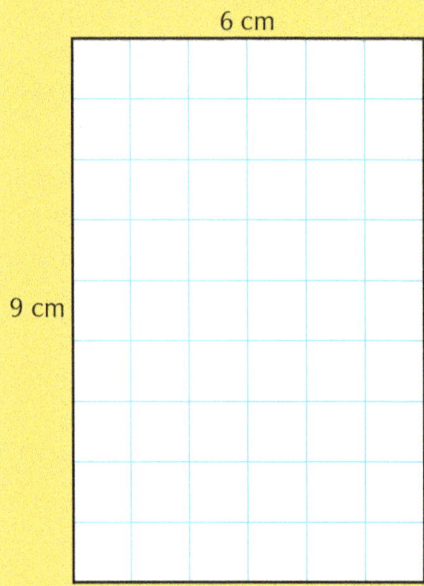

Example 25·5 Reduce this shape by a scale factor of $\frac{1}{2}$.

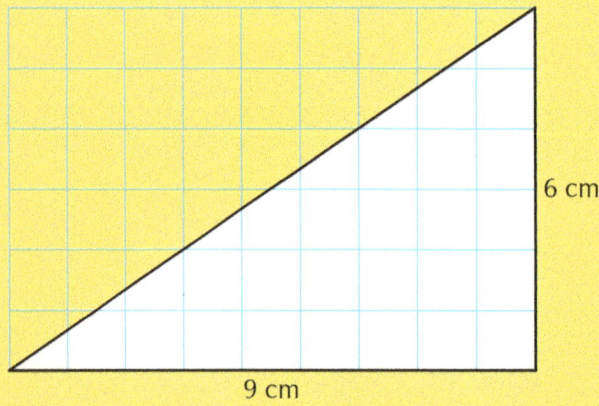

The scale factor is $\frac{1}{2}$ so this means that the length of each side of the triangle is multiplied by the scale factor of $\frac{1}{2}$. The angles remain the same so the reduced base and height must meet at a right angle.

Original base = 9 cm   Reduction base = 9 cm × $\frac{1}{2}$ = 9 ÷ 2 = 4·5 cm

Original height = 6 cm   Reduction height = 6 cm × $\frac{1}{2}$ = 6 ÷ 2 = 3 cm

Angle, symmetry and transformation

Example 25·5

Continued

So the reduction looks like this:

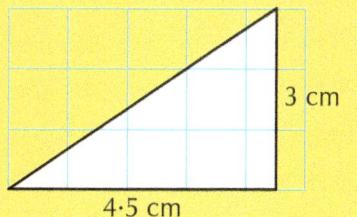

Exercise 25B

1 Enlarge these shapes by a scale factor of 2.

a 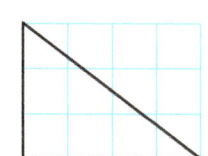   b 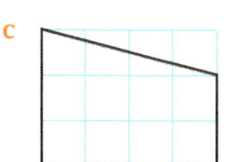   c   d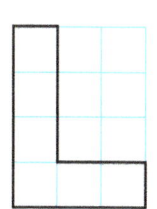

2 Reduce these shapes by a scale factor of $\frac{1}{2}$.

a    b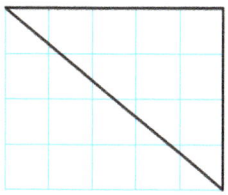

c    d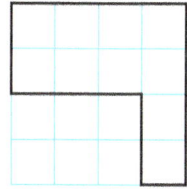

3 Enlarge these shapes by a scale factor of 3.

a    b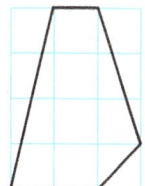

4 Reduce these shapes by a scale factor of $\frac{1}{4}$.

a 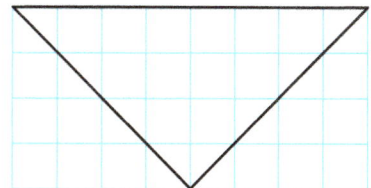   b

243

Shape, position and movement

 5 a Enlarge the shape below by a scale factor of 2.

b Reduce the shape below by a scale factor of $\frac{1}{2}$.

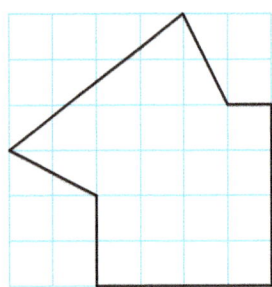

Challenge

The position of an enlargement is described by the centre of enlargement.

The diagram below shows triangle ABC enlarged to give triangle A'B'C'.

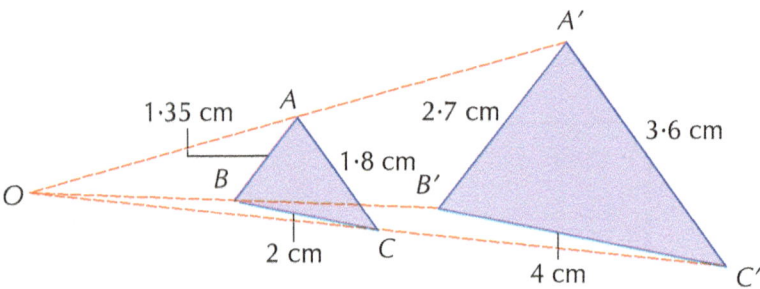

All the sides of triangle A'B'C' are twice as long as the sides of triangle ABC. Notice also that OA' = 2 × OA, OB' = 2 × OB and OC' = 2 × OC.

We say that triangle ABC has been enlarged by a scale factor of 2 about the centre of enlargement O to give the image triangle A'B'C'. The dotted lines are called the guidelines, or rays, for the enlargement. To enlarge a shape we need: a centre of enlargement and a scale factor.

Enlarge the triangle XYZ by a scale factor of 2 about the centre of enlargement O.

Draw rays OX, OY and OZ. Measure the length of the three rays and multiply each of these lengths by 2. Extend each of the rays to these new lengths measured from O and plot the points X', Y' and Z'. Join X', Y' and Z'.

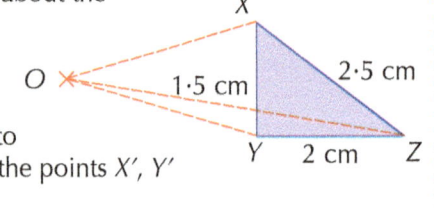

Triangle X'Y'Z' is the enlargement of triangle XYZ by a scale factor of 2 about the centre of enlargement O.

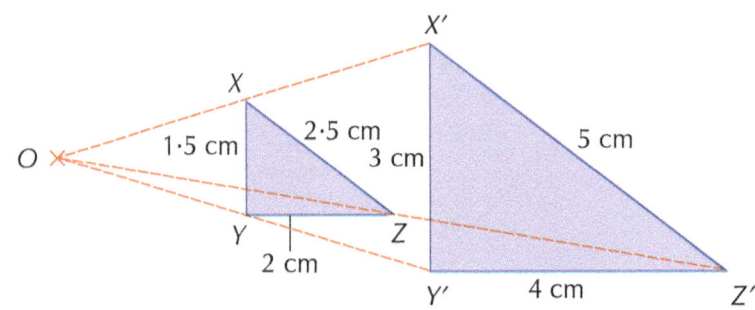

Continued

Continued

Angle, symmetry and transformation

1. Draw copies of the following shapes and enlarge each one by the given scale factor about the centre of enlargement O.

   a Scale factor 2

   b Scale factor 3

   c Scale factor 2

   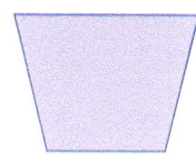

   d Scale factor 3

   (**Note:** × is the centre of square)

Using technology

In this section you will use technology to draw enlargements and reductions of shapes.

Example 25·6   Follow these instructions to use technology to enlarge a rectangle by a scale factor of 2.

- Open up a new Word document.
- Draw a rectangle of any size.

- Make a copy of the rectangle.

- Right-click on the copy of the rectangle and click on 'Format AutoShape...'.

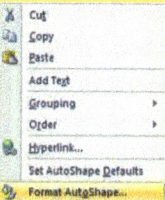

Continued

Shape, position and movement

Example 25·6
Continued

- Click on the Size tab and go to the Scale section. The scale factor is written as a percentage.

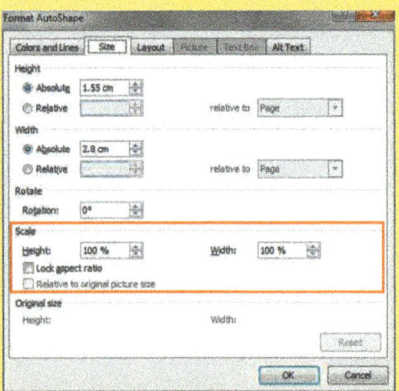

To make the enlargement change both Height and Width to 200%. Then click OK.

Exercise 25C

1. Follow the instructions given in Example 25·6 and enlarge a triangle by a scale factor of 3.

2. Follow the instructions given in Example 25·6 and reduce a parallelogram by a scale factor of $\frac{1}{4}$.

★ 3. Follow the instructions given in Example 25·6 and reduce a hexagon by a scale factor of $\frac{1}{6}$.

Challenge

Investigate what happens to the area of a shape when it is enlarged, using the shapes in Example 25·6.

- Right-click on the original rectangle and click on 'Format AutoShape...'.
- Click on the Size tab and note down the height and width measurements.

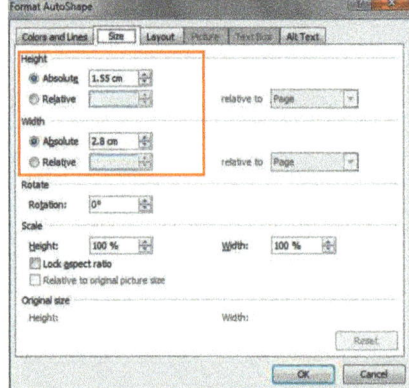

Continued

Continued

- Repeat these steps for the enlarged rectangle.

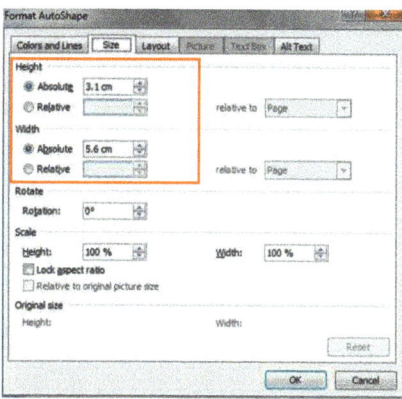

The height of the original rectangle is 1·55 cm and the width is 2·8 cm.

To calculate area, multiply height and width together.

So the area of the original rectangle = 1·55 × 2·8 = 4·34 cm².

The height of the enlargement is 3·1 cm and the width is 5·6 cm.

So the area of the enlarged rectangle = 3·1 × 5·6 = 17·36 cm².

So, enlarging a shape by scale factor of 2 doesn't double its area, that is, the shape isn't twice as big.

1. Carry out the same investigation using the scale factor of 2 but different rectangles.
2. Calculate the area scale factor – what factor is the original area multiplied by?
3. Carry out the same investigation using different scale factors.
4. Calculate the area scale factors. What type of numbers are the area scale factors?

- By working on this topic I can explain how to enlarge or reduce 2D shapes using scale factors.
- I can explain the significance of whether a scale factor is greater than 1 or less than 1.
- I can find the scale factor when a shape is enlarged or reduced. ★ Exercise 25A Q4
- I can use a scale factor to draw an enlargement or reduction. ★ Exercise 25B Q5
- I can use technology to enlarge or reduce a shape. ★ Exercise 25C Q3

Shape, position and movement

26
I can use my knowledge of the coordinate system to plot and describe the location of a point on a grid.

MTH 3-18a

This chapter will show you how to:
- identify 2D shapes by plotting their vertices on a coordinate grid
- use your knowledge of 2D shapes to find the coordinates of missing vertices
- use coordinates to help find the area of 2D shapes plotted on a coordinate grid.

You should already know:
- how to accurately plot a point on a coordinate grid
- how to read the coordinates of a point on a coordinate grid.

Coordinates

Example 26·1

Look at the grid on the right.

You can describe the position of a point on the grid using coordinates.

For example, the point A has the coordinates (2, 3).

This is found by starting at the origin (0, 0), reading from the horizontal axis first and the vertical axis second.

What are the coordinates of these points?

a B   b C   c D

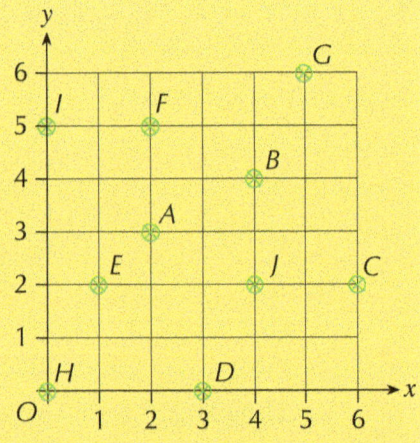

By reading first from the horizontal axis and then from the vertical axis, the coordinates are:

a (4, 4)   b (6, 2)   c (3, 0)

Exercise 26A

1 From the grid above, write the coordinates of the points E, F, G, H, I and J.

2 Draw six copies of the grid above without the points marked on. Plot the following sets of points on separate grids and join them together to make a 2D shape. Write down the mathematical name of each shape.

   a (5, 2), (1, 4), (0, 2) and (4, 0)
   b (6, 5), (6, 2), (1, 0), (3, 5)
   c (0, 2), (2, 6), (5, 2)
   d (3, 6), (6, 4), (4, 1), (1, 3)
   e (2, 4), (4, 2), (3, 0), (1, 0), (0, 2)
   f (3, 0), (1, 3), (3, 6), (5, 3)

248

Angle, symmetry and transformation

3 Look at the grid on the right.

   a Write down the coordinates of *A*, *B* and *C*.

   b *ABC* are three corners of a square.

   Write down the coordinates of the fourth corner.

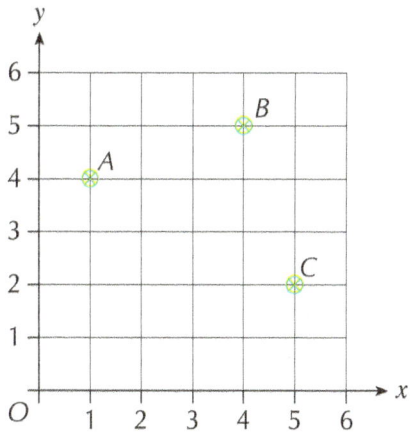

4 Look at the grid on the right.

   a Write down the coordinates of *L*, *M* and *N*.

   b *LMN* are three corners of a parallelogram.

   Write down two possible coordinates for the position of the fourth corner.

5 a Make a copy of the grid in Question 4. Plot the points *W* (1, 0), *X* (6, 1) and *Y* (5, 6).

   b The points form three vertices of a square *WXYZ*. Plot the point *Z* and draw the square.

   c What are the coordinates of point *Z*?

   d Draw in the diagonals of the square. What are the coordinates of the point of intersection of the diagonals?

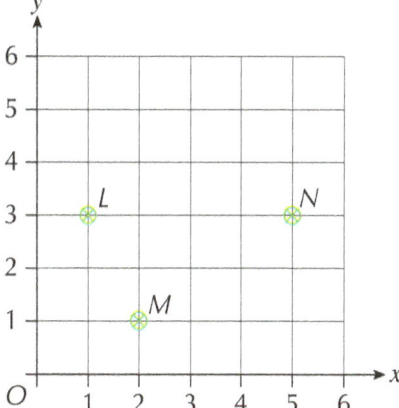

★ 6 The points (1, 5), (4, 8) and (7, 5) form three vertices of a quadrilateral *ABCD*. Plot these points on a coordinate grid with axes for *x* and *y* from 0 to 8.

   a Find the coordinates of the other vertex if the shape is:

   i a square   ii an arrowhead (or delta)   iii a kite

   b What happens if you plot the vertex at (4, 5)?

7 The points (4, 1) and (1, 4) form two vertices of a square *ABCD*. Plot these points on a coordinate grid with axes for *x* and *y* from 0 to 8 and find the other two vertices if these points are:

   a *A* and *C*   b *A* and *B*

249

Shape, position and movement
Using coordinates

You can use coordinates to help find the areas of shapes and the image of a shape after a transformation.

Example 26·2

Triangle A has been mapped onto triangle B by a translation 3 units right, followed by 2 units up.

Points on triangle A are mapped by the same translation onto triangle B, as shown by the arrows.

When a shape is translated onto its image, every point on the shape moves the same distance, so the coordinates of triangle A that were (0, 0), (1, 2) and (2, 0) move to (3, 2), (4, 4) and (5, 2) for triangle B.

The area of both triangles is $A = \frac{1}{2} \times 2 \times 2 = 2$ units².

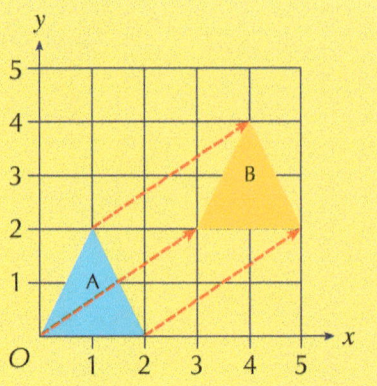

Exercise 26B

For Questions 1 to 3, draw axes for *x* and *y* from 0 to 8 on centimetre squared paper, then plot the coordinates and find the area of each shape.

MNU 3-11a

1 a △ABC with A (2, 0), B (5, 0) and C (4, 4)
   b △DEF with D (1, 1), E (6, 1) and F (3, 5)
   c △PQR with P (2, 1), Q (2, 5) and R (5, 3)
   d △XYZ with X (0, 5), Y (6, 5) and Z (4, 1)

2 a Parallelogram ABCD: A (2, 0), B (6, 0), C (8, 5) and D (4, 5)
   b Parallelogram EFGH: E (1, 2), F (4, 2), G (7, 7) and H (4, 7)
   c Parallelogram PQRS: P (1, 8), Q (7, 5), R (7, 1) and S (1, 4)

3 a Trapezium ABCD: A (1, 1), B (6, 1), C (5, 7), D (2, 7)
   b Trapezium EFGH: E (0, 5), F (6, 5), G (5, 3), H (3, 3)
   c Trapezium PQRS: P (2, 3), Q (2, 5), R (7, 6), S (7, 1)

4 Copy the triangle ABC onto squared paper. Label it P.

   a Write down the coordinates of the vertices of triangle P.
   b Translate triangle P 6 units left and 2 units down. Label the new triangle Q.
   c Write down the coordinates of the vertices of triangle Q.
   d Translate triangle Q 5 units right and 4 units down. Label the new triangle R.
   e Write down the coordinates of the vertices of triangle R.
   f Describe the translation which translates triangle R onto triangle P.

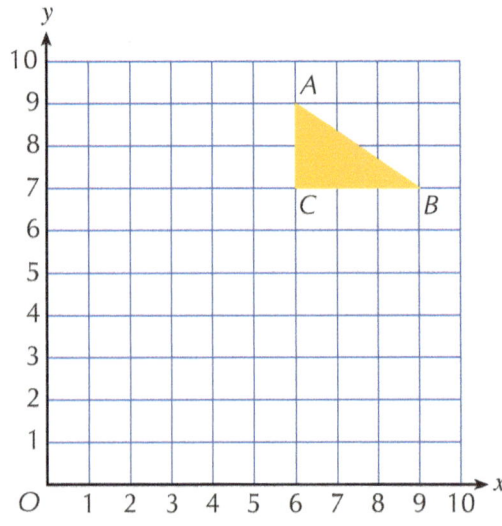

Angle, symmetry and transformation

MTH 3-19a

5 a Copy the grid onto squared paper and draw the triangle ABC. Write down the coordinates of A, B and C.

b Reflect the triangle in mirror line A. Label the vertices of the image P, Q and R and write down their coordinates.

c Reflect triangle PQR in mirror line B. Label the vertices of this image S, T and V and write down their coordinates.

d Reflect triangle STV in mirror line A. Label the vertices of the image W, X and Y and write down their coordinates.

e Describe the reflection that maps triangle WXY onto triangle ABC.

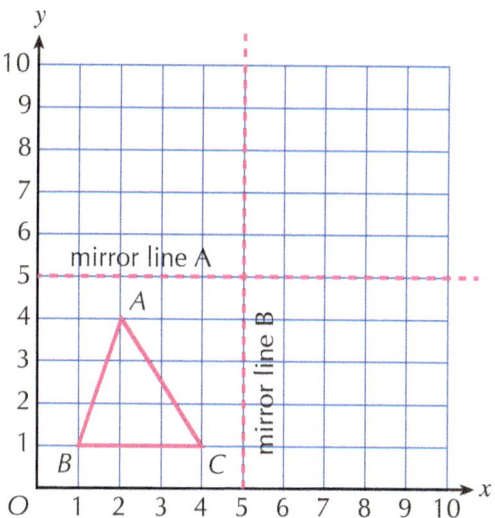

MTH 3-17c

6 Copy the diagrams below onto centimetre squared paper and enlarge each one by the given scale factor, using the origin O as the centre of enlargement.

a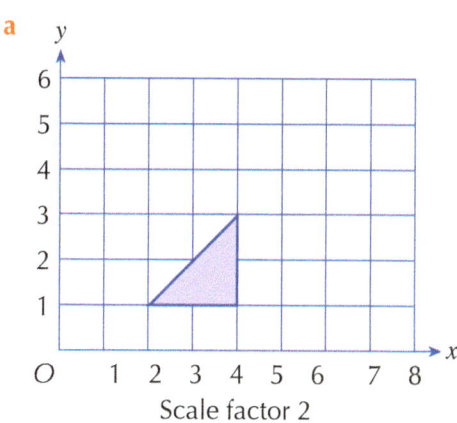
Scale factor 2

b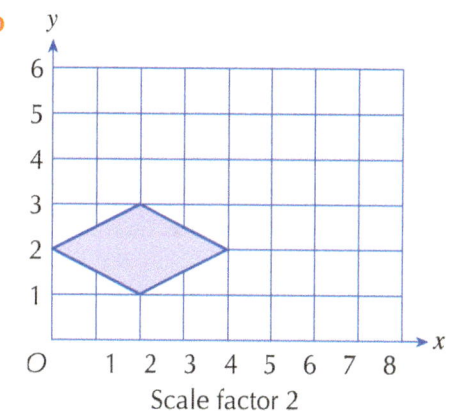
Scale factor 2

c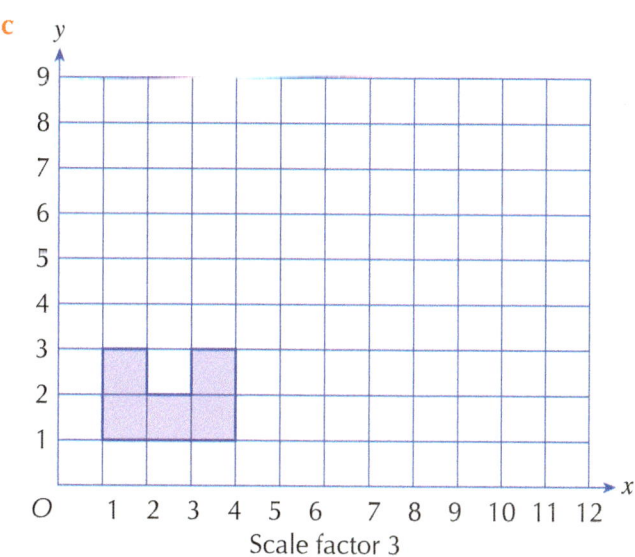
Scale factor 3

Shape, position and movement

7 Draw axes for *x* and *y* from 0 to 10 on centimetre squared paper. Plot the points A (4, 5), B (5, 4), C (4, 1) and D (3, 4) and join them together to form the kite ABCD. Enlarge the kite by a scale factor of 2 using O as the centre of enlargement.

★ 8 a Draw axes for *x* and *y* from 0 to 12 on centimetre squared paper. Plot the points A (1, 3), B (3, 3), C (3, 1) and D (1, 1) and join them together to form the square ABCD.

   b Write down the area of the square.

   c Enlarge the square ABCD by a scale factor of 2 using the origin as the centre of enlargement. What is the area of the enlarged square?

   d Enlarge the square ABCD by a scale factor of 3 using the origin as the centre of enlargement. What is the area of the enlarged square?

   e Enlarge the square ABCD by a scale factor of 4 using the origin as the centre of enlargement. What is the area of the enlarged square?

   f Write down anything you notice about the increase in area of the enlarged squares. Can you write down a rule to explain what is happening?

   g Repeat the above using your own shapes. Does your rule still work?

Challenge

1 Battleships

Play the game of 'Battleships' against a partner, using a coordinate grid.

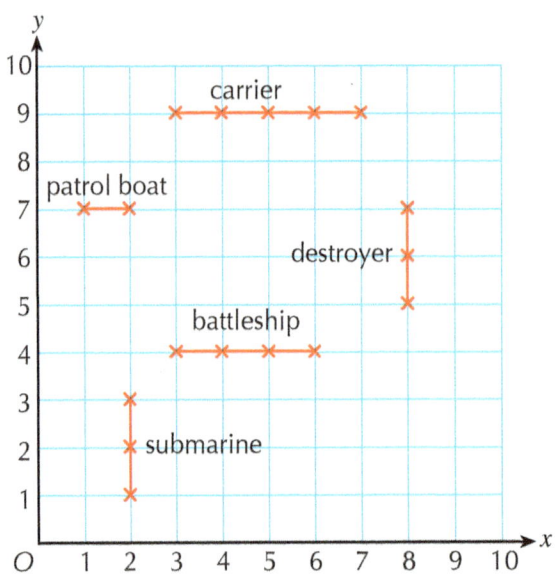

- Each player draws axes for *x* and *y* from 0 to 10 on squared paper.
- You have 5 ships to place anywhere on your grid:
  - patrol boat – 2
  - submarine – 3
  - destroyer – 3
  - battleship – 4
  - carrier – 5

Continued

Angle, symmetry and transformation

Continued

- Once you have both placed your ships, take turns to tell your opponent the coordinates you want to fire at.
- Your opponent will tell you if you have hit or missed and if you have sunk one of their ships.
- The winner is the first to sink all their opponent's ships.

You could make up an alternative version of this game using shapes instead of ships, for example, square, rectangle, right-angled triangle, parallelogram or trapezium.

2 Four in a row

The aim of this game is to be the first to plot four adjacent points in a straight line, whether horizontally, vertically or diagonally, on a coordinate grid.

- This time you and your opponent play using the same grid.
- Draw axes for x and y from 0 to 8 on squared paper.
- Take turns to pick a point and mark it with your colour.
- Try to get four points in a row, while blocking your opponent.
- The winner is the first to get four in a row.

Challenge the winner to find and describe a connection between the coordinates of the points on their line.

- By working on this topic, I have learnt how to apply my knowledge of coordinates to problems involving 2D shapes, their areas and transformations.

- I can explain how to find the missing vertices in a 2D shape. ★ Exercise 26A Q6

- I can plot a 2D shape on a coordinate grid and then calculate its area. ★ Exercise 26B Q8

- I can transform a 2D shape by reflection, translation or enlargement and find the coordinates of its image. ★ Exercise 26B Q8

Shape, position and movement

27

I can illustrate the lines of symmetry for a range of 2D shapes and apply my understanding to create and complete symmetrical pictures and patterns.

MTH 3-19a

This chapter will show you how to:
- recognise lines of symmetry in complex shapes, pictures and patterns
- complete symmetrical patterns with two or more lines of symmetry
- use your knowledge to create more complex symmetrical shapes, pictures and patterns.

You should already know:
- how to recognise symmetry in shapes, pictures and patterns
- how to draw lines of symmetry on shapes, pictures and patterns
- how to complete symmetrical patterns with up to two lines of symmetry
- how to draw your own symmetrical shapes, pictures and patterns.

Line symmetry

A 2D shape has a **line of symmetry** when one half of the shape fits exactly over the other half when the shape is folded along that line.

A mirror or tracing paper can be used to check whether a shape has a line of symmetry. Some shapes have no lines of symmetry while others have more than one.

A line of symmetry is also called a **mirror line**.

Example 27·1 This T-shape has one line of symmetry, as shown.

Put a mirror on the line of symmetry and check that the image in the mirror is half the T-shape.

Next, trace the T-shape and fold the tracing along the line of symmetry to check that both halves of the shape fit exactly over each other.

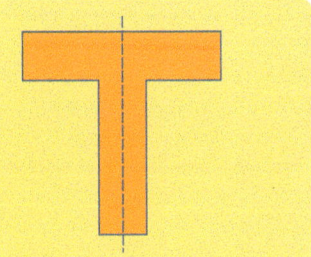

Example 27·2 This cross has four lines of symmetry, as shown.

Check that each line drawn here is a line of symmetry. Use either a mirror or tracing paper.

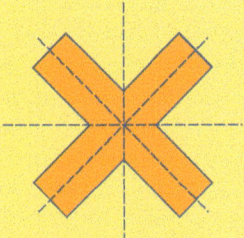

Angle, symmetry and transformation

Example 27·3 This L-shape has no lines of symmetry.

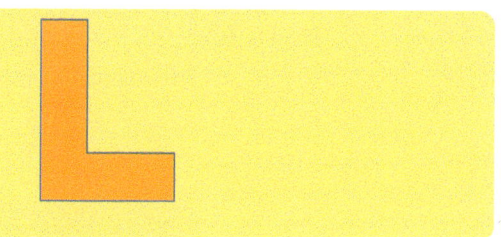

Exercise 27A

For this Exercise, you may find tracing paper or a mirror helpful.

1. Copy each of these shapes and draw its lines of symmetry. Write below each shape the number of lines of symmetry it has.

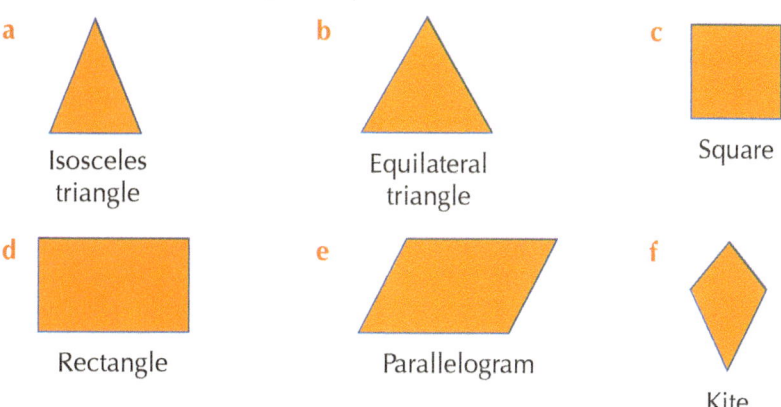

a Isosceles triangle b Equilateral triangle c Square
d Rectangle e Parallelogram f Kite

2. Write down the number of lines of symmetry for each of the following shapes.

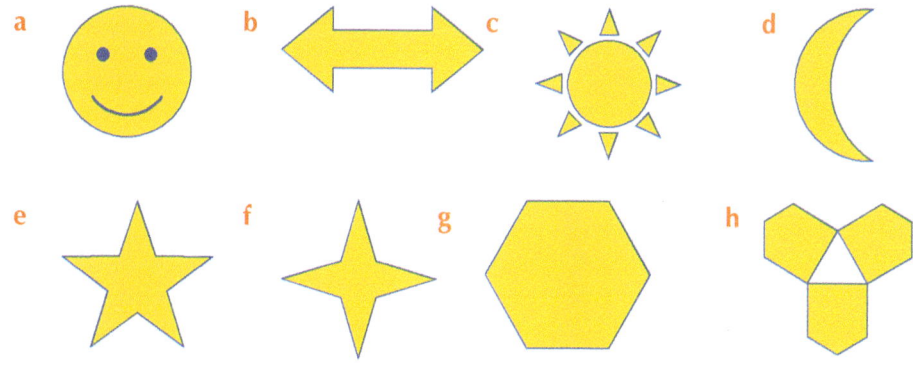

3. Write down the number of lines of symmetry for each of these road signs.

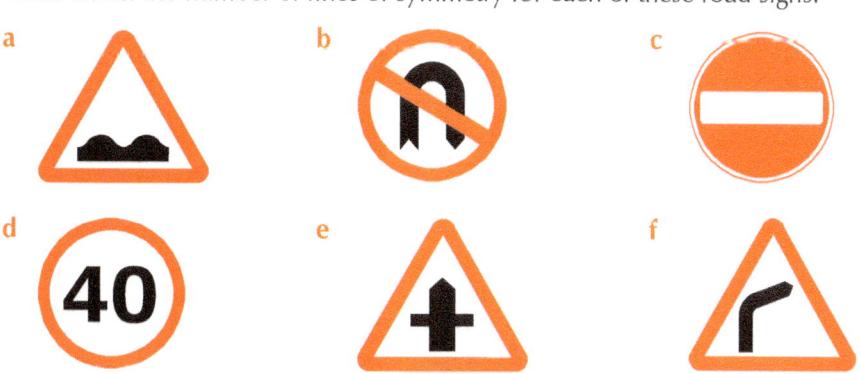

Shape, position and movement

★ 4 Look at these famous landmarks. How many lines of symmetry does each picture have? Draw sketches to show the lines of symmetry.

a The Angel of the North

b Notre Dame Cathedral

c O2 Arena

d Taj Mahal

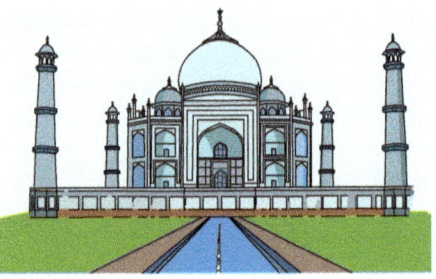

e Eiffel Tower

f Petronas Towers

Angle, symmetry and transformation

Challenge

1 Sports logo

Design a logo for a new sports and leisure centre that is due to open soon. Your logo should have four lines of symmetry.

2 Symmetry squares

Two squares can be put together along their sides to make a shape that has line symmetry.

Three squares can be put together along their sides to make two different shapes that have line symmetry.

Investigate how many different symmetrical arrangements there are for four squares. What about five squares?

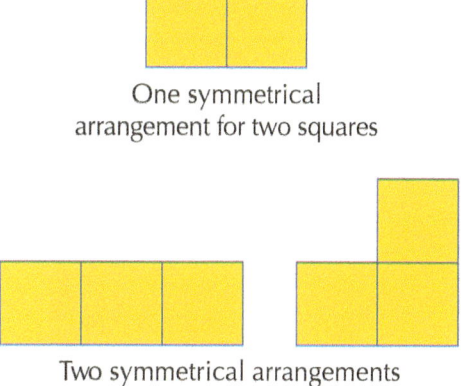

One symmetrical arrangement for two squares

Two symmetrical arrangements for three squares

Reflections

The picture shows an L-shape reflected in a mirror.

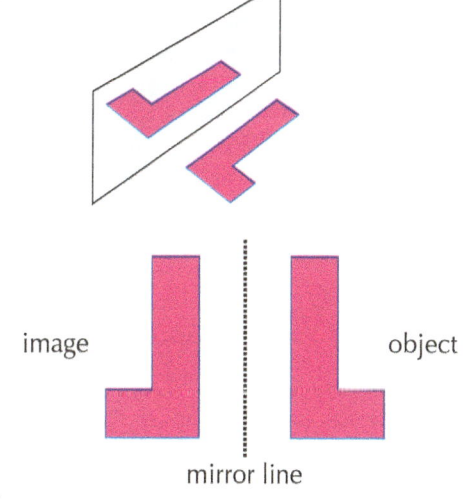

You can draw the picture without the mirror, as follows.

The **object** is reflected in the mirror line to give the **image**. The mirror line becomes a line of symmetry. So, if the paper is folded along the mirror line, the object will fit exactly over the image. The image is the same distance from the mirror line as the object is.

A **reflection** is an example of a transformation. A **transformation** is a way of changing the position or the size of a shape.

Example 27·4

Reflect this shape in the given mirror line.

Notice that the image is the same size as the object, and that the mirror line becomes a line of symmetry.

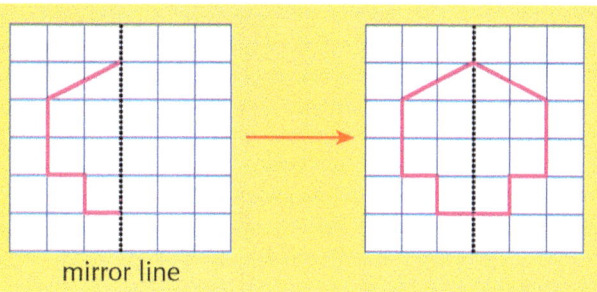

mirror line

257

Shape, position and movement

Example 27·5 Triangle A'B'C' is the reflection of triangle ABC in the given mirror line.

Notice that the point A and the point A' are the same distance from the mirror line, and that the line joining A and A' crosses the mirror line at 90°. This is true for all corresponding points on the object and its image.

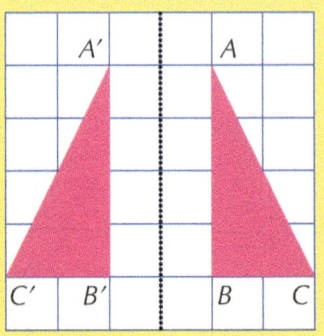

Example 27·6 Reflect this rectangle in the mirror line shown.

Use tracing paper or a mirror to check the reflection.

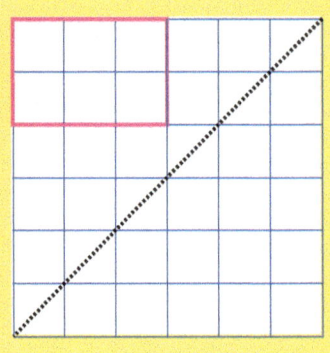 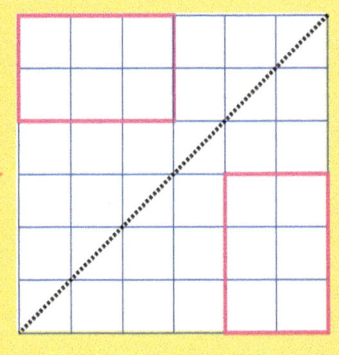

Exercise 27B For this Exercise, you may find tracing paper or a mirror helpful.

 1 Copy each of these diagrams onto squared paper and draw its reflection in the given mirror line.

a   b   c   d

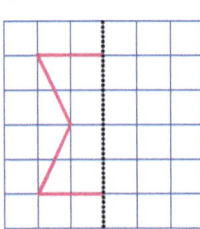 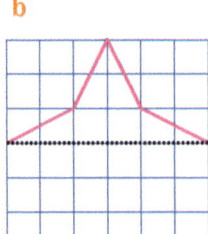 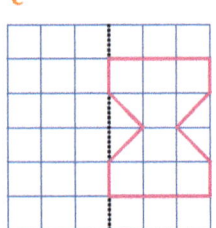 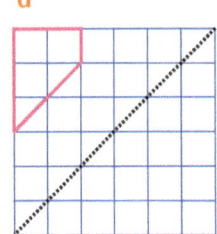

2 Copy each of these shapes onto squared paper and draw its reflection in the given mirror line.

a   b   c   d

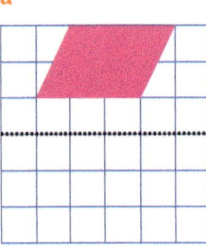 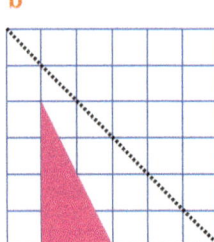 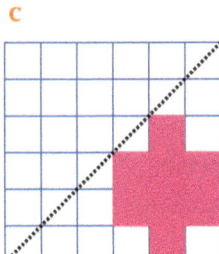 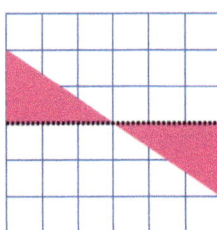

Angle, symmetry and transformation

3 Copy each shape onto squared paper and draw its reflection in the given mirror line.

a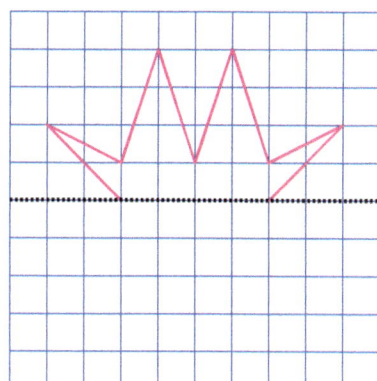

b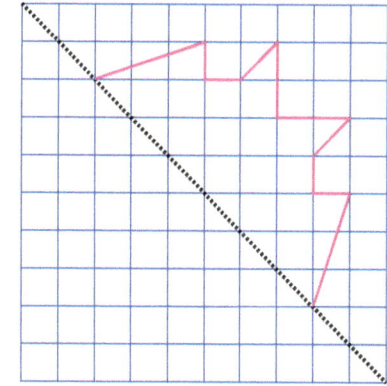

4 Copy this diagram. Create a shape on one side and get your partner to complete the diagram.

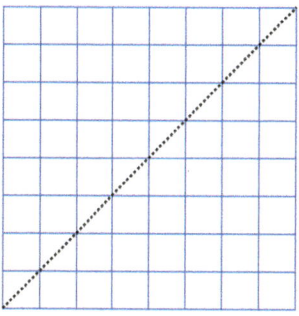

Challenge

1 a Copy the diagram onto squared paper and reflect the triangle in mirror line *A*. Now reflect the image you got in mirror line *B*. Finally, reflect this image in mirror line *C*.

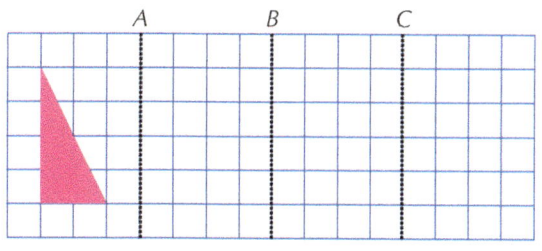

b Make up your own patterns using a series of parallel mirror lines.

2 a Copy the grid on the right onto centimetre squared paper.

b Reflect the shape in mirror line 1.

Then reflect the new shape in mirror line 2.

c Make up your own patterns using two perpendicular mirror lines.

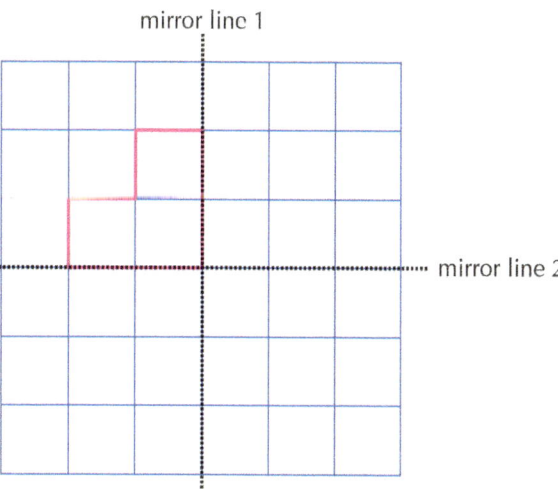

259

Shape, position and movement
Reflections in two mirror lines

A shape can be reflected in two perpendicular mirror lines as Example 27·7 shows.

Example 27·7 Reflect the L-shape in mirror line 1, then in mirror line 2.

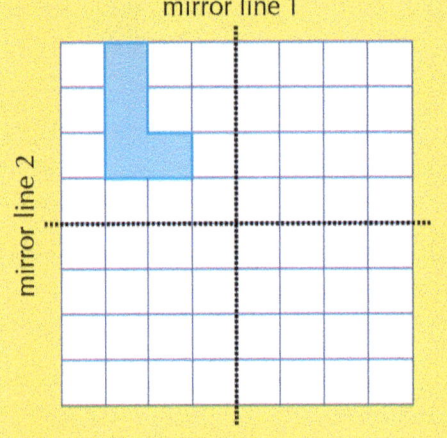

Reflecting in mirror line 1 gives:

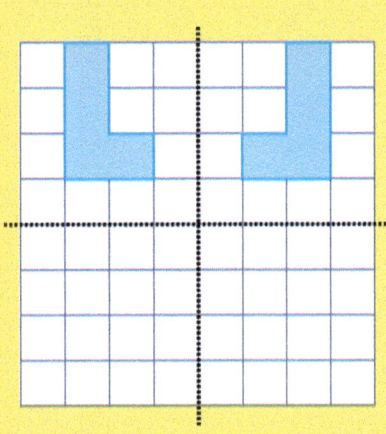

Reflecting both shapes in mirror line 2 gives:

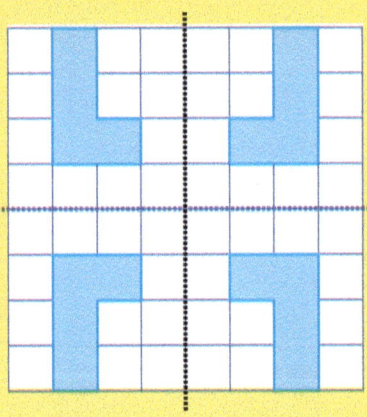

Example 27·8 Reflect the shape in mirror line 1, and then reflect both shapes in mirror line 2.

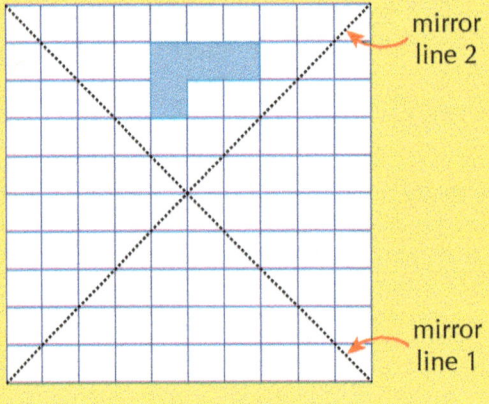

Continued

260

Angle, symmetry and transformation

Example 27·8

Continued

Reflecting in mirror line 1 gives:

Reflecting both shapes in mirror line 2 gives:

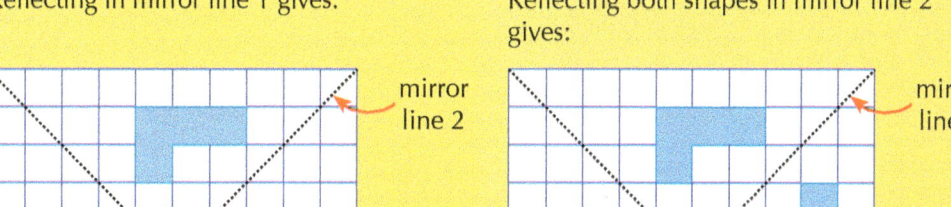

Exercise 27C

1   Copy each of the following shapes onto centimetre squared paper and reflect it in both mirror lines shown.

a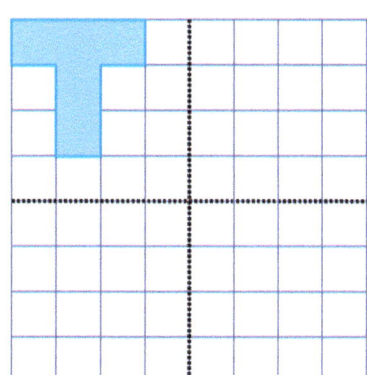

b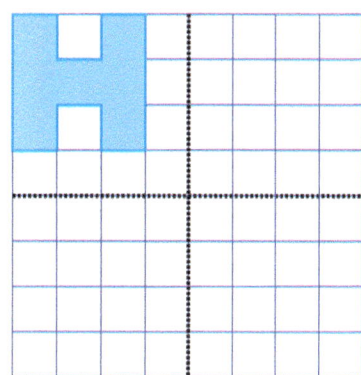

c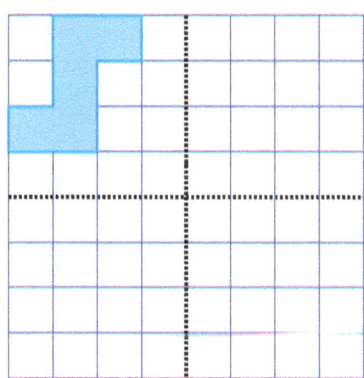

d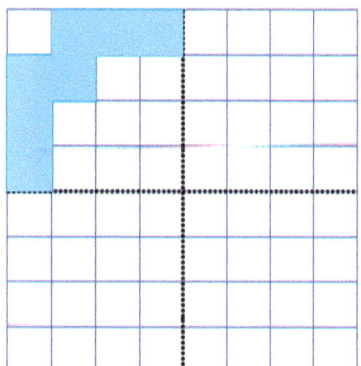

Shape, position and movement

2 Copy each of the following diagrams onto centimetre squared paper and reflect the shape in both mirror lines.

a   b

★ **3** Copy the diagram onto centimetre squared paper and reflect the shape in the two diagonal mirror lines.

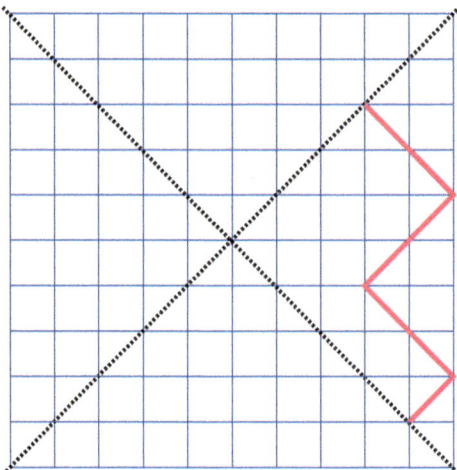

Angle, symmetry and transformation

4 Copy each of the following diagrams onto centimetre squared paper. Reflect the shape in one mirror line and then reflect both shapes in the second mirror line.

a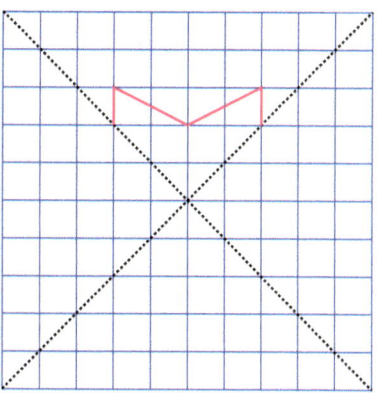

b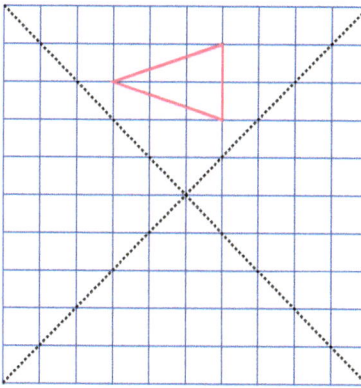

c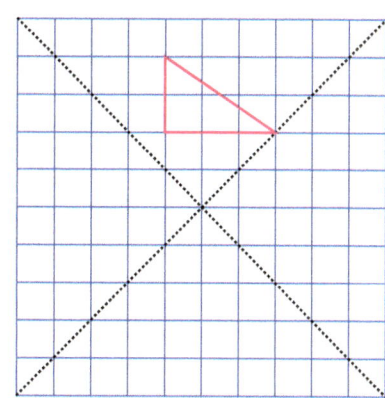

d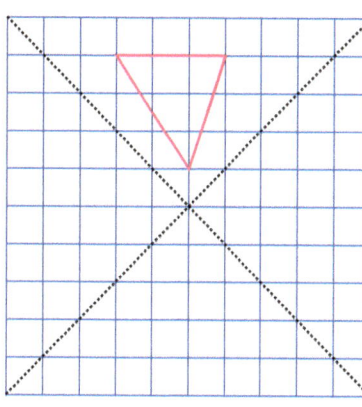

e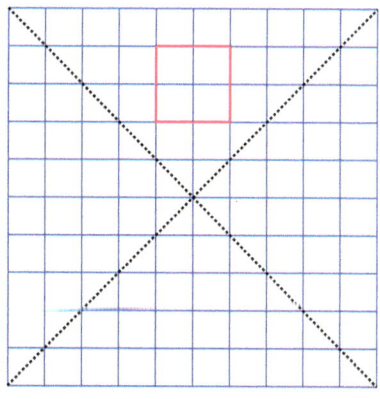

f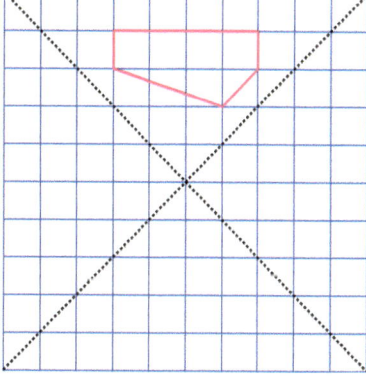

5 Copy this diagram. Create a shape on one side and get your partner to complete the diagram by reflecting your shape in the mirror lines.

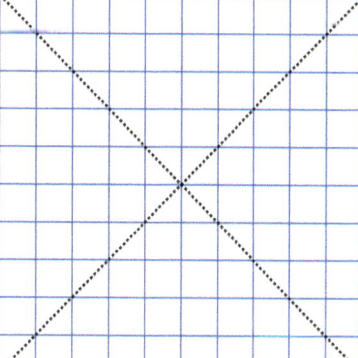

Shape, position and movement

Challenge   Rangoli patterns

Rangoli patterns are symmetrical designs used by Sikh and Hindu families. They are traditionally drawn on the floor near the entrance to a house to welcome guests.

1 Draw a 10 × 10 grid on centimetre squared paper and draw four diagonal lines.

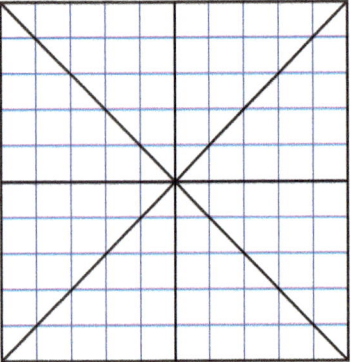

2 In the top right corner, design a pattern, for example, like the one shown.

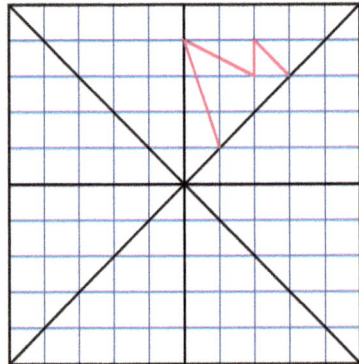

3 Moving round in a clockwise direction, reflect the shape in the diagonal line.

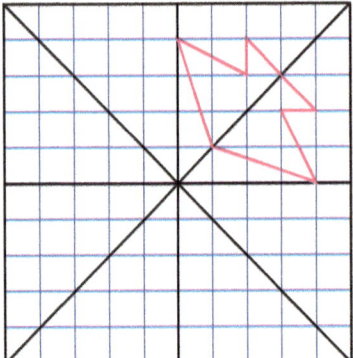

4 Continue reflecting the shape in each diagonal, horizontal or vertical line until the shape is complete.

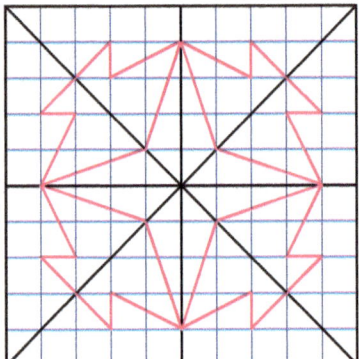

- By working on this topic I can explain how symmetry and reflection are related.
- I can use my knowledge of symmetry and reflection to create symmetrical pictures and patterns.
- I can find the lines of symmetry of a shape, picture and pattern. ★ Exercise 27A Q4
- I can reflect shapes in one mirror line. ★ Exercise 27B Q1
- I can reflect shapes in more than one mirror line. ★ Exercise 27C Q3

28

I can work collaboratively, making appropriate use of technology, to source information presented in a range of ways, interpret what it conveys and discuss whether I believe the information to be robust, vague or misleading.

MNU 3-20a

This chapter will show you how to:
- recognise that statistics are presented in different types of media
- understand that statistics can be shown in different pictorial ways
- interpret data on a given topic and explain whether the information gathered is reliable or misleading.

You should already know:
- how to interpret and draw conclusions from a range of data displays
- how to check that the data display is a reliable source of information and can be used to make accurate statements
- how to check that statements made about data are correct when compared with the data given in a data display.

Information from charts

We often see charts in magazines, papers, websites and on TV giving us a lot of different types of information.

Data and information

The words **data** and **information** are used a lot when we talk about charts and statistics. **Data** refers to the original facts or numbers that are collected. When we interpret the data to look for a meaning or pattern we get **information**. For example, if you collected the dates of birth of all the people in your class you would have a list of days, months and years, which is your data. If you plotted this onto a bar chart you might see that more people were born in October than any other month. By drawing the graph you have organised and interpreted the data and created information about the class.

Information handling

Example 28·1

Here are the results of a girls' long-jump competition.

The table shows how high they jumped in centimetres.

	1st jump	2nd jump	3rd jump	4th jump
Amy	218	105	233	297
Donna	154	108	287	176
Gaynor	202	276	95	152
Joy	165	197	240	295

a How far did Gaynor jump on her 3rd jump?

b Who improved with every jump?

c Who won the competition?

All the information is found by reading the table.

a Gaynor jumped 95 cm on her 3rd jump.

b Joy improved with every jump, the only girl to do that.

c Amy had the longest jump of 297 cm, so she won the competition.

Example 28·2

An internet company charges delivery for goods based on the type of delivery – normal delivery (taking 3 to 5 days) or next-day delivery – and also on the cost of the order. The table shows how it is calculated.

Cost of order	Normal delivery (3 to 5 days)	Next-day delivery
£0–£10	£1·95	£4·95
£10·01–£30	£2·95	£4·95
£30·01–£50	£3·95	£6·95
£50·01–£75	£2·95	£4·95
Over £75	Free	£3·00

a Comment on the difference in delivery charges for normal and next-day delivery.

b Two items cost £5 and £29. How much would you save by ordering them together using: **i** normal delivery? **ii** next-day delivery?

a It always costs more using next-day delivery but for goods costing between £10·01 and £30, or between £50·01 and £75, it is only £2 more. It is £3 more for all other orders.

b Using normal delivery and ordering the items separately, it would cost £1·95 + £2·95 = £4·90, but ordering them together would cost £3·95. The saving would be £4·90 − £3·95 = 95p.

Using next-day delivery and ordering the items separately, it would cost £4·95 + £4·95 = £9·90, but ordering them together would cost £6·95. The saving would be £9·90 − £6·95 = £2·95.

Data and analysis

Exercise 28A

1. The chart below shows the distances, in kilometres, between certain towns in the UK.

Aberdeen				
652	Birmingham			
169	787	Inverness		
377	331	430	Newcastle	
904	275	958	529	Portsmouth

   a How far apart are Birmingham and Newcastle?
   b How far apart are Inverness and Aberdeen?
   c Which town is 430 km from Inverness?
   d Which two towns are 377 km from each other?
   e How much further away from Aberdeen is Portsmouth than Birmingham?

2. The table shows who is able to babysit on which nights.

	Mon	Tue	Wed	Thu	Fri	Sat	Sun
Ali		✓		✓			✓
Davinder		✓	✓	✓	✓		
Kylie	✓					✓	
Marco				✓			✓
Teresa	✓	✓	✓				

   a On which days can Ali babysit?
   b Who is available to babysit on Wednesday?
   c Who can babysit the most nights?
   d Which days are best for finding a babysitter?
   e On which days would someone have least choice for a babysitter?
   f Who could babysit on Thursday if Ali and Marco are unavailable?

Information handling

3. This table shows some information about the subjects taken by five S5 pupils:

	Maths	Geology	History	Literature
Eli	✓		✓	✓
Dan	✓	✓	✓	
Gill	✓	✓		✓
Eve	✓		✓	✓
Fynn	✓	✓		

This table shows some information about their families:

	Dad	Mum	Brothers	Sisters
Eli		✓	1	0
Dan	✓		0	2
Gill	✓	✓	2	1
Eve		✓	1	1
Fynn	✓	✓	0	0

a How many of the pupils took geology?

b Which subjects did Eli take?

c One pupil said, 'Last night my brother helped me with my geology homework'. Who was this pupil?

d In the literature lesson, one pupil was talking about their brother and sister fighting the previous night. Which pupils could this have been?

4. The table shows some information about pupils in an S2 class.

a How many pupils are in the class?

b How many of the pupils have brown eyes?

Eye colour	Number of boys	Number of girls
Blue	7	8
Brown	5	6
Green	2	4

Data and analysis

5 Ali and Padmini go to their local supermarket and record data on the cars parked there.

		Colour of cars				
		Blue	White	Red	Black	Other
Make of cars	Toyota	6	2	6	2	3
	Peugeot	8	1	3	3	1
	Vauxhall	4	2	5	1	2
	Ford	9	0	4	3	1
	Other	5	2	3	4	2

a How many blue Toyotas are there?
b How many Peugeots are not blue?
c How many more blue Fords are there than red Vauxhalls?
d Which make of car is the most common at this car park?

6 Reha, Jake, Colin and Celina had a games competition. They played two games, 'Noughts and Crosses' and 'Boxes'. Each played each of the others at both games. Colin recorded how many games each person won:

Reha	I I I
Jake	I I I
Colin	I I
Celina	I I I I

Celina recorded who won each game:

| Noughts and Crosses | Jake, Colin, Jake, Celina, Reha, Jake |
| Boxes | Celina, Celina, Reha, Celina, Colin |

a Celina has missed one name out from her table. Use Colin's table to say which name is missing.
b Who won most games of Noughts and Crosses?
c Give a reason why Colin's table is a good way of recording the results.
d Give a reason why Celina's table is a good way of recording the results.

Information handling

7 The table shows the number of pupils who have school lunches in S1, S2 and S3.

	Have school lunch	Do not have school lunch
S1	120	64
S2	97	87
S3	80	104

a How does the number of pupils who have school lunch change as they get older?

b Between which two years are the greatest changes? Explain your answer.

c By looking at the changes in the table, approximately how many pupils would you expect to not have a school lunch in S4?

8 The table shows the percentage of boys and girls by age group who have mobile phones.

a Work out the differences in the percentages between boys and girls at age 10 and age 15.

b Write down what you notice about the differences in the percentages for boys and girls.

c Comment on any other trends that you notice.

Age	Boys	Girls
10	18%	14%
11	21%	18%
12	42%	39%
13	53%	56%
14	56%	59%
15	62%	64%

9 The heights of 70 S3 pupils are recorded. Here are the results given to the nearest centimetre.

Height (cm)	Boys	Girls
130–139	3	3
140–149	2	4
150–159	10	12
160–169	14	11
170–179	6	5

Use the results to examine the claim that boys are taller than girls in S3. You may use a bar chart to help you.

Data and analysis

10 A running club wants to know the age profile of its members and prints this chart showing the numbers of members in four age ranges.

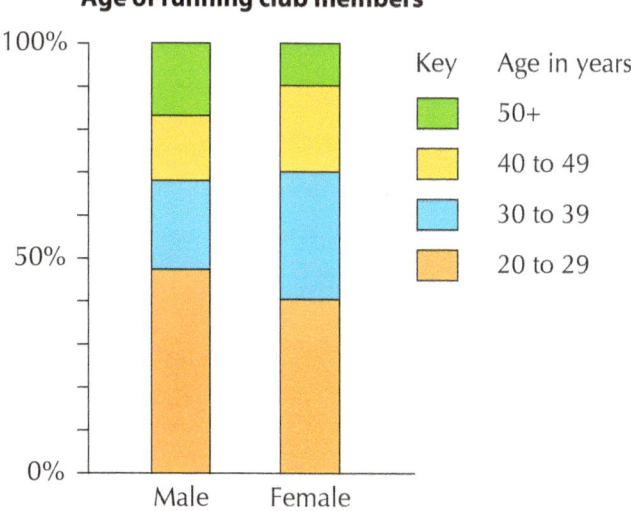

a Describe the age profiles for male and female runners in the club.

b List the percentages for each age group for male and female runners.

c There are approximately 200 male runners in the club. Calculate how many there are in each age group.

11 The diagrams show the amounts of rainfall in January and February in Crieff.

a On how many days in January was there less than 10 mm of rain?

b On how many days in February was there between 26 mm and 50 mm of rain?

c Estimate which month had more rainfall.

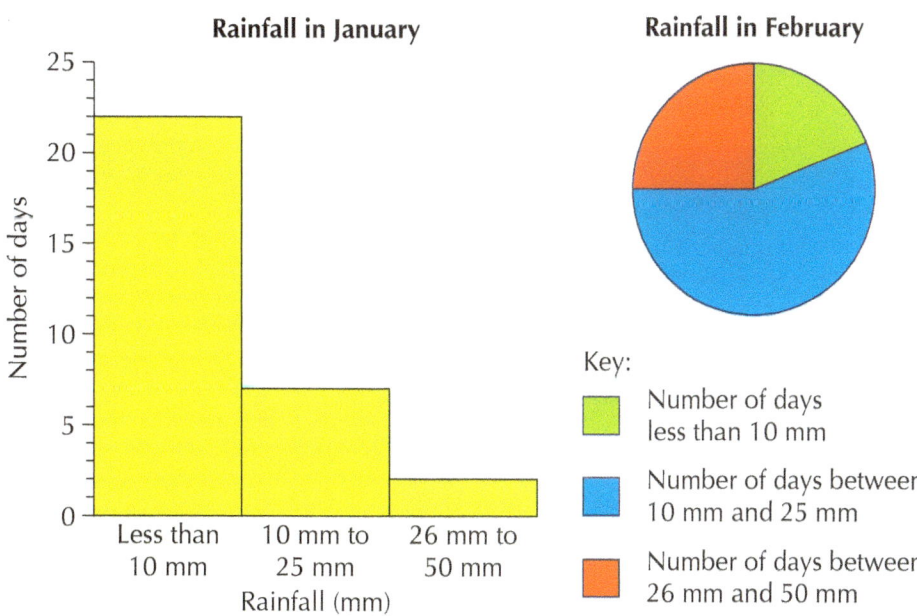

Information handling

12 This label shows the percentage of protein, carbohydrates, fat, fibre and sodium in a toffee cake.

TOFFEE CAKE	
Typical values	per 50 g
Energy	710 kJ/170 kcal
Protein	1·4 g
Carbohydrates	19·5 g
Fat	9·5 g
Fibre	0·15 g
Sodium	0·1 g

The pie chart shows the percentage of each part of the toffee cake given in the label above.

Match the labels to the key.

The first one is done for you.

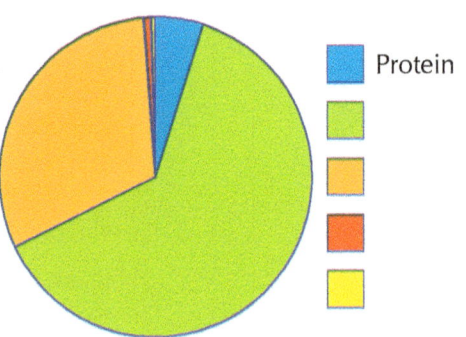

Fat

Carbohydrates

Sodium

Fibre

Challenge A school analyses the information on the month of birth for 1000 pupils. The results are shown in the table.

Month	Jan	Feb	Mar	Apr	May	Jun	Jul	Aug	Sep	Oct	Nov	Dec
Boys	34	36	43	39	47	50	44	39	55	53	42	35
Girls	37	31	36	35	44	43	36	40	52	49	43	37

a On the same grid plot a multiple bar chart to compare the data for boys and girls.

b Use the graphs to examine the claim that more children are born in the summer than in the winter.

Data and analysis

Interpreting graphs and diagrams

In this section you will learn how to interpret graphs, charts and diagrams to see if the information they present is robust, vague or misleading. Sometimes people make statements or decisions based on misleading information; you can help them by reading the information correctly and spotting any mistakes.

Example 28.3

The diagram shows how a group of pupils say they spend their time per week.

Matt says: 'The diagram shows that pupils spend too much time at school and doing homework.' Give two arguments to suggest that this is not true.

The diagram represents a group of pupils, so the data may vary for individual pupils. It could also be argued, for example, that pupils spend longer watching TV than doing homework.

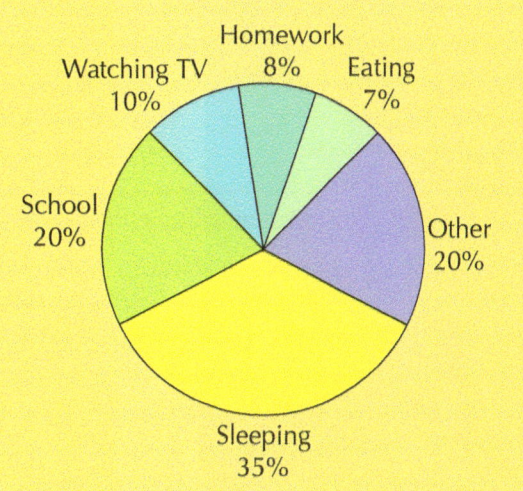

Exercise 28B

 1 The results of a ball throwing competition are shown in the bar chart.

   a How many pupils threw between 3 and 4 metres?

   b How many pupils threw between 1 and 2 metres?

   c How many pupils threw 2 metres or more?

   d Mr Jarvis the PE teacher claims that the chart shows that the total distance of all the throws is 17 m, a new school record. Is he correct? Explain your answer.

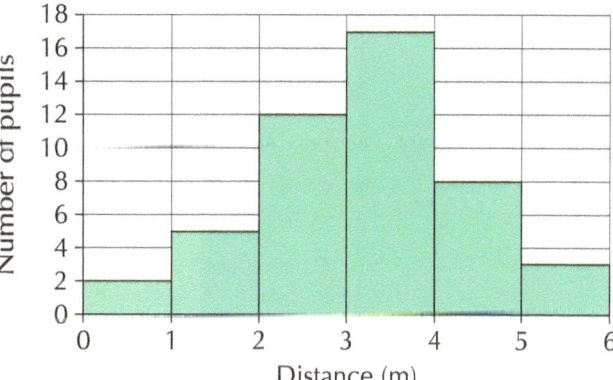

273

Information handling

2 The pie chart shows how crimes were committed in a town over a weekend.

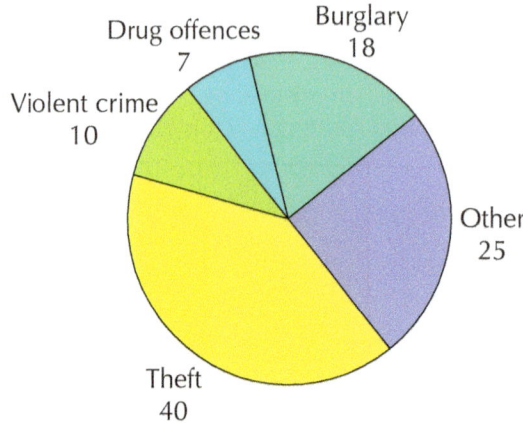

a How many drug offences were committed?

b How many offences were committed altogether?

c It is claimed that most crime involves theft. Explain why this is incorrect.

d Write down two statements using the information in the pie chart.

3 The table shows the total animal populations on four small farms in the years 2005 and 2010.

Farm	Animal population 2005	Animal population 2010
A	275	529
B	241	205
C	75	65
D	63	40
Total	654	839

a Which farm has increased the number of animals between 2005 and 2010?

b Which farm has shown the largest decrease in animal population from 2005 to 2010?

c A newspaper headline says that farm animal populations are increasing. Using the information in the table, explain if you agree or not with this headline.

Data and analysis

4 The pie charts show how the economies of the UK and Nepal are made up. Mr Dee the economist says, 'Nepal is manufacturing the same amount as the UK.' Ms Archibold thinks that the data presented is vague and is sure that the UK manufactures more than Nepal.

  a Whom do you agree with? Explain your answer.

  b How can the pie charts be improved?

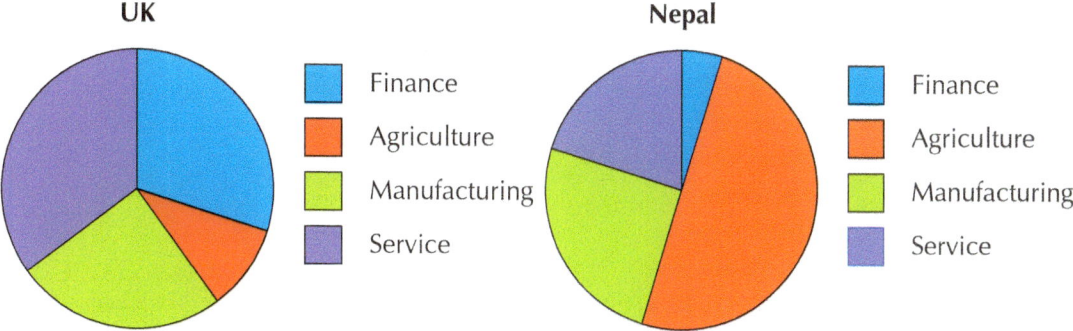

5 Sheena and Jools are working on a presentation together. They want to show how profits are increasing in the company they work for. They both create a graph. Advise them on which graph they should use in their presentation, giving clear reasons for your choice.

Sheena's graph

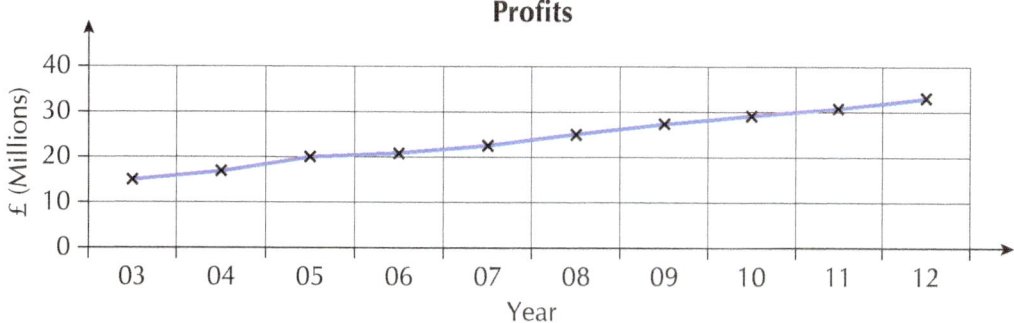

Jools's graph

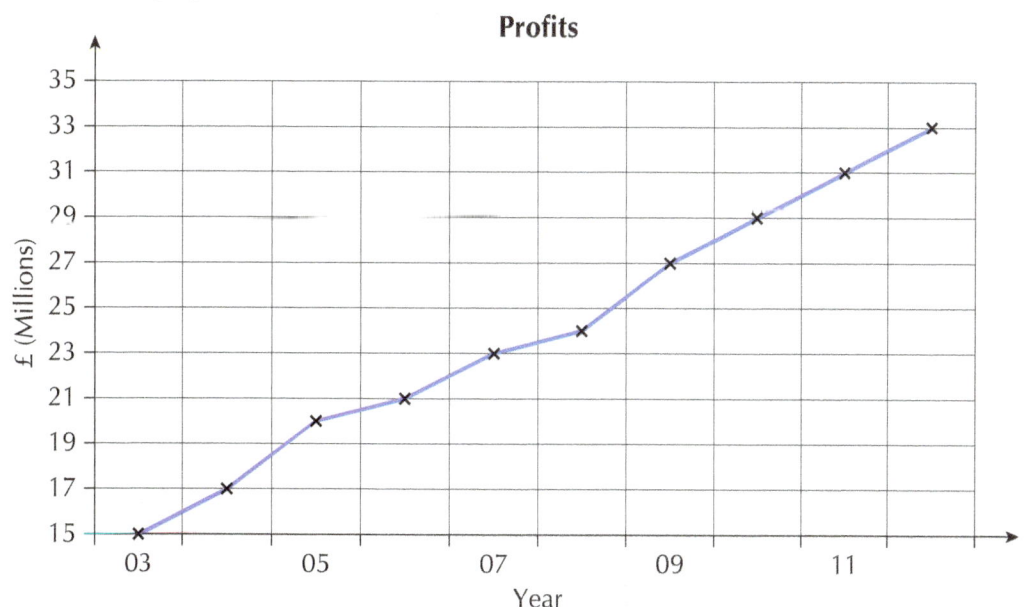

275

Information handling

6 Mrs Best uses this bar chart to summarise the sales of video games in her store. She claims, 'Sales increased dramatically in 2011.' Do you agree with her? Explain your answer fully.

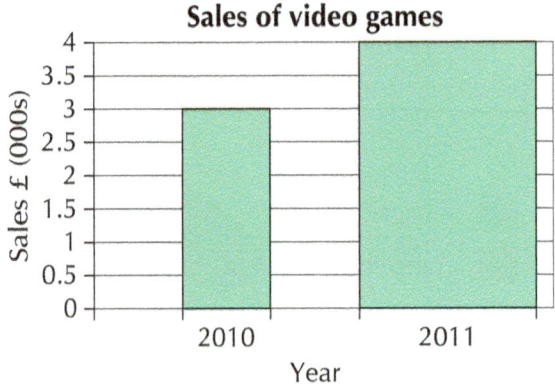

7 George wants to find out which is the most successful football team of all time. He intends to use the internet for his research. Give him some advice on how to start his search. How can he ensure the information he finds is robust and not subject to other people's personal preference? (If you are not a football fan, imagine trying to work out the most successful song, book, video game or movie.)

Challenge Find a graph or chart from a newspaper. Write down the facts that the newspaper article is claiming that the graph or chart shows. Use different arguments, referring to the graph or chart, to cast doubt on or to agree with the facts given.

- By working on this topic I can work with others to get information and explain why I think it is reliable or not.

- I have learnt how to read and interpret information from a range of different graphs and diagrams. ★ Exercise 28A Q10

- I can analyse graphs and diagrams and say if statements made about the data are true or misleading. ★ Exercise 28B Q2 ★ Exercise 28B Q5

29

When analysing information or collecting data of my own, I can use my understanding of how bias may arise and how sample size can affect precision, to ensure that the data allows for fair conclusions to be drawn.

MTH 3-20b

This chapter will show you how to:
- understand the terms sampling, reliability and bias
- understand how the size of samples can affect reliability
- justify your data collection strategy by choosing an appropriate sample.

You should already know:
- how to gather information for surveys and investigations using questionnaires, observations, experiments, etc.
- how to collate, organise and communicate the results of surveys and investigations
- how to draw conclusions from investigations and surveys and communicate them to others.

Sample size

When you have a problem to solve or a decision to make, it is usually best to gather as much information as you can to work out what is the right decision for you. This is also true of businesses, public bodies (such as schools), football teams and even entire countries.

When you carry out a survey or experiment you need to decide how many people you are going to question or how many observations you will make. It is usual practice to **sample** a small section of people rather then aiming to ask the entire **population**. For example, if you wanted to find out how many hours of TV young people watch it would be very difficult to ask every young person in Scotland (the population) but you could take a sample from your class or from the school. The bigger your sample, the more confidence you can have in your analysis, but the longer your investigation will take.

Do certain newspapers use more long words than the other newspapers?

There are many different newspapers. Can you list six different national newspapers?

Now consider the question in the speech bubble. The strict way to answer this would be to count, in each newspaper, all the words and all their letters. But this would take too long, so we take a sample. We count, say, 100 words from each newspaper to find the length of each word.

Exercise 29A

This whole exercise is a class activity.

 1 a You will be given either a whole newspaper or a page from one.

   b Create a data capture form (a tally chart) like the one below.

Information handling

Number of letters	Tally	Frequency
1		
2		
3		
4		
5		

c Choose a typical page from the newspaper. Then select at least two different articles. Next, count the number of letters in each word from each article (or paragraph), and fill in the tally chart. Do not miss out any words. Before you start to count, see part **f** below.

d Decide what to do with such things as:

Numbers – 6 would count as 1, *six* would count as 3.

Hyphenated words – Ignore the hyphen.

Abbreviations – Just count the letters that are there.

e Once you have completed this task, fill in the Frequency column. The frequency is the total of the Tally column. Now create a bar chart of the results.

f Each of you will have taken a different newspaper. So what about comparing your results with others'? To do this, you must all use the same number of words.

Using the correct data

Look at the following misleading conclusions that arise from not using like data.

Two different classes did a survey of how pupils travelled to school.

They both made pie charts to show their results.

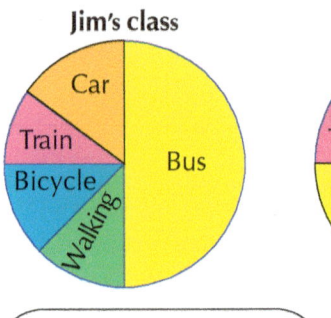

Look at both charts and ask yourself:

Which class had the most pupils using the bus?

Obvious – it's Jim's class!

BUT if you are told:

In Jim's class there were only 24 pupils, and in Sara's class there were 36 pupils.

Then you will see that:

In Jim's class, half of the 24 used the bus, which is 12.

In Sara's class, more than one third used the bus, which is about 14.

Clearly, you have to be very careful about interpreting data in its final form.

Here, you can say that a bigger proportion of Jim's class used the bus, but *not* more pupils used the bus than in Sara's class.

So, it is most important to try to use like numbers of items when you are going to make a comparison of data from different sources.

Data and analysis

> **Challenge** Use the newspaper articles for another experiment. Ask pupils in your class, and in other classes, to read the sample of 100 words and time them to see if the papers with longer words take any longer to read. Write a short report on your findings.

Bias

In the picture the teenagers are carrying out a **survey** about healthy eating. Do you think that their results will be fair if they only interview people who are buying burgers? How can they collect the **data** to avoid any **bias** and make sure that many different opinions are obtained? How many people do you think they should ask?

Example 29·1 Here are some questions that might be used in a survey. Give a reason why each question is not very good and then write a better question.

  a How old are you?
  b Do you eat lots of fruit or vegetables?
  c Don't you agree that exercise is good for you?
  d If you go to a sports centre with your friends and you want to play badminton, do you usually play a doubles match or do you just practise?

  a This is a personal question. If you want to find out about the ages of people, use answer boxes and group several ages together.

  ☐ 0–15   ☐ 16–30   ☐ 31–45   ☐ 46–50   ☐ More than 50

  b This question is asking about two different items, so the answers may be confusing. Also the word 'lots' can mean different amounts to different people. A better question would be: 'How many pieces of fruit did you eat last week?'

  c This question is trying to force you to agree. It is a leading question. A better question would be, 'Is exercise good for you?' You could then limit people to answers such as 'very' and 'not at all'.

  d This question is too long. There is too much information, which makes it difficult to answer. It should be split up into several smaller questions.

Exercise 29B

 1 Choose one of the following statements for a statistical survey. For the statement chosen:

- Make a list of the information you need to collect.
- Decide how you will collect your data and write any questions you need to ask. Make sure your questions avoid bias.
- Decide which groups of people you need to ask. How many people should you ask?

Statements

- Girls spend more on clothes than boys.
- Old people use libraries more than teenagers.
- Pupils who enjoy playing sports eat healthier foods.
- More men wear glasses than women.
- Families eat out more than they used to.

Information handling

> **Challenge** Take each problem statement from the Exercise and write down how you would collect the data required. Try to justify your chosen method for collecting the data and clearly state how you would avoid introducing bias so that any conclusions taken from the investigation are fair and balanced.

Collecting data for statistical surveys

MTH 3-20a
MTH 3-21a

If you are about to carry out your own statistical survey you will need to plan how you intend to carry out the survey and decide how you are going to collect your data. When you search for information online or in books or DVDs, this is called **secondary data**, because someone else collected the information. If you decide to conduct your own survey or experiment, the data you collect will be your own **primary data**.

When you collect your own data you need to consider the **sample size**, that is, the number of people you will collect data from or the number of times you run an experiment. Usually, the bigger the sample size, the more accurate the results will be. However, it is often impractical to ask large numbers of people because of time and cost constraints. For example, market research companies who are trying to predict the results of a UK election will ask around 1000 people which party they intend to vote for, but 1000 people is a tiny proportion of the population. If you are conducting a survey think carefully about how many people you will need to ask in order to make the results meaningful.

Your data may be obtained in one of the following ways.

- A survey of a sample of people. To collect data from your chosen sample, you will need to use a data collection sheet or a questionnaire.
- Carrying out experiments. You will need to think how you will record your observations. You could use an electronic recording device or spreadsheet; this could help guard against writing the wrong value or result by mistake.
- Searching secondary sources. Examples are reference books, newspapers, ICT databases and the internet.

If you are using secondary data you should check the following to ensure it is reliable and fair:

- Who conducted the survey and are they impartial?
- How big was the sample?
- Did the data collection process avoid bias?
- Do their conclusions agree with the data they collected?

If you plan to use a questionnaire, remember the following points:

- Make the questions short and simple.
- Choose questions that require simple responses. For example: Yes, No, Do not know. Or use tick boxes with a set of multi-choice answers.
- Avoid personal and embarrassing questions.
- Never ask a leading question designed to get a particular response.

When you have collected all your data, you are ready to write a report on your findings.

Your report should be short and based only on the evidence you have collected. Use statistical diagrams to illustrate your results and to make them easier to understand. For example, you might draw bar charts, line graphs or pie charts. Try to give a reason why you have chosen to use a particular type of diagram.

Data and analysis

To give your report a more professional look, use ICT software to illustrate the data.

Finally, you will need to write a short conclusion based on your evidence.

Example 29·2   Look at the different methods of collecting data shown, and then decide which is the most suitable method for each of the two tasks **a** and **b**. Briefly outline a plan for each task.

Methods of collecting data
Construct a questionnaire
Do research on the internet
Carry out an experiment
Use a software database (e.g. an encyclopaedia CD)
Visit a library for books and other print sources

a Girls have faster reaction times than boys.

b People in the UK live longer than people in Brazil.

a This task would require an experiment. The plan would consider:

- the detailed design of the reaction time experiment
- what results to expect (a hypothesis)
- the number of pupils in each group
- how the pupils would be selected to ensure that there was no bias
- how to record the data
- how to analyse the results and report the findings.

b This task would use data obtained from the internet, a software database or a library. The plan would consider:

- which countries and what data to look at
- what results to expect (a hypothesis)
- the most efficient way of obtaining the data
- how many years of data to compare
- how to record the data
- how to analyse the results and report the findings.

Exercise 29C

1 Look at the different methods of collecting data in Example 29·2. Decide which is the most suitable method for each of the following tasks. Briefly outline a plan for each task including the points mentioned in Example 29·2.

   a Comparing how easy two newspapers are to read.

   b Testing someone's memory in a game.

   c Finding out people's opinions on smoking.

2 What are the possible problems for the following methods of data collection? How can you make them better?

   a Recording long jump data for a whole school in one set of data

   b Collecting data about the age of the population of a country and putting it into groups of five years, i.e. 0–5, 6–10, etc.

Information handling

c Giving a questionnaire about fitness to a small sample of members of a sports club

d Testing boys' reaction times in the morning and girls' reaction times in the afternoon using a different test.

★ 3 Write your own statistical report on one or more of these statements:

- Young people watch more TV than older people.
- Schoolchildren like the same music as their teachers.
- More people play for football teams than any other sport.
- Holidays to Europe are more expensive than holidaying in Scotland.
- Taller people have a larger head circumference.

Challenge

1 Write your own statistical report on one or more of the following problems. For these problems, you will need to use secondary sources to collect the data.

- Do football teams in the Scottish Premier League score more goals than teams in Division One of the League?
- Compare the frequency of letters in the English language to the frequency of letters in the French language.
- Compare the prices of second-hand cars using different motoring magazines.

2 **Statistics in the press**

Look through newspapers and magazines to find as many statistical diagrams as you can. Make a display to show the variety of diagrams used in the press.

What types of diagram are most common? How effective are the diagrams in showing the information? Are any of the diagrams misleading? If they are, explain why.

- By studying this chapter I can describe the terms sampling, reliability and bias.
- I can choose an appropriate sample size when conducting a survey. ★ Exercise 29A Q1
- I have learned that the composition of my sample can introduce bias and affect how reliable my survey will be. ★ Exercise 29B Q1
- When carrying out a survey I know how to choose and justify my data collection strategy. ★ Exercise 29C Q3

30 Data and analysis

I can display data in a clear way using a suitable scale, by choosing appropriately from an extended range of tables, charts, diagrams and graphs, making effective use of technology.

MTH 3-21a

This chapter will show you how to:
- choose how to display data, explaining and justifying your choices
- create bar charts, pie charts, line graphs and tables accurately, including all key features
- interpret scales used on data displays and choose appropriate scales when displaying data
- use technology effectively when displaying data.

You should already know:
- how to display data using a range of tables, charts, diagrams and graphs
- how to construct data displays correctly, using grids, labels and titles where necessary
- how you can use technology such as spreadsheets, graphing calculators or websites to create tables, charts, diagrams and graphs.

In this chapter you will practise drawing a range of tables, charts, diagrams and graphs. When you draw any sort of diagram to communicate information it is important that it is easily understood by your intended audience. Your diagram should have a title, labels on the axes, a key if required and you should choose a suitable scale so that your diagram does not become misleading.

In all of the exercises you may have access to a computer. If you use a software package to draw your graph remember to check its title, labels and scale.

Creating graphs, charts and diagrams

A **frequency diagram** is a means of displaying raw data in a chart to make it easier to interpret. Any chart that is produced from data in a frequency table can be called a frequency diagram. Examples of frequency diagrams can include bar charts, histograms, line graphs and pie charts.

When choosing what type of diagram to draw you need to consider the type of data you are using: is it **discrete** or **continuous**? A set of data is **discrete** if the values are separate and distinct, for example, if the data is based on counting, like the number of apples on an apple tree or the types of cars in a car park. **Continuous** data can take any value within a range; examples of continuous data include temperature, weight, length or time.

Example 30·1 Construct a frequency diagram for the data about eye colour given in the table.

The eye colour of pupils in class 2B

Eye colour	Frequency
Brown	12
Blue	9
Hazel	2
Green	3

Continued

Information handling

Example 30·1
Continued

The data in the table is discrete. This means that a bar chart is a suitable diagram to draw. It is important that the diagram has a title and labels as shown.

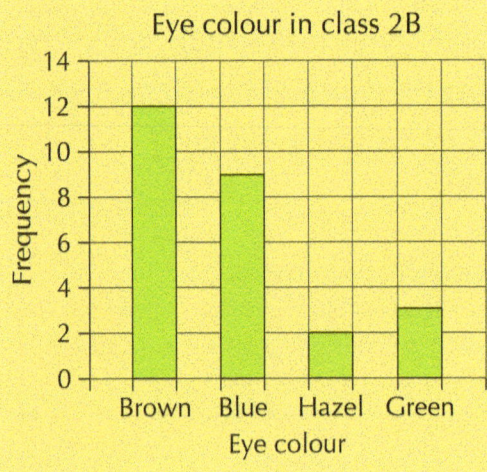

Example 30·2

Construct a frequency diagram to show how lawn-mower sales at a shop vary throughout a year.

Jan	Feb	Mar	Apr	May	Jun	Jul	Aug	Sep	Oct	Nov	Dec
0	25	63	75	92	68	53	32	76	15	0	12

Write down why you think the sales are high in September and why there are some sales in December.

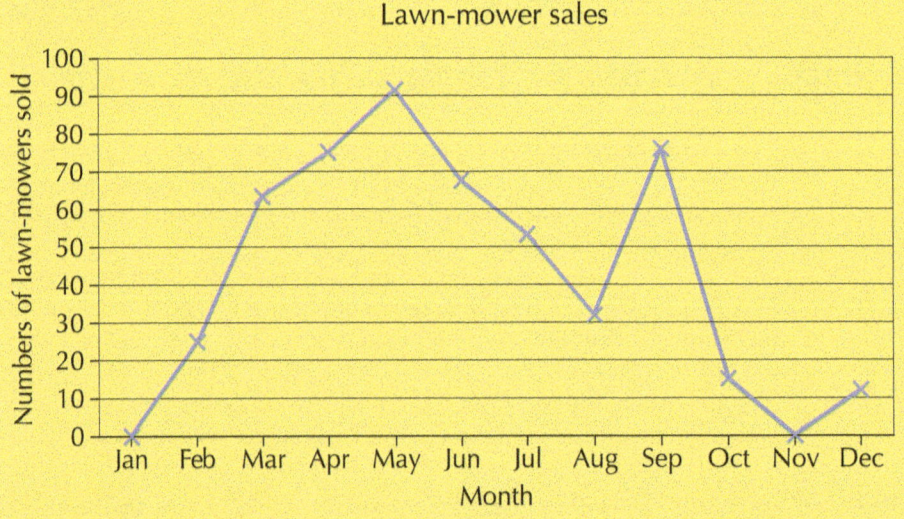

The reasons for the sales in September could be that the shop reduces the prices in an attempt to sell off stock before winter. The December sales could be Christmas presents.

Data and analysis

Exercise 30A

1 For each frequency table, construct a bar chart.

a The number of restaurants in a particular town

Meal	Frequency
Chinese	3
Italian	7
Indian	15
Turkish	12
Greek	5
English	2

b The number of types of worker at a factory

Type of worker	Frequency
Trainee	38
Skilled	20
Manager	5
Director	1

c The number of tomatoes on a truss

Number of tomatoes	Frequency
1	3
2	5
3	12
4	18
5	4
6 or more	2

2 For each frequency table, construct a frequency diagram. Choose between a bar chart or line graph and explain your choice.

a The number of extra minutes added on to the end of 44 football matches

Minutes	Frequency
0	1
1	7
2	15
3	12
4	6
5 or more	3

285

Information handling

  b The stock market totals in London

Month	Index value
Jan	4550
Feb	5100
Mar	5400
Apr	5250
May	4700
June	5450

  c Dougie's height, in cm, from age 6 to 12

Age (years)	Height (cm)
6	69
7	87
8	98
9	110
10	118
11	127
12	140

★ 3 The following table shows the rainfall, in millimetres, for a town in the north of England.

Month	Jan	Feb	Mar	Apr	May	Jun	Jul	Aug	Sep	Oct	Nov	Dec
Rainfall (mm)	45	36	44	47	51	54	49	55	50	44	51	50

  a Construct a line graph of this data.
  b Which month had the greatest rainfall?
  c Which month had the least rainfall?
  d For how many months was the amount of rainfall below 45 mm?
  e What is the difference between the amount of rainfall in July and in August?
  f Between which two consecutive months is there the greatest difference in rainfall?

4 The table shows population forecasts for the UK and Afghanistan.

Year	2015	2020	2025	2030	2035	2040	2045	2050
Population of UK (millions)	62	63	64	64	65	65	64	64
Population of Afghanistan (millions)	37	40	44	48	53	58	61	66

  a Choose a frequency diagram to display the data in the table. Plot both diagrams on the same axes.
  b Estimate the year when the populations of the two countries will be equal.
  c Estimate the year when the population of the UK is at its maximum. State what this maximum population could be.
  d Justify your choice of frequency diagram. Do you think you made the best choice or is there a better way to display the data?

Data and analysis

5 Baby Nieve was 4 kg when she was born in April. Use the data in the table to construct a frequency diagram to show how Nieve's weight changes in the first months of her life.

Month	April	May	June	July	August	September
Weight (kg)	4	5	5	6	6.5	7.5

Challenge

For this activity you will need access to a software package that can draw graphs and charts, such as a spreadsheet program.

For each of the following sets of data time yourself drawing the chart by hand and then drawing the same chart using a spreadsheet package.

Write a short report to answer the following questions:

- Which method was quicker?
- How easy was it to include axes, title, labels, etc. in your electronic chart?
- What are the advantages and disadvantages of using a software package?

a Choose a frequency diagram for the number of pupils in each year group. Experiment with the different graph types to find which graph makes the data easiest to interpret.

Year	Number of pupils
S1	120
S2	140
S3	105
S4	90
S5	65
S6	35

b Display the data for the maximum temperatures measured in Elgin for one year. In your report describe the trend you notice for the temperatures.

Month	Jan	Feb	Mar	Apr	May	Jun	Jul	Aug	Sep	Oct	Nov	Dec
Temperature (°C)	4	3	7	18	22	26	30	24	16	10	8	6

Throughout the rest of this chapter you can repeat this exercise to see if it is quicker to draw different charts by hand or by computer. Remember your titles and labels!

Information handling
Grouped frequencies

Example 30·3

A class was asked this question and the replies, in minutes, were:

"How long does it take you to get to school in the morning?"

6 min, 3 min, 5 min, 20 min, 15 min,
11 min, 13 min, 28 min, 30 min, 5 min,
2 min, 6 min, 8 min, 18 min, 23 min,
22 min, 17 min, 13 min, 4 min, 2 min,
30 min, 17 min, 19 min, 25 min, 8 min,
3 min, 9 min, 12 min, 15 min, 8 min

There are too many different values here to make a sensible bar chart. So we group them to produce a **grouped frequency table**, as shown below. The different groups the data has been put into are called **classes**. Where possible, classes are kept the same size as each other.

Time (minutes)	1–5	6–10	11–15	16–20	21–25	26–30
Frequency	7	6	6	5	3	3

Notice how we use 6–10, 11–15, … to mean 'over 5 minutes up to 10 minutes', and so on.

A frequency diagram has been drawn from this data, and information put on each bar about the method of transport.

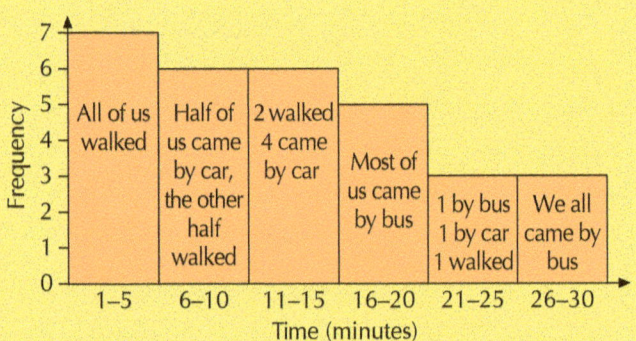

Data, such as time, is continuous data, so we can create a continuous scale. This new diagram is called a **histogram**.

Time to travel to school (minutes)	Frequency
$0 < \text{Time} \leq 5$	7
$5 < \text{Time} \leq 10$	6
$10 < \text{Time} \leq 15$	6
$15 < \text{Time} \leq 20$	5
$20 < \text{Time} \leq 25$	3
$25 < \text{Time} \leq 30$	3

The diagram would now look like this.

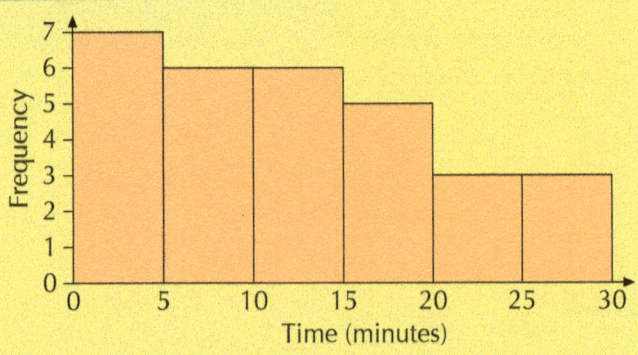

Data and analysis

Exercise 30B

1. A class did a survey on how many pencils each pupil had with them in school. The results of this survey are:

   4, 7, 2, 18, 1, 16, 19, 15, 13, 0, 9, 17, 4, 6, 10, 12, 15, 8, 3, 14, 19, 14, 15, 18, 5, 16, 3, 6, 5, 18, 12

   a Put this data into a grouped frequency table with a class size of 5: that is, 0–4, 5–9, 10–14 …

   b Draw a frequency diagram from the data.

★ 2. A teacher asked her class: 'How many hours a week do you spend on a computer?' She asked them to give a rounded figure in hours. This was their response:

   3, 6, 9, 2, 23, 18, 6, 8, 29, 27, 2, 1, 0, 5, 19, 23, 20, 21, 7, 4, 23, 8, 7, 1, 0, 25, 24, 8, 13, 18, 15, 16

   These are some of the reasons pupils gave for the length of time they spent on a computer:

   - 'I haven't got one.'
   - 'I play games on mine.'
   - 'I always try to do my homework on the computer.'
   - 'I can't use it when I want to, because my brother's always on it.'

   a Put the above data into a grouped frequency table with a class size of five.

   b Draw a frequency diagram with the information. Can you match any of the pupils' responses to any parts of the frequency diagram?

3. The table shows the amount of call time used by 50 S5 pupils.

Call time used (min)	Frequency
$0 <$ Time ≤ 20	14
$20 <$ Time ≤ 40	9
$40 <$ Time ≤ 60	5
$60 <$ Time ≤ 80	4
$80 <$ Time ≤ 100	18

   a Draw a histogram to show the data in the table.

   b How many pupils used more than 60 minutes of call time?

   c Describe what the graph tells you about the pupils' use of call time.

4. The table shows the times of goals scored in football matches played on one weekend in November.

Time of goals (minutes)	Frequency
$0 <$ Time ≤ 15	3
$15 <$ Time < 30	6
$30 <$ Time ≤ 45	8
$45 <$ Time ≤ 60	4
$60 <$ Time ≤ 75	2
$75 <$ Time ≤ 90	5

   a One goal was scored after exactly 75 minutes. In which class was it recorded?

   b Five teams scored in the last five minutes of their games. Write down what other information this tells you.

   c Draw a frequency diagram to represent the data in the table.

Challenge Collect data about the heights of the pupils in your class. Design a frequency table to record the information. Draw a frequency diagram from your table of results. Comment on your results.

Information handling
Conversion graphs

Example 30·4

When you fill your car with petrol, both the amount of petrol you have taken and its cost are displayed on the pump. In February 2012, one litre of petrol cost about £1·30, but this rate does change from time to time.

The table below shows the costs of different quantities of petrol as displayed on a petrol pump.

Petrol (litres)	5	10	15	20	25	30
Cost (£)	6·50	13·00	19·50	26·00	32·50	39·00

This information can also be represented by the following ordered pairs:

(5, 6·5)   (10, 13)   (15, 19·5)   (20, 26)   (25, 32·5)   (30, 39)

An **ordered pair** links two items of data together and can be plotted as coordinates. For example, (5, 6·5) represents '5 litres of petrol costs £6·50'.

On the right is the graph which relates the cost of petrol to the quantity bought.

This is an example of a **conversion graph**. You can use it to find the cost of any quantity of petrol, or to find how much petrol can be bought for different amounts of money.

Conversion graphs are usually straight-line graphs.

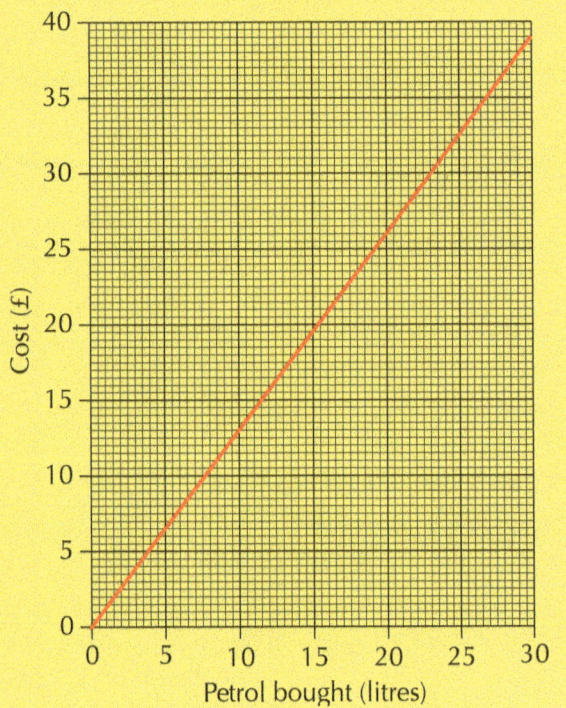

290

Exercise 30C

1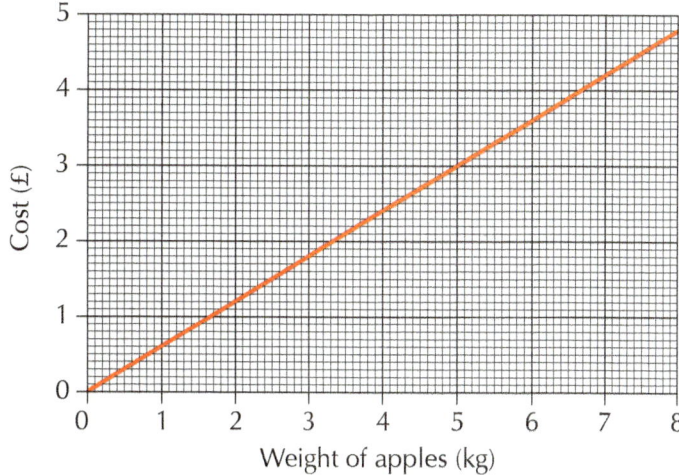

Use the graph to answer these questions.

a Find the cost of each quantity of apples.

   i 3 kg   ii 7 kg

b What weight of apples can be bought for:

   i £3   ii £2·40

2 The graph below shows the distance travelled by a car over a period of 5 minutes.

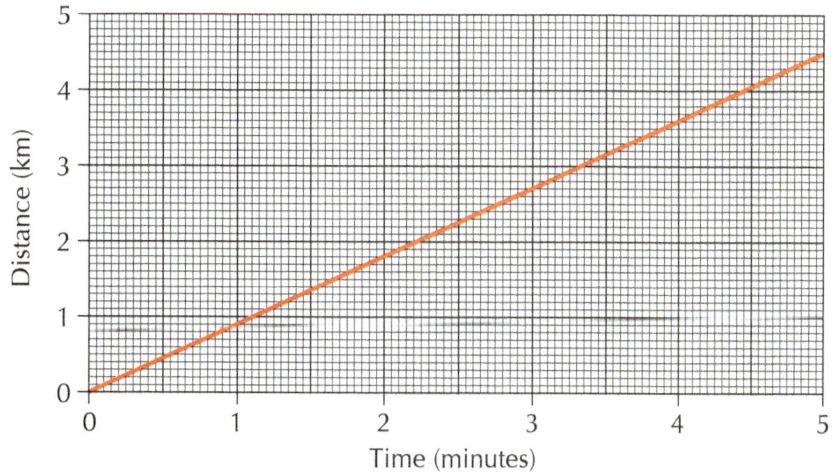

a Find the distance travelled during the second minute of the journey.

b Find the time taken to travel 3 km.

Information handling

3 Here is a kilometre–mile conversion graph.

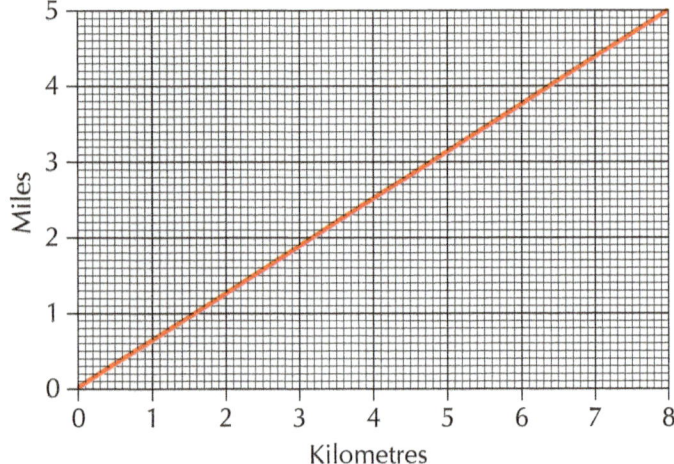

a Express each of the following distances in km.

  i   3 miles
  ii 4·5 miles

b Express each of the following distances in miles.

  i   2 km
  ii   4 km
  iii 6 km

4 a Copy and complete the following table for the exchange rate of the euro.

Euros (€)	1	5	10	15	20
Pounds (£)	0·80	4·00			

b Use the data from this table to draw a conversion graph from pounds to euros.

c Use your graph to convert each of the following to pounds.

  i €7   ii €16   iii €17·50

d Use your graph to convert each of the following to euros.

  i £9   ii £12   iii £10·80

5 A box weighs 2 kg. Packets of juice, each weighing 425 g, are packed into it.

a Draw a graph to show the weight of the box plus the packets of juice and the number of packets of fruit juice put into the box.

b Find, from the graph, the number of packets of juice that make the weight of the box and packets as close to 5 kg as possible.

6 A taxi firm charges a basic charge of £2 plus £1 per kilometre.

Draw a graph to show how much the firm charges for journeys up to 10 kilometres.

7 Draw a graph to show the cost of gas from a supplier who charges a fixed fee of £18 plus 3p per unit of gas. Draw the horizontal axis from 0 to 500 units and the vertical axis from £0 to £35.

Pie charts

Data and analysis

A **pie chart** is a good way to summarise data and to compare proportions. The area of each sector is proportional to the quantity it represents. However, pie charts are not as good as other graph types at showing totals, and it can be difficult to compare data between different pie charts.

Example 30·5 Draw a pie chart to represent the following set of data showing how a group of people travel to work.

Type of travel	Walk	Car	Bus	Train	Cycle
Frequency	24	84	52	48	32

It helps to set out your workings in a table.

Type of travel	Frequency	Calculation	Angle
Walk	24	$\frac{24}{240} \times 360 = 36°$	36°
Car	84	$\frac{84}{240} \times 360 = 126°$	126°
Bus	52	$\frac{52}{240} \times 360 = 78°$	78°
Train	48	$\frac{48}{240} \times 360 = 72°$	72°
Cycle	32	$\frac{32}{240} \times 360 = 48°$	48°
Total	240		360°

We work out the angle for each sector using:

$$\frac{\text{Frequency}}{\text{Total frequency}} \times 360°$$

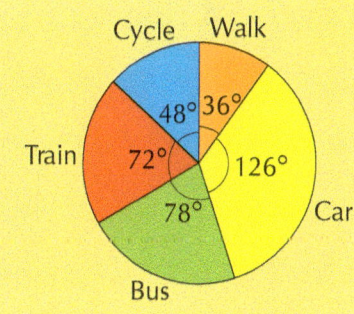

Exercise 30D

 1 Draw pie charts to represent the following data.

a The favourite subject of 36 pupils.

Subject	Maths	English	Science	Languages	Other
Frequency	12	7	8	4	5

b The type of food that 40 people usually eat for breakfast.

Food	Cereal	Toast	Fruit	Cooked	Other	None
Frequency	11	8	6	9	2	4

Information handling

c The number of goals scored by an ice hockey team in 24 matches.

Goals	0	1	2	3	4	5 or more
Frequency	3	4	7	5	4	1

d The favourite colour of 60 S1 pupils.

Colour	Red	Green	Blue	Yellow	Other
Frequency	17	8	21	3	11

2 A teacher asked her class of 24 pupils:

'What type of book do you like most?'

The table shows the results.

Type of book	Frequency
Crime	6
Romance	8
Science fiction	3
Sport	7

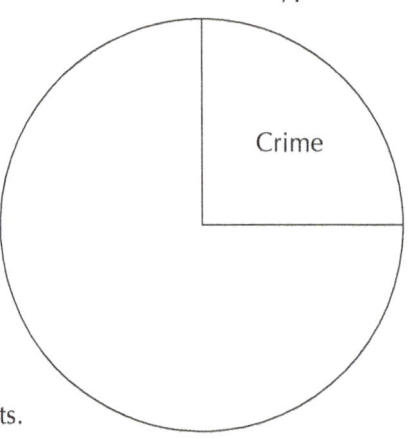

Favourite book type

Copy and complete the pie chart to show the results.
Show your working and draw your angles carefully.

3 a Tom Watson lives on his own and earns £480 per week. The pie chart shows how he spends his money. How much does Tom Watson spend on travel each week?

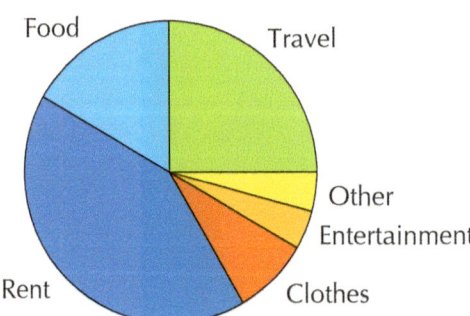

b Jacqui Wilson lives with her husband and their two children. Jacqui and her husband earn £800 per week. The table shows how they spend their money.

Rent	£300
Food	£200
Clothes	£100
Travel	£100
Entertainment	£50
Other	£50

Data and analysis

Draw a pie chart to show how the Wilson family spend their money.

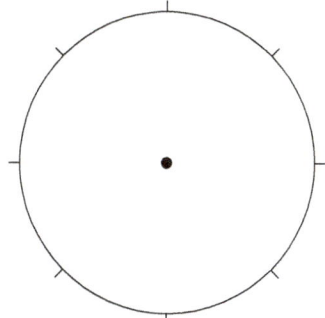

4 A group of 60 rail commuters were asked how they travelled to the rail station. The table shows the results.

Mode of travel	Number of people
Walk	36
Cycle	4
Car	20

Create a pie chart to show this information.

5 Design a poster showing information about the pupils in your class. Include pie charts that you have drawn yourself or use a spreadsheet to produce the pie charts. Make sure that any pie chart you produce has labels and is easy to understand.

Challenge

1 The label below is from a packet of porridge oats.

a Work out the percentages of protein, carbohydrates, fat, fibre and sodium.

PORRIDGE OATS

Typical values	per 100 g
Energy	1555 kJ/ 372 kcal
Protein	7·5 g
Carbohydrates	71 g
Fat	6·0 g
Fibre	6·0 g
Sodium	0·3 g

b Draw a pie chart to show the percentage of each constituent of the porridge oats.

2 Obtain labels from a variety of cereals and other food items. Draw a pie chart for each of them.

3 a What types of food have the most fat? What types of food have the most energy?

b Is there a connection between the energy of food and the fat and carbohydrate content?

Information handling
Choosing data displays and communicating findings

In Chapter 29 you will have collected data through a survey. To help your audience understand your conclusions you can present the data in a range of graphs and charts. This exercise will help you practise choosing the best diagrams to display and communicate your survey findings.

Exercise 30E

 1 A class surveyed some S1, S2 and S3 pupils about their preferences for the school disco. They asked questions about cost, start and finish times, and what they would like to eat. Their data collection chart is shown below.

Year group	Boy or girl	How much to charge	Time to start	Time to finish	What would you like to eat?
S1	B	£1	7 pm	11 pm	Crisps, beefburgers, chips
S1	G	50p	7 pm	9 pm	Chips, crisps, ice pops
S2	G	£2	7:30 pm	10 pm	Crisps, hot dogs
S3	B	£3	8:30 pm	11:30 pm	Chocolate, pizza
S3	G	£2	8 pm	10 pm	Pizza
S3	B	£2·50	7:30 pm	9:30 pm	Hot dogs, Chocolate
S2	G	£1	8 pm	10:30 pm	Crisps
S1	B	75p	7 pm	9 pm	Crisps, beefburgers
S1	B	£1	7:30 pm	10:30 pm	Crisps, ice pops
S2	B	£1·50	7 pm	9 pm	Crisps, chips, hot dogs
S3	G	£2	8 pm	11 pm	Pizza, chocolate
S3	G	£1·50	8 pm	10:30 pm	Chips, pizza
S3	G	£2	8 pm	11 pm	Crisps, pizza
S1	G	£1·50	7 pm	9 pm	Crisps, ice pops, chocolate
S2	B	£2	7:30 pm	9:30 pm	Crisps, ice pops, chocolate
S2	B	£1	8 pm	10 pm	Chips, hot dogs
S3	B	£1·50	8 pm	11 pm	Pizza
S1	B	50p	7 pm	9:30 pm	Crisps, hot dogs
S2	G	75p	8 pm	10:30 pm	Crisps, chips
S3	B	£2	7:30 pm	10:30 pm	Pizza
S2	G	£1·50	7:30 pm	10 pm	Chips, hot dogs, chocolate
S2	B	£1·25	7 pm	9:30 pm	Chips, hot dogs, ice pops
S3	G	£3	7 pm	9:30 pm	Crisps, pizza
S3	B	£2·50	8 pm	10:30 pm	Crisps, hot dogs
S1	G	25p	7:30 pm	10 pm	Crisps, beefburgers, ice pops
S1	G	50p	7 pm	9 pm	Crisps, pizza
S1	G	£1	7 pm	9:30 pm	Crisps, pizza
S2	B	£2	8 pm	10 pm	Crisps, chips, chocolate
S2	G	£1·50	7:30 pm	9:30 pm	Chips, beefburgers
S1	B	£1	7:30 pm	10 pm	Crisps, ice pops

Data and analysis

   a   Create and complete a tally chart for each option for each year.
   b   Choose a graph or chart to display the findings of the food suggestions survey.
   c   Comment on the differences between the year groups.
   d   Did your chart help you to interpret that data? Could you use a better chart?

2 The table shows how much pupils are prepared to pay for lunch each day.

Year group	Price		
	under £1	£1–£2	over £2
S1	30	45	12
S2	25	50	18
S3	18	42	25
S4	11	52	27
S5	8	55	22

   a   How many of each year group were asked in the survey?
   b   Draw a diagram to help you interpret the data in the table.
   c   Explain the differences between the year groups.
   d   Combine the information into one table and create a single diagram to illustrate the whole school price choices.

★ 3 a   Find a sample of people and ask them the following question, keeping a tally of the answers.

      Approximately how many hours do you watch TV over a typical weekend?

Time (hours)	Tally	Frequency
2 hours or less		
Over 2 but less than 4 hours		
Between 4 and 8 hours		
Over 8 hours		

   b   Create a chart illustrating your collected data.

4 For this question, change some of the named sports if you wish.

   a   Find a sample of boys and ask them the following question, keeping a tally of the answers.

      Which of the following sports do you play outside school?

Sport	Tally	Frequency
Football		
Cricket		
Tennis		
Badminton		
Something else		

Information handling

    b Now find a sample of girls and ask them the same question, keeping a tally of the results.

    c Create charts illustrating your data and also illustrating any differences between the two groups.

5 For this question, change some of the bands if you wish.

    a Find a sample of pupils and ask them the following question, keeping a tally of the answers.

Which of the following bands would you most want to go and listen to at a concert?

Band	Tally	Frequency
Red Hot Chili Peppers		
Mika		
Ordinary Boys		
Kaiser Chiefs		
Arctic Monkeys		

    b Create charts illustrating your data.

6 a Ask the following sequence of questions to a sample of people:

      i Before this year, have you normally gone abroad for your holiday?   Yes/No

      ii Are you intending to go abroad on holiday this year?   Yes/No

    b Use your results to complete the following table:

	Going abroad this year	Not going abroad this year
Normally go abroad		
Do not normally go abroad		

    c From your results, is it true to say 'More people are taking holidays abroad this year'?

- By working on this topic I can explain the choices I make when displaying data.

- I can draw a range of charts, graphs, diagrams and tables including frequency diagrams, bar charts, conversion graphs and pie charts. ★ Exercise 30A Q3 ★ Exercise 30B Q2 ★ Exercise 30C Q4 ★ Exercise 30D Q1 ★ Exercise 30E Q3

- I have learnt to choose the best diagram to communicate the results of a survey. ★ Exercise 30E Q1

- By drawing a range of diagrams I know to label my axes, give the chart a title and I can choose an appropriate scale. ★ Exercise 30A Q3 ★ Exercise 30B Q2 ★ Exercise 30C Q4 ★ Exercise 30D Q1 ★ Exercise 30E Q3

31

Ideas of chance and uncertainty

I can find the probability of a simple event happening and explain why the consequences of the event, as well as its probability, should be considered when making choices.

MNU 3-22a

This chapter is going to show you how to:
- list all the possible outcomes of an event
- understand that the probability of an event occurring can be placed on a scale from 0 to 1
- use a 0–1 probability scale to show the probability of an event occurring
- calculate an expected outcome for a given probability
- work out the probability of an event occurring
- understand that other (non-numerical) factors in real-life situations can affect outcomes
- understand why such factors need to be taken into account when making decisions.

You should already know:
- how to communicate your predictions and findings of chance experiments using the vocabulary of probability
- how to give a numerical value to the likelihood of simple events.

Probability and the probability scale

Probability is the way of describing and measuring the chance or likelihood that an **event** will happen.

The chance of an event happening can be shown on a **probability scale**:

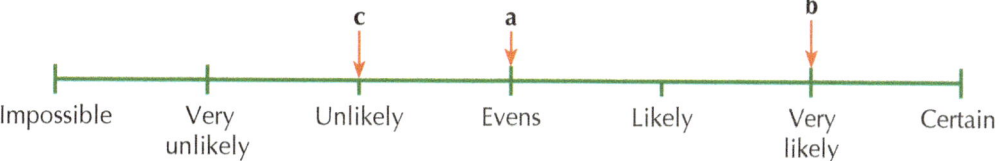

An evens chance is often referred to as 'a 50–50 chance'. Other everyday words used to describe probability are: uncertain, possible, probable, good chance, poor chance.

Example 31·1 The following events are shown on the probability scale above.

a The probability that a new-born baby will be a girl.
b The probability that a person is right-handed.
c The probability of rolling a 5 or 6 on a fair dice.

To measure probability, we use a scale from 0 to 1. So probabilities are written as fractions or decimals, and sometimes as percentages, as in the weather forecast.

The probability scale is now drawn as:

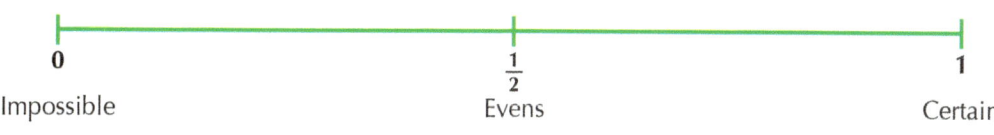

Information handling

We define the probability of an event happening as:

$$P(\text{event}) = \frac{\text{Number of outcomes in the event}}{\text{Total number of all possible outcomes}}$$

Example 31·2

A bag contains five counters. Two are red and three are blue.

When you take a counter out of the bag without looking, there are five to choose from and each one is equally likely to be taken. There are five **outcomes**, all **equally likely**.

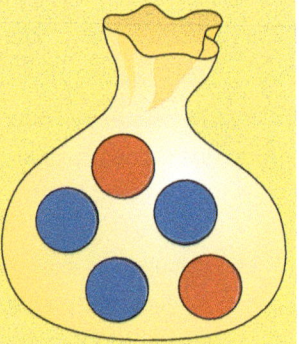

The probability of choosing a red counter is 2 out of 5. This is written as:

$$P(\text{red}) = \frac{2}{5}$$

Probabilities in examples like this one are usually written as fractions.

In the same way,

$$P(\text{blue}) = \frac{3}{5}$$

The probability of choosing a green counter is:

$$P(\text{green}) = \frac{0}{5} = 0$$

since there are no green counters in the bag.

Example 31·3

When throwing a fair dice, there are six equally likely outcomes: 1, 2, 3, 4, 5, 6.

So, for example, the chance of throwing 6 is:

$$P(6) = \frac{1}{6}$$

And the chance of thowing 1 or 2 is:

$$P(1 \text{ or } 2) = \frac{2}{6} = \frac{1}{3}$$

Fractions are *always* cancelled down in probability.

Example 31·4

Which is more likely to happen: tossing a tail on a coin or rolling a number less than 5 on a dice?

A coin can only land two ways (head or tail). So, provided the coin is fair, there is an even chance of landing on a tail and $P(\text{tail}) = \frac{1}{2}$.

On a dice there are four numbers less than 5 and two numbers that are not. So there are four chances out of six of rolling a number less than 5; the probability is $\frac{4}{6} = \frac{2}{3}$. There is a more than even chance of rolling a number less than 5.

So, rolling a number less than 5 is more likely than getting a tail when a coin is tossed:

$$\frac{2}{3} > \frac{1}{2}.$$

Ideas of chance and uncertainty

Exercise 31A

1 Copy the probability scale shown below:

Place the following events onto the above probability scale:

A You will be Prime Minister one day.
B You will have some homework this week.
C You can swim to the Moon.
D If you flip a coin, you will get a head.
E You will sit down today.
F It will snow in May.
G If you roll a dice, you will get a number bigger than 2.

2 Cards numbered 1 to 10 are placed in a box. A card is drawn at random from the box. Find the probability that the card drawn is:

a 5
b an even number
c a number in the 3 times table
d 4 or 8
e a number less than 12
f a prime number

3 Adam picks a card at random from a normal pack of 52 playing cards. Find each of the following probabilities.

a P(a Jack)
b P(a Heart)
c P(a picture card)
d P(the Ace of Spades)
e P(a 9 or a 10)
f P(an Ace)

4 A bag contains five red discs, three blue discs and two green discs. Linda takes out a disc at random. Find the probability that she takes out:

a a red disc
b a blue disc
c a green disc
d a yellow disc
e a red or blue disc

 5 Syed is using a fair, eight-sided spinner in a game.

a Find the probability that the score he gets is:

i 0   ii 1   iii 2   iv 3

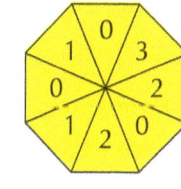

b Show each of these probabilities on a probability scale.

6 a Write down five events that have a probability of 0.
b Write down five events that have a probability of 1.

7 Sandhu rolls a fair dice. Find each of the following probabilities.

a P(3)
b P(odd number)
c P(5 or 6)
d P(even number)
e P(6)
f P(1 or 6)

8 Mr Evans has a box of 25 calculators, but 5 of them do not work very well.

What is the probability that the first calculator taken out of the box at random does not work very well?

Write your fraction as simply as possible.

Information handling

9 At the start of a tombola, there are 300 tickets inside the drum. There are 60 winning tickets available.

What is the probability that the first ticket taken out of the drum is a winning ticket? Write your fraction as simply as possible.

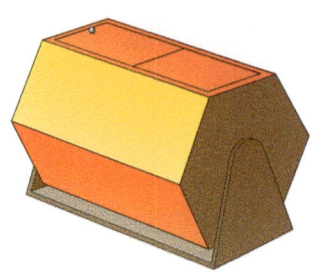

10 Bag A contains 10 red marbles, 5 blue marbles and 5 green marbles. Bag B contains 8 red marbles, 2 blue marbles and no green marbles. A girl wants to pick a marble at random from a bag.

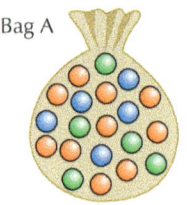 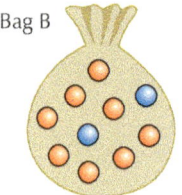

   a Which bag should she choose to have the better chance of:

      i a red marble   ii a blue marble   iii a green marble?

      Explain your answers.

   b Show your answers on a probability scale.

11 The letters which make up the word

are put into a bag.

A letter is taken out of the bag. Write down, as a fraction, the probability that the letter chosen is:

   a T   b B   c I   d a vowel

12 Mitchell has a box of coloured squares with shapes drawn on them.

He takes out a square at random. Find the probability that he takes out:

   a a red square

   b a square with a circle on it

   c a green square with a triangle on it

   d a blue square with a rectangle or a circle on it.

Ideas of chance and uncertainty

13 Seonaid did a survey of how all the pupils in her class travelled to school. The table shows her results.

	Number of boys	Number of girls
Walked	10	4
Bus	4	6

  a How many pupils are in Seonaid's class?

  b How many boys in Seonaid's class walked to school?

  c How many girls are there in Seonaid's class?

  d What is the probability of selecting a pupil at random who:

   i is a boy who walks to school

   ii is a girl?

14 Chi did a survey of hair colour of all the pupils in her class.
The table shows her results.

	Number of boys	Number of girls
Dark hair	12	6
Light hair	4	9

  a How many pupils are in Chi's class?

  b How many girls in Chi's class had dark hair?

  c How many pupils in Chi's class had dark hair?

  d What is the probability of selecting a pupil at random who:

   i is a girl with dark hair

   ii is dark-haired?

15 Look at this table. If a car is chosen at random, what is the probability that it is one of the following?

  a Peugeot   b Red   c Red Peugeot

  d Not blue   e Not a Ford

		Colour of cars				
		Red	White	Blue	Black	Other
Make of cars	Peugeot	8	1	4	1	4
	Ford	11	2	4	2	6
	Vauxhall	5	4	0	0	2
	Citroen	1	2	2	0	3
	Other	6	3	3	4	2

Information handling

Challenge

1. You will need a set of cards numbered 1 to 10 for this question.

   Line up the cards, face down and in random order.

   a Turn over the first card.

   b Work out the probability that the second card will be higher than the first card.

   c Work out the probability that the second card will be lower than the first card.

   d Turn over the second card.

   e Work out the probability that the third card will be higher than the second card.

   f Work out the probability that the third card will be lower than the second card.

   g Carry on the process. Write down all your results clearly and explain any patterns that you notice.

   Repeat the experiment. Are your results the same?

2. Get the necessary data to find the following probabilities, selecting at random a pupil from your class who is:

   a a blue-eyed boy

   b a girl who can swim

   c a dark-haired boy

   d a girl who has been abroad for her holiday

   e a girl who can ride a bike

Calculating the probability of an event not happening

Before we go outside, we might think about rain. There are only two possible events: raining, and not raining. The probabilities of all the possible events must add up to 1. So if the probability of it raining is p, then the probability of it not raining is $1 - p$.

The weather is very complex. We could look out of the window and think it looks like it might rain. The weather forecast may say the chance of rain is 30% or 0·3. That may sound high but the chance of it **not** raining can be calculated as $1 - 0·3 = 0·7$ or a 70% chance of no rain. Now that sounds a lot more promising.

Example 31·5

The probability that a woman washes her car on Sunday is 0·7. What is the probability that she does not wash her car on Sunday?

These two events are opposites of each other, so the probabilities add up to 1. The probability that she does not wash her car is $1 - 0·7 = 0·3$.

Example 31·6

A girl plays a game of tennis. The probability that she wins is $\frac{2}{3}$. What is the probability that she loses?

The probability of not winning (losing) is $1 - \frac{2}{3} = \frac{1}{3}$.

Ideas of chance and uncertainty

Example 31·7

A number of discs marked 1, 2, 3 and 4 are placed in a bag. The probabilities of randomly drawing out discs marked with a particular number are given as:

$P(1) = 0·2$ $P(2) = 0·3$ $P(3) = 0·25$

What is the probability of drawing a disc marked: **a** 1, 2 or 3   **b** 4

a The probability of drawing a disc marked 1, 2 or 3 is
$0·2 + 0·3 + 0·25 = 0·75$

b The probability of drawing a disc marked 4 is
$1 - 0·75 = 0·25$

Exercise 31B

1 Tony has a bag of jelly babies. Seven are red, two are green and one is black.

  a What colour jelly baby is Tony most likely to pick at random?

  b What is the probability of Tony, picking at random:

   i a black jelly baby   **ii** a green jelly baby   **iii** a red jelly baby?

2 Mrs Bradshaw has 20 felt-tipped pens. Three of these do not work.

  a How many pens do work?

  b What is the probability of taking a pen at random and it:

   i working   **ii** not working?

   iii Show these two possibilities on a probability scale.

3 Ten cards are numbered 0 to 9.

A card is picked at random. Work out the probability that it is:

  a 2   **b** not 2   **c** odd   **d** not odd

  e 7, 8 or 9   **f** less than 7   **g** 4 or 5   **h** not 4 or 5

4 In a bus station there are 24 red buses, six blue buses and 10 green buses. Calculate the probability that the next bus to arrive at the bus station is:

  a green   **b** red   **c** red or blue   **d** yellow   **e** not green

  f not red   **g** neither red nor blue   **h** not yellow

5 The cellar of a café was flooded in the great floods of 2007. All the labels came off the tins of soup and were floating in the water.

The café owner knew that she had:

  17 cans of mushroom soup   10 cans of vegetable soup

  15 cans of tomato soup   8 cans of pea soup.

  a How many tins of soup were there in the flooded cellar?
  b After the flood, what is the probability that a tin of soup chosen at random is:

   i mushroom   **ii** tomato   **iii** vegetable   **iv** pea?

  c Show your answers on a probability scale.

Information handling

6 A bag contains 32 counters that are either black or white. The probability that a counter is black is $\frac{1}{4}$.

How many white counters are in the bag? Explain how you worked it out.

7 Joe has 1000 tracks on his MP3 player, which comprises the following.

250 tracks of Soft rock

200 tracks of Blues

400 tracks of Country & western

100 tracks of Heavy rock

50 tracks of Quiet romantic

He sets the player to play tracks at random.

What is the probability that the next track to play is:

a Soft rock
b Blues
c Country & western
d Heavy rock
e Quiet romantic?

8 In a city taxi fleet there are 25 black cabs, 8 yellow cabs and 7 blue cabs.

a What is the probability that the first cab to come along for Mr Adams is not yellow?

b What is the probability that the first cab to arrive is either yellow or blue?

9 A bag contains many counters that are red, blue, green or yellow. The probabilities of randomly drawing out a counter of a particular colour are given as:

$P(\text{red}) = 0 \cdot 4$

$P(\text{blue}) = 0 \cdot 15$

$P(\text{green}) = 0 \cdot 3$

Calculate the probability that a counter drawn out is:

a red or blue.
b red, blue or green.
c yellow.

10 In a game there are three types of prize: jackpot, runners-up and consolation. The probability of winning the jackpot is $\frac{1}{100}$, the probability of a runners-up prize is $\frac{1}{10}$, and the probability of a consolation prize is $\frac{1}{2}$.

a Calculate the probability of winning a prize.

b Calculate the probability of not winning a prize.

c Which event is more likely to happen: winning a prize or not winning a prize?

d After many games the jackpot had been won three times. How many games would you expect to have been played?

Ideas of chance and uncertainty

Listing all the possible outcomes

This Exercise will help you think about how to work out all the possible outcomes of an event. This is an important skill for working out the chance of something happening.

Look at the pictures of the yachts. Can you spot the differences? Some yachts have a flag on the top of the mast. The sails and hulls are different colours.

Exercise 31C

Look at the 20 yachts in the picture and answer the questions.

1 How many yachts have each of these features?

 a Round sails
 b A blue hull
 c Yellow sails
 d A flag at the top of the mast
 e A pointed hull
 f Red sails and a curved hull
 g A yellow hull but no flag
 h Straight green sails
 i A curved red hull
 j A red hull, blue sails and a flag at the top of the mast

★ **2** A sailor takes a yacht at random. What is the probability that it has each of the following?

 a Round sails
 b A blue hull
 c Yellow sails
 d A flag at the top of the mast
 e A pointed hull
 f Red sails and a curved hull
 g A yellow hull but no flag
 h Straight green sails
 i A curved red hull
 j A red hull, blue sails and a flag at the top of the mast

3 All 20 boats were in a race. Three friends were on the boats.

 Peter was on a boat with green sails.
 Ali was on a boat with a blue hull.
 Andy was on a boat with straight sails.

 a A boat with a blue hull won the race.
   What was the probability that Ali was on the boat that won the race?

 b A boat with green sails was the only boat to capsize.
   What is the probability that Peter was on a boat that capsized?

 c A boat with straight sails came last.
   What is the probability that Andy was on the boat that came last?

307

Information handling

> **Challenge**
>
> 1 Draw your own set of 10 boats so that:
>
>   a there are more boats with blue hulls than any other colour
>
>   b there are fewer boats with red hulls than any other colour
>
>   c there are the same number of boats with straight sails as curved sails.
>
> 2 If one of your boats is chosen at random, what is the probability that it will have:
>
>   a blue sails   b a red hull
>
>   c straight sails   d a straight sail and a blue hull?

Making choices and decisions based on chance and uncertainty

If you can calculate the probability of an event happing (or not happening), then you can use your understanding to make decisions about events that are uncertain.

Now look closely at the statements in Examples 31·8, 31·9 and 31·10, and the comments on them. In Exercise 31D, you will have to decide which given statements are sensible.

Example 31·8

Daniel says: 'There is a 50–50 chance that the next person to walk through the door of a supermarket will be someone I know because I will either know them or I won't.'

The next person who walks through the door may be someone whom he knows, but there are far more people whom he does not know. So, there is more chance of she/he being someone whom he does not know. Hence, the statement is incorrect.

Example 31·9

Clare says: 'If I buy a lottery ticket every week, I am bound to win sometime.'

Each week, the chance of winning is very small (1 chance in 13 983 816), so it is highly unlikely that Clare would win in any week. Losing one week does not increase your chances of winning the following week.

Example 31·10

Matthew and Keira are tossing a coin and trying to predict the next result. Matthew says, 'The next throw will be a tail because the last throw was a head.' Keira says, 'The chances of a head or a tail are equally likely, so we don't know what the next throw will be.' Who is right?

Keira is correct. The probability of getting a head or a tail on a fair coin is 50–50 for every throw – it doesn't matter what the last throw was. If you try an experiment with a coin there will be many times when you get a head several times in a row, but in the long run you will get each result 50% of the time. Try it!

Example 31·11

The probability of rain falling on any day in June is 0.25. How many days would you expect it to rain in June?

There are 30 days in June. Multiply the number of days by the probability.
30 × 0.25 = 7.5 days. Round this up to 8 days.

Ideas of chance and uncertainty

Exercise 31D

1. Write a comment on each of the following statements, explaining why the statement is incorrect.

   a. A game for two players is started by rolling a six on a dice. Ashad says: 'I never start first because I'm unlucky.'

   b. It will rain tomorrow because it rained today.

   c. There is a 50% chance of snow tomorrow because it will either snow or it won't.

   d. There are mint, chocolate and plain sweets in the packet, so the probability of picking out a chocolate sweet is $\frac{1}{3}$.

 2. Decide whether each of the following statements is correct or incorrect.

   a. I fell down yesterday. I don't fall down very often, so it could not possibly happen again today.

   b. I have just tossed a coin to get a head three times in succession. The next time I throw the coin, the probability that I will get a head is still $\frac{1}{2}$.

   c. My bus is always on time. It will probably be on time tomorrow.

   d. There is an equal number of red and blue counters in a bag. My friend picked a counter out and it was blue. She then put it back. It is more likely that I will get red when I pick one out.

3. Here are three coloured grids. The squares have either a winning symbol or a losing symbol hidden.

   a. If you pick a square from each grid, is it possible to know on which you have the greatest chance of winning?

   b. You are now told that on Grid 1 there are three winning squares, on Grid 2 there are five winning squares and on Grid 3 there are four winning squares. Which grid gives you the least chance of winning?

   c. Helen says that there are more winning squares on Grid 2, which means that there is more chance of winning using Grid 2. Explain why she is wrong.

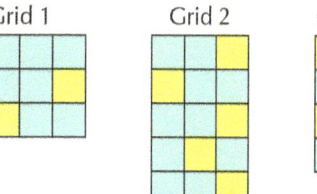

4. Fiya plays the lottery twice a week, every week. Fiya says, 'When I play the lottery I don't expect to win the jackpot, but I do expect to win a small prize regularly.' The probability of winning a small prize on a national lottery is 0·0095. Do you think Fiya is correct? Explain your answer.

5. The odds of being injured by a firework are 1 in 20 000. Write a statement saying whether or not you think it is safe to go to a firework display and explain why.

6. Dougie needs a new kettle. He has a choice between a £10 'Hotstuff' kettle with a 3-year warranty and a £3 'Boiler' kettle that has a 1-year guarantee. Which kettle do you think he should buy and why?

Information handling

7 Choose one (or more) of the following statements. Write a brief paragraph explaining if you agree or disagree with the statement. Include reasons for your decision. If you have access to the internet, see if you can find some facts and figures to back up your decision or you could conduct an experiment.

- An outdoor birthday party will always be spoiled by rain.
- It is always sunny in Spain.
- If the first-born child is a boy the second is always a girl.
- The chance that my football team will win is one third because they can win, lose or draw.
- I am unlucky, I never win when I enter a raffle.

8 Calum gets the bus to school every school day for 4 weeks. The probability of him getting a seat on the bus is 0.4. How many days would you expect him to get a seat?

9 There are 60 children in Primary 2. The probability of a child having blue eyes is $\frac{1}{3}$. How many children in Primary 2 would you expect to have blue eyes?

10 In April in Iceland, there is a 50% chance it will snow. How many days would you expect it to snow in Iceland in April?

11 The probability that a hockey player will score a goal in a game is 0.25. How many goals would you expect her to score if she plays 40 games in a season?

12 At the school fete, the probability of winning a prize in the Lucky Dip is 5%. 140 people play the Lucky Dip game. How many prizes would you expect to be won?

★ **13** The probability of seeing the snow leopards at the Highland Wildlife Park on any day is 0.3. On one day in August, there were 4180 visitors to the park. How many visitors would you expect to see the snow leopards?

- By working on this topic I can explain how likely things are to happen and use this to make choices.
- I can calculate the probability of an event happening. ★ Exercise 31A Q5
- I have learned that probabilities can be expressed as fractions and put on a scale from 0 to 1. ★ Exercise 31A Q5
- I can calculate the probability of an event not happening. ★ Exercise 31B Q4
- I can list all the possible outcomes of an event. ★ Exercise 31C Q2
- I can use my understanding of chance and uncertainty to make informed decisions. ★ Exercise 31D Q2
- I can use a given probability to calculate an expected outcome. ★ Exercise 31D Q13

Answers

Chapter 1

Exercise 1A

1. a i 5000   ii 4600   iii 4560
   b i 3000   ii 3200   iii 3250
   c i 6000   ii 6000   iii 5990
   d i 35 000   ii 35 200   iii 35 230
   e i 0   ii 500   iii 460
   f i 10 000   ii 10 000   iii 9980
   g i 43 000   ii 43 300   iii 43 260
   h i 5000   ii 5200   iii 5240
   i i 9000   ii 9000   iii 9040
   j i 5000   ii 5000   iii 5010

2. a i 5   ii 4·7   b i 3   ii 3·1
   c i 3   ii 2·6   d i 2   ii 1·9
   e i 1   ii 0·8   f i 1   ii 0·9
   g i 4   ii 4·0   h i 3   ii 2·6
   i i 3   ii 3·2   j i 3   ii 3·5
   k i 1   ii 1·5   l i 2   ii 1·9

3. a 4·72   b 3·10   c 6·23   d 4·94   e 0·78
   f 1·00   g 4·00   h 2·60   i 3·19   j 6·50

4. a 6·326 b 8·122 c 1·002 d 9·871
   e 5·555 f 1·999 g 0·003 h 5·421
   i 3·782 j 0·995

Exercise 1B

1. a 20 × 10 = 200   b 50 × 10 = 500   c 40 × 20 = 800
   d 40 × 40 = 1600 e 20 × 70 = 1400

2. a 1200 ÷ 30 = 40   b 200 ÷ 40 = 5
   c 1400 ÷ 20 = 70   d 1400 ÷ 70 = 20
   e 1600 ÷ 40 = 40

3. a $\frac{1000}{200} = 5$ b $\frac{280}{40} = 7$
   c $\frac{220}{110} = 2$ d $\frac{900}{30} = 30$

4. a 70 × 0·6 = 42   b 60 × 0·7 = 42   c 40 × 0·8 = 32
   d 40 × 0·2 = 8   e 60 × 0·3 = 18   f 70 × 0·7 = 49
   g 40 × 0·2 = 8   h 20 × 0·9 = 18   i 40 × 0·9 = 36
   j 30 × 0·8 = 24   k 30 × 0·5 = 15   l 120 × 0·2 = 24

5. a 3000 − 400 = 2600   b 230 × 20 = 4600
   c 800 ÷ 40 = 20   d 60 ÷ 15 = 4
   e 400 × 400 = 160 000 f 160 ÷ 40 = 4
   g 60 ÷ 10 = 6   h $\frac{40 \times 60}{20} = 120$

6. Must be less than £3 since 6 × 46p is less than 6 × 50p (£3).
7. No; 8 × 60p = £4·80 so £4·50 is not enough.
8. 53·00 + 1·47 = 54·47. The prices were entered incorrectly into till.
9. b
10. d
11. d
12. £4·99 < £5 so Delroy had at least £5 spare. £2·65 + £1·92 < £5, so the goods cost less than £10. Using complements to 100: 1p + 35p + 8p = 44p, so Delroy cannot afford the chocolate bar as well.

Challenge

1. a The total area of grid is 64 so must be less.
      Length of side of square is greater than 6, so area must be greater than 36.
   b 40 square units.

2. a 600 ÷ 6 = 100   b 1400 ÷ 7 = 200   c 400 ÷ 8 = 50
   d 400 ÷ 2 = 200   e 600 ÷ 3 = 200   f 700 ÷ 7 = 100
   g 400 ÷ 2 = 200   h 1800 ÷ 9 = 200   i 2700 ÷ 9 = 300
   j 400 ÷ 8 = 50   k 300 ÷ 5 = 60   l 1200 ÷ 2 = 600

Exercise 1C

1. a 5·0   b 6·5   c 4·4   d 26·7
   e 98·8   f 214·0 g 2·5   h 3·3

2. $\frac{230 + 170}{80 - 30} = \frac{400}{50} = 8$ Calculator answer to 1 dp = 8·7

3. a 19·1   b 2·5   c 1·6
4. a 42·83 b 22·28

Challenge

Worker 1 rounded off after the first part of his calculation. You should only round off at the end.
Worker 1's method could make the company buy more metal then needed.

Chapter 2

Exercise 2A

1. a 18   b 18   c 42   d 8
   e 25   f 16   g 20   h 0
   i 60   j 7   k 28   l 100
   m 72   n 54   o 27   p 45
   q 35   r 63   s 48   t 56

2. a 240   b 300   c 200   d 300
   e 420   f 2400 g 1800 h 5600
   i 270   j 2000 k 7200 l 7200

3. a 340   b 8900   c 700
   d 4000   e 3·4   f 0·89
   g 0·07   h 0·004 i 0·058

4. a 30   b 100   c 0·3   d 100

5. a 45   b 6   c 530   d 3
   e 5800   f 700   g 0·45   h 0·06
   i 0·053   j 0·0003   k 0·0058 l 0·004
   m 0·501   n 637·8   o 210

6. a 3   b 1000   c 0·03
   d 100   e 3   f 300
   g 30   h 30   i 300

7. a ×100   b ÷100   c ×1000
   d ÷10   e ÷100   f ÷1000

8. £40, £37, £9·50; Total = £86·50

311

Answers

Exercise 2B

1. a)
×	30	4
7	210	28

   238

   b)
×	4
20	80
6	24

   104

   c)
×	50	2
7	350	14

   364

2. a) 406 b) 280 c) 68 d) 152
   e) 192 f) 294 g) 336 h) 99

3. a) 27, 140 b) 52, 100

4. a) £14·97 b) £40
   c) 8; 5 for £20, then an additional 3 costing £14·97.

5. a) 29, 30, 31 b) 11, 12
   c) All consecutive integers have a difference of 1.

6. 6, 7
7. 14, 15
8. 18, 19
9. 5, 6
10. 9, 10
11. 7, 8
12. a) 11, 13 b) 31 × 4 = 124

13.
4	3	8
9	5	1
2	7	6

2	9	4
7	5	3
6	1	8

4	9	2
3	5	7
8	1	6

Exercise 2C

1. a) 51 b) 128 c) 95 d) 336
   e) 692 f) 1623 g) 1029 h) 1917

2. 272 miles
3. 574
4. 2436
5. £2·25
6. £1467
7. a) i) 11 r 2 ii) 11·3
   b) i) 18 r 1 ii) 18·3
   c) i) 24 r 4 ii) 24·8
   d) i) 18 r 4 ii) 18·7
   e) i) 17 r 3 ii) 17·8
   f) i) 35 r 5 ii) 35·8
8. 260
9. 13
10. 11
11. a) 10 b) £711

Exercise 2D

1. £36
2. £3·15
3. 5 × 5 + 5, 6 × 6 − 6
4. a) 2 + 3 = 1 + 4 b) 3 × 4 = 12 c) 12 ÷ 4 = 3
5. 356·3 litres
6. a) 70 cm b) 130 cm
7. £5·10
8. £14·40
9. £426
10. 56p
11. 22 miles
12. There are two chains of length 21 starting with 18 and 19.

Exercise 2E

1. a) 5 b) 30 c) 7
2. a) (17 + 8) ÷ (7 − 2) b) (53 − 8) ÷ (3·5 − 2)
   c) (19·2 − 1·7) ÷ (5·6 − 3·1)
3. a) 5 b) 6·5 c) 4·4 d) 26·7
   e) 98·8 f) 214 g) 2·5 h) 3·3
4. Estimate = 14; actual answer = 12·7

Exercise 2F

1. a) 240 b) 396 c) 1161 d) 1922
   e) 4097 f) 7608 g) 11 565 h) 29 232
2. a) 240 b) 396 c) 1161 d) 1922
   e) 4097 f) 7608 g) 11 565 h) 29 232
3. a) 391 b) 1344 c) 855 d) 2576
   e) 4152 f) 17 312 g) 3969 h) 8307

Exercise 2G

1. a) 32 b) 21 c) 23 d) 24
   e) 24 f) 35 g) 41 h) 32
2. a) 32 b) 21 c) 23 d) 24
   e) 24 f) 35 g) 41 h) 32
3. a) 36 b) 42 c) 27 d) 25
   e) 24 r 18 f) 26 r 18 g) 28 r 26 h) 19 r 2

Exercise 2H

1. 5916
2. 39
3. 576
4. 3825
5. 15
6. a) 16 b) £12 288
7. £22·75
8. £3480
9. 192
10. 13

Exercise 2I

1. a) 686·2 b) 26·19 c) 33·21 d) 12·772
   e) 1·4178 f) 1·89 g) 17·095 h) 0·2856
2. 1·6482 m²
3. a) 2·4 b) 0·65 c) 1·58 d) 3·9
   e) 3·42 f) 0·85 g) 0·25 h) 9·6
4. 1·45 cm

Exercise 2J

1. 23, 25
2. 42
3. 4
4. a) 5 b) 5 c) 7 or 23 d) 19 or −11
5. Multipack.
6. 53, 57
7. The weight of the odd numbered counters ⟶ 5 × 6;
   the total weight ⟶ 10 × 6;
   the weight of the counters that are blue or red ⟶ (5 + 3) × 6

Answers

8 60

9 Olivia 24, Jack 12.

10 125

11 a £7·20   b £86·34

12 £383·40

Challenge

16	2	3	13
5	11	10	8
9	7	6	12
4	14	15	1

Chapter 3

Exercise 3A

1 a 20   b 200   c 700   d 1400
   e 1200 f 21 000 g 800   h 3000
   i 1400 j 1800   k 3200   l 90 000
   m 42 000 n 40 000 o 2400   p 2500
   q 2400 r 4900

2 a 1800   b 12 000 c 20 000 d 420

3 a 30   b 5   c 70   d 20
   e 30   f 16   g 60   h 100
   i 50   j 40   k 30   l 50
   m 200   n 70   o 50   p 60

4 a i 25 × 10 = 250   ii 253
   b i 45 × 10 = 450   ii 516
   c i 70 × 20 = 1400   ii 1296
   d i 30 × 40 = 1200   ii 1344
   e i 30 × 70 = 2100   ii 2139
   f i 20 × 40 = 800   ii 684
   g i 50 × 60 = 3000   ii 2784
   h i 70 × 20 = 1400   ii 1584
   i i 60 × 30 = 1800   ii 2079
   j i 40 × 80 = 3200   ii 3159
   k i 55 × 20 = 1100   ii 1134
   l i 35 × 50 = 1750   ii 1764

5 a i 1200 ÷ 30 = 40   ii 36·8
   b i 200 ÷ 40 = 5   ii 5·4
   c i 1400 ÷ 20 = 70   ii 64·2
   d i 1400 ÷ 70 = 20   ii 20·3
   e i 1500 ÷ 50 = 30   ii 32·0
   f i 1600 ÷ 100 = 16   ii 16·1
   g i 1800 ÷ 30 = 60   ii 57
   h i 2000 ÷ 40 = 50   ii 54
   i i 2000 ÷ 40 = 50   ii 48·7
   j i 2400 ÷ 60 = 40   ii 38·3
   k i 3500 ÷ 50 = 70   ii 72·0
   l i 2000 ÷ 80 = 25   ii 26·2

6 a i 1000 ÷ 200 = 5   ii 4·8
   b i 300 ÷ 50 = 6   ii 5·7
   c i 200 ÷ 50 = 4   ii 4·4
   d i 900 ÷ 30 = 30   ii 26·7

7 a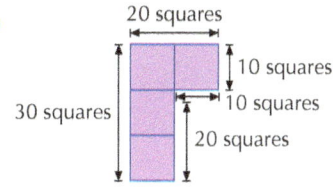

There are 200 little squares in the larger shape.

   b

There are 400 little squares in the larger shape.

   c 20 000

Challenge

3 42 000 000 000 000 000

4 210 000 000 000

5 4

Exercise 3B

1 a Last digit should be 2.
   b The answer is roughly 60 × 80 = 4800.
   c The answer is roughly 100 ÷ 10 = 10.
   d By the inverse operation 8 × 45 must equal 440, but 8 × 45 = 360.
   e Last digit should be 3.
   f Answer should have 2 dp ending in … ·33.

2 Estimated answers may vary depending on the numbers rounded.
   a 4800 − 700 = 4100   b 300 × 30 = 9000
   c 700 + 140 = 840   d 75 ÷ 15 = 5
   e 500 × 500 = 250 000   f 200 ÷ 50 = 4
   g 50 ÷ 5 = 10   h $\frac{(90 \times 20)}{60} = 30$

3 After DVD he has about £5 left.
   £2·75 + £1·82 is less than £5.
   He has enough for the 35p bag of crisps.

4 Should be less than 8 × £0·60 = £4·80.

5 No. 4 × 60p = £2·40 so £2 is not enough.

6 60 + 1·95 = 61·95 (60p entered incorrectly as 60·00)

7 a 2·5   b 11   c 0

Exercise 3C

1 a 800   b 0·8   c £3 million

2 a 150 kg   b 0·15 cm c 6000

3 140 + 60 = 200
   1·4 + 0·6 = 2·0
   14 000 + 6000 = 20 000
   0·020 − 0·006 = 0·014
   200 − 140 = 60
   0·14 + 0·06 = 0·20
   20 000 − 6000 = 14 000
   0·6 + 201·4 = 202·0
   2000 − 1400 = 600
   1·014 + 0·006 = 1·020

313

Answers

4 a 130 000 **b** 0·9 **c** 1100
  d 80 **e** 1·5 **f** 50
  g 11 000 **h** 0·011 **i** 1·16
  j 20, 120 **k** 1080 **l** 4·06
  m £0·4 million **n** 7·017 kg **o** 100·6 mm

5 a £50 000 **b** 12·2 seconds
  c 0·6 g **d** £80

Exercise 3D

1 a 0·8 **b** 0·6 **c** 4·2
  d 3·5 **e** 1·6 **f** 2·4
  g 0·9 **h** 1·6 **i** 5·6
  j 4·5 **k** 5·4 **l** 6·3

2 a 0·08 **b** 0·06 **c** 0·42
  d 0·35 **e** 0·16 **f** 0·24
  g 0·09 **h** 0·16 **i** 0·56
  j 0·45 **k** 0·54 **l** 0·63

3 a 24 **b** 12 **c** 30
  d 12 **e** 12 **f** 20
  g 14 **h** 18 **i** 40
  j 42 **k** 3 **l** 48

4 a 0·008 **b** 0·032 **c** 0·0028
  d 0·0054 **e** 0·004 **f** 0·003
  g 0·0007 **h** 0·0049

5 a 240 **b** 24 **c** 300 **d** 12
  e 40 **f** 3 **g** 3·6 **h** 320

6 £27   **7** 420 milligrams

Challenge

1 a 156·4 **b** 15·64 **c** 460 **d** 460
2 a 0·1824 **b** 18 240 **c** 1824 **d** 0·1824
3 a 1540 **b** 2·8 **c** 550 **d** 154

Exercise 3E

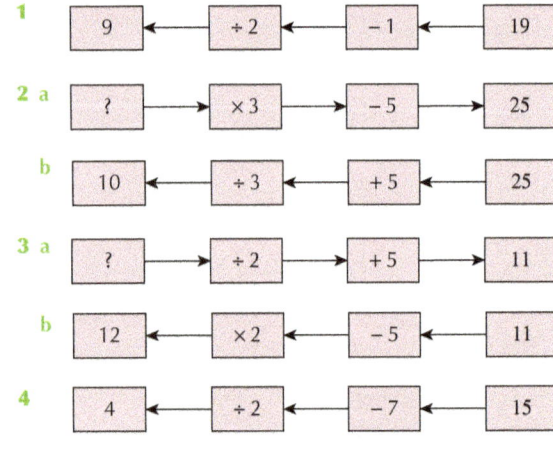

6 a 5 °C **b** 154·4 °F

Exercise 3F

1 a   32
      + 17

        49

  b   43
       + 33

         76

  c 132
       + 17

        149

  d   87
       − 42

         45

  e 238
       − 121

         117

  f 135
       ×   4

        540

2 3 + 5 = 8, 1 + 9 = 10…

3 2 + 4 = 6, 12 + 8 = 20…

4 5 + 4 = 9, 11 + 10 = 21…

5 7 × 9 = 63, 3 × 7 = 21…

6 7 × 8 = 56, 9 × 12 = 108…

7 Small: £1·25 per litre; medium: £1·20 per litre; large: £1·24 per litre. So medium is best value.

8 a 6 litres for £12
  b 8 kg for £8
  c 300 g for £5
  d Four bars for 90p

9 1350 g

Challenge

The answer always adds up to 1089.

If the end digits are the same, e.g. 121, 232 or 444, then you get 0. Otherwise you still get 1089.

Chapter 4

Exercise 4A

1 a 1, 2, 3, 4, 5, 6 **b** −3, −2, −1, 0, 1

2 a T **b** T **c** F
  d F **e** T

3 a < **b** < **c** > **d** >
  e = **f** >

4 a −5 < 4 **b** −7 > −10 **c** 3 > −3 **d** −12 < −2

5 a −5 **b** −1 **c** −5

6 a −3 **b** −5 **c** −2 **d** 0
  e 3 **f** −10 **g** 1 **h** −7
  i −1 **j** −2 **k** −6 **l** −4

7 a −4 **b** 3 **c** 4 **d** 6
  e 7 **f** 12 **g** 7 **h** 6
  i −7 **j** 0 **k** −2 **l** −6

8 a 17 m **b** −£10

Challenge

1 a 20 **b** 10 **c** 0 **d** 5
  All results are in 5 times table.

2 a 32 **b** 20 **c** 8 **d** 14
  Results are no longer all in 5 times table.

Exercise 4B

1 a 4 °C, −2 °C **b** −6 °C, −10 °C **c** −5 °C, −20 °C

Answers

2 a 7 °C   **b** −12 °C   **c** 9 °C

3 −1 °C

4

Deposits (£)	Withdrawals (£)	Balance (£)
100		100
	120	−20
30		10
	60	−50
	40	**−90**
10		**−80**
	30	**−110**

5 a

Deposits (£)	Withdrawals (£)	Balance (£)
100		100
	60	40
70		110
	160	−50
40		−10
30		20
	20	0

b

Deposits (£)	Withdrawals (£)	Balance (£)
500		500
	200	300
	400	−100
150		50
	50	0
	75	−75
65		−10

6 a 1 birdie   2 bogey   3 bogey
4 par   5 eagle   6 bogey
7 double bogey   8 birdie   9 par
b i −1   ii +1

7 Donald   −4
Woods   −2
McIlroy   −2

8 a In 2012, Archimedes 2299 years ago
Pythagoras 2582 years ago.
b 283
c Pythagoras 495 BC, Archimedes 212 BC
d

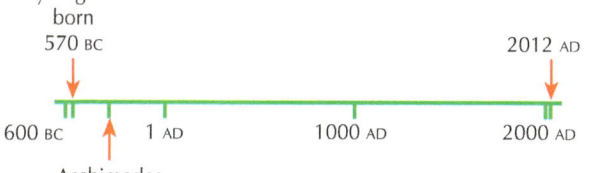

9 a 19·7 km   **b** 2·1 km

Exercise 4C

1 a 4, 3, 2   **b** 10, 9, 8, 7   **c** 3, 2, 1   **d** −4, −5, −6
2 a 4   **b** 2   **c** 20 − 7 = 13   **d** 16 − 5 = 11
e −6 − 7 = −13
3 a −2   **b** 6   **c** −1   **d** 0
e −5   **f** −8   **g** 6   **h** −9
i 5   **j** −14   **k** 2   **l** 0
m −1   **n** −10   **o** −8   **p** −8
4 a 15   **b** −25   **c** −40   **d** −30
e 4   **f** 0   **g** −25   **h** −120
i 5   **j** −120   **k** −9   **l** −28
5 a 3   **b** −3   **c** 0   **d** −23
6 a −2   **b** 20
7 a

−6	−1	−8
−7	−5	−3
−2	−9	−4

b

0	−7	−2
−5	−3	−1
−4	+1	−6

c

3	−4	1
−2	0	2
−1	4	−3

Challenge

−9	−2	0	5
2	3	−7	−4
−3	−8	6	−1
4	1	−5	−6

Exercise 4D

1 a 4, 5, 6   **b** 9, 10, 11   **c** −4, −3, −2   **d** −9, −8, −7
2 a 8   **b** 16   **c** 20 + 8 = 28
d −14 + 5 = −9   **e** −6 + 7 = 1
3 a 11   **b** 9   **c** −1   **d** 15
e 1   **f** 8   **g** 6   **h** −1
i 12   **j** −2   **k** −1   **l** 0
m −8   **n** 10   **o** 8   **p** 4
4 a 28   **b** −5   **c** 130   **d** 0
e 40   **f** 60   **g** 25   **h** −60
i 24   **j** −50   **k** 41   **l** 11
5 a 21   **b** −2   **c** 33   **d** 3
6 biggest = 16   smallest = −16
7 a −3   **b** −3   **c** 2   **d** 10
e −11   **f** −2   **g** −18   **h** 4
i −7   **j** −14   **k** 40   **l** −4
m 7   **n** −16   **o** 0   **p** −21
q 19   **r** −100   **s** −98   **t** 98
8 a −1   **b** −2   **c** 12   **d** −12

315

Answers

Chapter 5

Exercise 5A

1. a 4, 8, 10, 18, 72, 100
   b 18, 33, 69, 72, 81
   c 10, 65, 100
   d 10, 100

2. a 4, 8, 12, 16, 20, 24, 28, 32, 36, 40
   b 5, 10, 15, 20, 25, 30, 35, 40, 45, 50
   c 8, 16, 24, 32, 40, 48, 56, 64, 72, 80
   d 9, 18, 27, 36, 45, 54, 63, 72, 81, 90
   e 10, 20, 30, 40, 50, 60, 70, 80, 90, 100

3. 42, 75, 786

4. a 40   b 20   c 36   d 40

5. a 45   b 25   c 24   d 12
   e 24   f 60   g 63   h 77

6. 28 December

7. a 9:36 am   b 10 times

Exercise 5B

1. a 30   b 24   c 56
   d 210 e 42   f 420

2. 6 pm

3. 120 minutes (2 hours)

4. 72 weeks

Exercise 5C

1. a No   b Yes   c Yes   d No

2. 54

3. a 1, 3, 5, 15   b 1, 2, 4, 5, 10, 20
   c 1, 2, 4, 8, 16, 32 d 1, 2, 3, 4, 6, 12
   e 1, 5, 25

4. a 5   b 5   c 4   d 4

5. a 3   b 4   c 2   d 4
   e 2   f 2   g 9   h 1

6. a 6, 8   b No

7. 1 by 18, 2 by 9, 3 by 6, 6 by 3, 9 by 2, 18 by 1

Exercise 5D

1. a 6, 8   b 6, 9   c 15, 20

2.

x	y	Product	HCF	LCM
4	14	56	2	28
9	21	189	3	63
12	21	252	3	84
18	24	432	6	72

HCF × LCM = Product

3.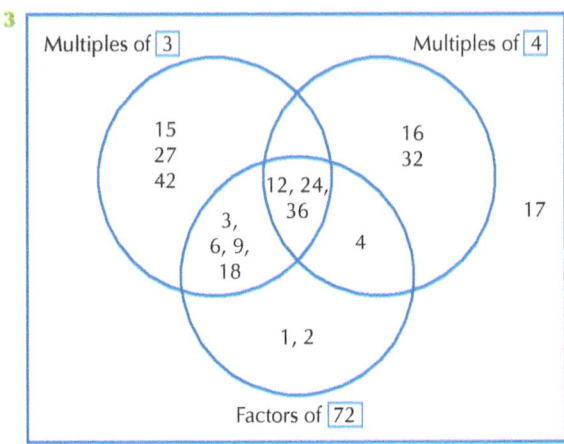

4. 5, 6, 7

5. a 4   1, 2, 4
   b 12   1, 2, 3, 4, 6, 12

6. 24

Challenge

1. a HCF = $2b$,   LCM = $20abc$
   b HCF = $3m$,   LCM = $30mn$
   c HCF = xz,   LCM = $21wxyz$

2. a i 1, 35   ii 1, 12   iii 1, 22
   b xy

3. a i 5, 10   ii 3, 18   iii 4, 20
   b HCF = x,   LCM = y

Chapter 6

Exercise 6A

1. a 2, 2, 2   b 2, 5   c 2, 2, 2, 2   d 2, 2, 5
   e 2, 2, 7   f 2, 17   g 5, 7   h 2, 2, 13
   i 2, 2, 3, 5   j 2, 2, 3, 3, 5

2. a 2, 3, 7   b 3, 5, 5   c 2, 2, 5, 7
   d 2, 5, 5, 5   e 2, 2, 2, 2, 2, 3, 5

3. a 2, 2, 2, 5, 5   b 2, 2, 2, 2, 2   c 2, 2, 2, 3, 3
   d 3, 5, 7   e 2, 2, 5, 11

4. a 200 = 2 × 2 × 2 × 5 × 5   b 32 = 2 × 2 × 2 × 2 × 2
   c 72 = 2 × 2 × 2 × 3 × 3   d 105 = 3 × 5 × 7
   e 220 = 2 × 2 × 5 × 11

Exercise 6B

1. a HCF = 6   LCM = 360
   b HCF = 10   LCM = 450
   c HCF = 12   LCM = 336

2.

HCF = 30   LCM = 600

Answers

3. HCF = 30, LCM = 630 (Venn diagram: 210 and 90; left circle 7, intersection 2,3,5, right circle 3)

4. a HCF = 25, LCM = 1400 b HCF = 8, LCM = 2520
   c HCF = 21, LCM = 210

Challenge

1. a 11 b 12 c 2
   d Prime numbers greater than 3 are either 1 more than a multiple of 6, or 1 less than a multiple of 6.
2. a No, e.g. 2 + 3 = 5
   b No. The smallest prime is 2. The smallest number you can make from adding three primes is 6 (2 + 2 + 2). So you can't make 2, 3 or 5 by adding three primes together. So cannot do for any prime.
3. a 15 b 16 c 13 d 21

Chapter 7

Exercise 7A

1. a 3^4 b 5^3 c 7^5 d 2^2
   e 1^4 f 10^2 g 4^3 h 3^6
   i 2^4 j 4^7 k f^3 l m^5
2. a 8 b 9 c 125 d 1 000 000
   e 32 f 1 g 27 h 64
   i 16 j 216 k 16 l 81
3. a $3^3 \times 4$ b $5^3 \times 7^2$ c $2^3 \times 3^4$
   d $5^2 \times 6^3 \times 7^2$ e $3^3 \times 4^2 \times 9^2$ f $3 \times 7^3 \times 10^3$
   g $m^4 \times n^3$ h $t^2 \times v \times w^2$ i $a^4 \times b^3 \times c^2$
4. a 41 b 121 c 25 d 116
   e 14 f 39 g 9909 h 243
5. a $25\,cm^2$ b $49\,cm^2$ c $81\,cm^2$ d $144\,cm^2$
6. a $27\,cm^3$ b $216\,cm^3$ c $1000\,cm^3$ d $64\,cm^3$

Exercise 7B

1. a 15625 b 243 c 7776 d 2401
   e 2187 f 3125 g 512
2. a 41 b 121 c 25 d 116
   e 14 f 39 g 9909 h 243
3. a 77.63 b 2.47 c 18

Exercise 7C

1. a 3^2 b 3^4 c 3^5
2. a 6^2 b 8^2 c 10^2 d 7^2
3. a i 2^4 ii 4^2 b i 2^6 ii 4^3
   c i 2^8 ii 4^4
4. The power of 4 is always half the power of 2.

Challenge

1. a 5^9 b 6^{12} c 4^7 d 8^9
   e 9^9 f 10^2 g 4^4 h 6^3

2. a 5^2 b 2^8 c 4 d 7^4
3. To multiply terms with the same base you add the powers. To divide terms with the same base you subtract the powers.
4. a m^9 b t^{11} c y^3
   d m^5 e w^3 f f^4

Chapter 8

Exercise 8A

1. a 12.9 b 3.4 c 9.1 d 18.3
   e 13.1 f 4 g 1 h 2.6
   i 15.2 cm j 1.1 m
2. a 17.59 b 32.755 c 14.163 d 74.73
   e 30.983 f 10.114 g 22.657 h 241.718
   i 14.482 j 15
3. 0.979 kg
4. a 5.604 kg b 2.227 kg
5. a 8.361 km b 4.711 km c 8.331 km
   d 14.876 km e 11.854 km
6. 5.56 m   7. 2.44 m   8. 10.98 litres

Exercise 8B

1. a 199.2 b 199.2 c 19.92 d 0.1992
2. a 80 b 800 c 8 d 0.8
3. a 0.6 b 0.8 c 3.6 d 1.4
   e 0.8 f 3.2 g 0.6 h 0.9
   i 5.6 j 4.0 k 2.7 l 5.4
4. a 0.06 b 0.08 c 0.36 d 0.14
   e 0.08 f 0.32 g 0.06 h 0.09
   i 0.56 j 0.40 k 0.27 l 0.54
5. a 0.008 b 0.032 c 0.0056 d 0.0054
   e 0.004 f 0.0018 g 0.0002 h 0.0049

Challenge

1. a 0.49 b 0.09 c 0.4 d 0.4
2. The same

Exercise 8C

1. a 5 b 7 c 30 d 6
   e 7 f 80 g 10 h 40
   i 24 j 8 k 100 l 35
   m 4 n 240 o 120 p 200
2. a 13 b 20.4 c 19.64 d 30.6
   e 220.5 f 179.2 g 87.6 h 76.4
3. £12 (1200p)   4. 300 mg

Exercise 8D

1. a 0.2 b 0.3 c 0.6 d 0.7
   e 0.3 f 0.6 g 0.6 h 0.8
2. a 0.18 b 0.09 c 0.04 d 0.05
   e 0.04 f 0.09 g 0.09 h 0.07

Answers

3 a 1·21 **b** 0·65 **c** 1·61 **d** 0·95
e 0·87 **f** 0·89 **g** 0·61 **h** 0·32

4 a 9·01 **b** 7·59 **c** 5·76 **d** 5·46
e 3·06 **f** 11·02 **g** 2·36 **h** 1·28

5 1·08 cm

6 £0·87 (87p)

Challenge

1 a 30 **b** 800 **c** 3 **d** 120
e 17 **f** 20 **g** 140 **h** 61
i 40 **j** 118

2 50

3 125

Exercise 8E

1 a £18 **b** 45 kg
c 12 metres **d** £8
e £13·80 **f** 525 houses
g 9 litres **h** 18 minutes
i £52·50 **j** 600 loaves
k 20 km **l** 420 crows
m £0·28 **n** 240 children
o 84 miles

2 780 g

3 60 kg

4 13 litres

5 24 000

6 48 pages

Exercise 8F

1 a $\frac{2}{5}$ **b** $\frac{5}{8}$ **c** $\frac{3}{14}$ **d** $\frac{3}{8}$
e $\frac{8}{63}$ **f** $\frac{2}{5}$ **g** $\frac{3}{7}$ **h** $\frac{3}{4}$
i $\frac{14}{45}$ **j** $\frac{4}{15}$

2 a $2\frac{2}{5}$ **b** $4\frac{4}{7}$ **c** $4\frac{2}{3}$ **d** $2\frac{4}{7}$
e $1\frac{2}{3}$ **f** 9 **g** 21 **h** $16\frac{2}{3}$
i $12\frac{6}{7}$ **j** $5\frac{1}{10}$

3 a $\frac{1}{15}$ **b** $\frac{3}{20}$ **c** $\frac{35}{64}$ **d** $\frac{5}{36}$
e $\frac{3}{16}$ **f** $\frac{8}{105}$ **g** 4 **h** $2\frac{6}{7}$

4 $4\frac{4}{5}$ **5** $6\frac{2}{3}$ m **6** $\frac{1}{6}$ **7** $\frac{3}{20}$

8 $\frac{1}{14}$ **9** $\frac{3}{8}$

Exercise 8G

1 a 64% **b** 85% **c** 40% **d** 64%
e 60% **f** 30% **g** 64% **h** 72%
i 66% **j** 16% **k** 8% **l** 24%
m 15% **n** 60% **o** 30% **p** 48%

2 78% maths; 80% English; 76% science. So English was the best mark.

3 48% electricity; 36% gas; 16% oil

4 28% text messages; 48% long distance calls; 24% local calls

5 a 55% **b** 23% **c** 33% **d** 40%
e 4% **f** 56% **g** 18% **h** 37%

6 14% number; 43% algebra; 29% geometry; 15% statistics. Total = 101%, because of rounding.

Challenge

Across
10 102 out of 200

Down
2 300 out of 400
6 87 out of 300
9 19 out of 20

Exercise 8H

1 a 24 **b** 3·5 **c** 90 **d** 12·8
e 81 **f** 144 **g** 161 **h** 210
i 4·5 **j** 19 **k** 13·5 **l** 104

2 a 38·4 **b** 22·05 **c** 80·6 **d** 12·24
e 1·32 **f** 78·4 **g** 18·72 **h** 203·94

3 a £72·80 **b** 29 books **c** 139 chairs
d £72·08 **e** 637 pupils **f** 68 buses
g 40 plants **h** 269 bottles **i** 226 days
j 2385 people **k** 246 fish **l** £20·25
m 238 eggs **n** £47·88 **o** 51 chickens

4 a 34% of 92 **b** 82% of £26
c 28% of 79 **d** 43% of 325

5 a 17 **b** 66% **c** 33

6 a 208 British, 68 American, 48 French, 76 German
b 19%

Exercise 8I

1 a 12 **b** 52

2 a 8000 **b** 12 000 **c** 60%

3 a £49·50 **b** £43·20
c £144 **d** £72
e £68·25 **f** £104·50
g £312·50 **h** £225
i £7·48 **j** £4·86

4 a i £15·30 **ii** £22·10
iii £42·50 **iv** £59·50
b i £210 **ii** £136·50
iii £399 **iv** £105

5 a 552 **b** 2952

6 a 99 **b** 121 **c** 55%

318

Answers

7 a €72·80   **b** €55·90
  c €153·72   **d** €381·60
  e €125·40   **f** €22·08
  g €320·85   **h** €345·22
  i €9·59   **j** €3·89

8 a i £20·56   **ii** £57·75
  iii £49·70   **iv** £108·50
  b i £268·75   **ii** £193·50
  iii £305·30   **iv** £213·93

Challenge

1 20   **2** 24   **3** £120   **4** £96
5 Pupil's own answer

Exercise 8J

1 TV Mania £200 (TV Bargains £224)
2 The Gravel King
3 a $\frac{1}{4}$ off   **b** $\frac{1}{3}$ off
4 Vince   **5** Michelle
6 David's car, by £400
7 They are the same.
8 No. In the second shop they cost £48.

Chapter 9

Exercise 9A

1 a $\frac{2}{3}$ **b** $\frac{3}{5}$ **c** $\frac{3}{7}$ **d** $\frac{1}{2}$
  e $\frac{4}{5}$ **f** $\frac{3}{4}$ **g** $1\frac{2}{3}$ **h** $1\frac{1}{2}$

2 a $\frac{1}{3}$ **b** $\frac{1}{5}$ **c** $\frac{1}{7}$ **d** $\frac{1}{2}$
  e $\frac{2}{5}$ **f** $\frac{1}{2}$ **g** $\frac{2}{3}$ **h** $\frac{1}{3}$

3 a $\frac{5}{7}$ **b** $\frac{7}{11}$ **c** $\frac{4}{5}$ **d** $\frac{10}{13}$
  e $\frac{5}{11}$ **f** $\frac{2}{9}$ **g** $\frac{1}{7}$ **h** $\frac{2}{5}$

4 a $\frac{1}{2}$ **b** $\frac{1}{2}$ **c** $\frac{2}{5}$ **d** $\frac{3}{5}$
  e $\frac{1}{3}$ **f** $\frac{1}{5}$ **g** $\frac{2}{3}$ **h** $\frac{1}{6}$

5 a $\frac{2}{3}$ **b** $1\frac{2}{3}$ **c** $\frac{3}{5}$ **d** $\frac{1}{3}$
  e $1\frac{2}{15}$ **f** $1\frac{1}{3}$ **g** $\frac{1}{2}$ **h** $1\frac{3}{7}$

6 a $6\frac{5}{7}$ **b** $\frac{5}{6}$ **c** $4\frac{7}{10}$ **d** $2\frac{7}{9}$
  e $\frac{13}{15}$ **f** $1\frac{5}{9}$ **g** $5\frac{7}{12}$ **h** $4\frac{1}{3}$

7 a $\frac{1}{8}$ **8** $1\frac{2}{3}$ km   **9** $1\frac{2}{5}$ kg

Exercise 9B

1 a 12   **b** 30   **c** 15   **d** 6
  e 20   **f** 4   **g** 18   **h** 12

2 a $\frac{11}{12}$ **b** $\frac{17}{30}$ **c** $\frac{11}{15}$ **d** $\frac{5}{6}$
  e $\frac{9}{20}$ **f** $\frac{3}{4}$ **g** $\frac{17}{18}$ **h** $\frac{5}{12}$

3 a $\frac{1}{12}$ **b** $\frac{7}{30}$ **c** $\frac{1}{15}$ **d** $\frac{1}{6}$
  e $\frac{3}{20}$ **f** $\frac{1}{4}$ **g** $\frac{13}{18}$ **h** $\frac{1}{12}$

4 a $\frac{7}{12}$ **b** $\frac{1}{2}$ **c** $\frac{11}{20}$ **d** $\frac{23}{24}$
  e $\frac{17}{30}$ **f** $1\frac{17}{24}$ **g** $\frac{5}{6}$ **h** $1\frac{7}{12}$
  i $\frac{13}{24}$ **j** $\frac{1}{2}$ **k** $\frac{1}{20}$ **l** $\frac{13}{18}$
  m $\frac{1}{6}$ **n** $\frac{1}{24}$ **o** $\frac{1}{3}$ **p** $\frac{7}{12}$

5 a $\frac{29}{40}$ **b** $\frac{27}{28}$ **c** $\frac{33}{56}$ **d** $1\frac{7}{18}$
  e $\frac{43}{60}$ **f** $\frac{13}{24}$ **g** $\frac{17}{24}$ **h** $1\frac{1}{2}$
  i $\frac{19}{40}$ **j** $\frac{13}{28}$ **k** $\frac{1}{28}$ **l** $\frac{7}{18}$
  m $\frac{19}{60}$ **n** $\frac{29}{56}$ **o** $\frac{11}{24}$ **p** $\frac{33}{56}$

6 a $\frac{7}{12}$ **b** 70

7 a $\frac{2}{15}$ **b** 120

8 a $\frac{37}{56}$ **b** $\frac{19}{56}$

9 a $\frac{5}{6}$ **b** $\frac{1}{6}$

10 $\frac{49}{60}$ litres

Challenge

1 a i $\frac{1}{4} + \frac{1}{8}$ **ii** $\frac{1}{2} + \frac{1}{4}$
  iii $\frac{1}{2} + \frac{1}{12}$ **iv** $\frac{1}{3} + \frac{1}{3}$ or $\frac{1}{2} + \frac{1}{6}$
  b i $\frac{1}{2} + \frac{1}{4} + \frac{1}{8}$ **ii** $\frac{1}{3} + \frac{1}{3} + \frac{1}{6}$
  iii $\frac{1}{4} + \frac{1}{4} + \frac{1}{8}$ **iv** $\frac{1}{2} + \frac{1}{3} + \frac{1}{8}$

2 a 1, 2, 3, 4, 5, 6, 8, 9, 10, 12, 15, 18, 20, 24, 30, 36, 40, 45, 60, 72, 90, 120, 180, 360
  b $\frac{1}{2} = \frac{180}{360}$, $\frac{1}{3} = \frac{120}{360}$, $\frac{1}{4} = \frac{90}{360}$
  $\frac{1}{5} = \frac{72}{360}$, $\frac{1}{6} = \frac{60}{360}$, $\frac{1}{8} = \frac{45}{360}$
  $\frac{1}{9} = \frac{40}{360}$, $\frac{1}{10} = \frac{36}{360}$, $\frac{1}{12} = \frac{30}{360}$

Answers

c i $\frac{5}{6}$ ii $\frac{2}{3}$
iii $\frac{7}{10}$ iv $\frac{7}{12}$
v $\frac{7}{24}$ vi $\frac{2}{15}$
vii $\frac{1}{12}$ viii $\frac{1}{40}$
ix $\frac{47}{60}$ x 0

Challenge

a $1\frac{1}{5}$ kg in each bag   b $1\frac{1}{2}$ bars each

c $\frac{3}{4}$ metre long

Chapter 10

Exercise 10A

1 a 6   b 18   c 8   d 32
2 a 23   b 36   c 23   d 52
3 a $\frac{5}{4}$ b $\frac{5}{2}$ c $\frac{19}{6}$ d $\frac{30}{7}$
   e $\frac{41}{8}$ f $\frac{13}{5}$ g $\frac{16}{9}$ h $\frac{15}{4}$
   i $\frac{22}{5}$ j $\frac{25}{11}$ k $\frac{37}{8}$ l $\frac{29}{9}$
   m $\frac{7}{4}$ n $\frac{9}{5}$ o $\frac{8}{3}$ p $\frac{11}{4}$
   q $\frac{19}{6}$ r $\frac{11}{3}$
4 $\frac{5}{2} = 2\frac{1}{2}$, $\frac{7}{4} = 1\frac{3}{4}$, $\frac{7}{5} = 1\frac{2}{5}$, $\frac{11}{5} = 2\frac{1}{5}$

Exercise 10B

1 a $1\frac{1}{4}$ b $3\frac{1}{2}$ c $1\frac{5}{6}$
   d $4\frac{1}{2}$ e $1\frac{3}{8}$ f $1\frac{3}{7}$
   g $1\frac{2}{3}$ h $3\frac{1}{5}$
2 a $2\frac{1}{4}$ b $2\frac{1}{3}$ c $1\frac{4}{7}$ d $2\frac{2}{3}$
   e $1\frac{1}{7}$ f $1\frac{5}{8}$ g $3\frac{1}{3}$ h $4\frac{2}{3}$
   i $2\frac{4}{5}$ j $2\frac{5}{6}$ k $3\frac{2}{5}$ l $4\frac{1}{5}$
3 a $2\frac{1}{3}$ b $2\frac{2}{7}$ c $2\frac{2}{5}$ d $4\frac{1}{2}$
   e $2\frac{6}{7}$ f $4\frac{4}{5}$ g $4\frac{1}{3}$ h $2\frac{3}{8}$
   i $12\frac{1}{6}$ j $7\frac{4}{5}$ k $4\frac{1}{3}$ l $11\frac{1}{3}$
4 a $1\frac{1}{6}$ b $1\frac{2}{3}$ c $1\frac{1}{7}$ d $1\frac{3}{4}$
   e $1\frac{2}{5}$ f $1\frac{2}{5}$ g $1\frac{1}{6}$ h $2\frac{1}{5}$
   i $1\frac{5}{6}$ j $3\frac{1}{5}$ k $1\frac{1}{2}$ l $2\frac{1}{7}$

Chapter 11

Exercise 11A

1 a 60%, $\frac{3}{5}$,   0·6
   b 40%, $\frac{2}{5}$,   0·4
   c 80%, $\frac{4}{5}$,   0·8
   d 25%, $\frac{1}{4}$,   0·25
2 a $\frac{2}{5}$ b $\frac{1}{3}$ c $\frac{3}{10}$ d $\frac{2}{5}$
3 a $\frac{1}{5}$,   20%
   b $\frac{4}{5}$,   80%
   c

Number of black squares	Number of white squares
1	4
2	8
3	12
5	20
8	32
10	40

4 a 60 cl   b $\frac{1}{4}$
   c

Blackcurrant (cl)	Water (cl)	Total volume (cl)	Blackcurrant proportion
20	50	70	$\frac{2}{7}$
10	40	50	$\frac{1}{5}$
25	50	75	$\frac{1}{3}$
15	60	75	$\frac{1}{5}$

5 a £0·60   b £7·20   c £18·00
6 a £0·83   b £6·64   c £498
7 a 3 eggs, 9 ounces of plain flour, 15 fluid ounces of milk
   b $1\frac{1}{2}$ eggs, $4\frac{1}{2}$ ounces of plain flour, $7\frac{1}{2}$ fluid ounces of milk.
8 a 24% or $\frac{6}{25}$ b 76% or $\frac{19}{25}$

Answers

9 a £2·40 b 48 minutes c 70 seconds
  d 5 seconds

10 a Large shampoo (30 cl): 11·5 cl/£; small shampoo (20 cl): 13·3 cl/£. So, the small shampoo (20 cl) is the better buy.
   b Large roll: 0·33 metres/p; small roll: 0·27 metres/p. So, the large roll is the better buy.
   c Standard pad (120 pages): 109 pages/£; thicker pad (150 pages): 100 pages/£. So, the standard pad is the better buy.
   d Small tin: 7·35 grams/p; large tin: 6·49 grams/p. So, the small tin is the better buy.

Exercise 11B

1 a 1:2 b 2:1
2 a 6:4 b 4:6 c 8:2 d 2·5:7·5
3 a 4:6 b 4:8 c 6:14 d 10:25
4 8:7
5 Ratio is 1:4, so proportion is 1 in 5 (or $\frac{1}{5}$, not $\frac{1}{4}$) or 20%.

Exercise 11C

1 a 1:2 b 2:1 c 1:5 d 4:3
  e 1:4 f 4:5 g 3:2 h 2:3
  i 1:7 j 7:2 k 3:5 l 5:6
2 a 3:7 b 1:3 c 4:1
3 a White 40%; blue 20%; green 15%; yellow 25%.
  b i 2:1 ii 4:3 iii 3:5 iv 8:4:3:5
4 a 1:2 b 2:5 c 2:1 d 3:5
  e 3:2
5 a 2:5 b 1:10 c 4:5 d 5:1
  e 1:4 f 12:5

Exercise 11D

1 a

First element	Second element
5	3
15	9

b

First element	Second element
5	2
30	12

c

First element	Second element	Third element
4	3	1
20	15	5

2 a 20 cl b 18 cl
3 a i 20 ii 50 iii 60 iv 85
    v 200
  b i 6 ii 15 iii 21 iv 33
    v 180
4 250 grams

5 a 10 bags of sand and 2 bags of cement
  b 30 bags of sand and 18 bags of gravel

Exercise 11E

1 a £40, £60 b £10, £90 c £70, £30
  d £25, £75 e £55, £45
2 210 girls, 140 boys
3 50 pop CDs, 70 dance CDs
4 £57·50 (35 50p coins, 40 £1 coins)
5 20 ordinary calculators, 4 scientific calculators
6 Mental test 15 marks; non-calculator part 45 marks; calculator part 60 marks
7 a 15 cups strawberries, 5 cups sugar b 4 cups
  c $7\frac{1}{2}$ cups of strawberries
8 a 3 b 1:6
9 a 21 b 3:5

Challenge

a Jack gets £40, Jill gets £60.
b Jack gets £41·67, Jill gets £58·33.
c

Year	Jack	Jill
1	£42·86	£57·14
2	£43·75	£56·25
3	£44·44	£55·56
4	£45·00	£55·00
5	£45·45	£54·55
6	£45·83	£54·17
7	£46·15	£53·85
8	£46·43	£53·57

d Jack gets £441·58, Jill gets £558·42.

Chapter 12

Exercise 12A

1 Leisureways are cheaper at £62·99 (All sports cost £63·75).
2 Ordinary bag: 1·07 grams/p; large bag: 1·11 grams/p; jumbo bag: 1·09 grams/p. So, the large bag is the best value.
3 12-packet box: 4 packets/£; 18-packet box: 3·6 packets/£; 30-packet box: 3·75 packets/£. So, the 12-packet box is the best valve.
4 First card: 0·67 songs/£; Second card: 0·6 songs/£. So the first card is the better value.
5 a East Elect £45; North Energy £50. He should choose East Elect.
  b £40·50
  c £486
6 VAT = 20%
  Bookworm £318·75; Shelve Co £330; Leafy Wood £288·75. She should choose Leafy Wood.

Answers

7 a i Tent Hire £80; Party hut £104
  ii Tent Hire £110; Party hut £126
  iii Tent Hire £230; Party hut £214
  b Tent Hire at £140 (Party hut costs £148)

8 Pupil's own answer, reflecting own use of tests, calls and web access.

9 a Pupil's own answer
  b i Zoomweb (unlimited downloads)
    ii Blue (reasonable speed, only £8/month) or Zoomweb (cheap)
    iii Either Blue or Zoomweb
    iv Speedynet (fast)
  c

Company	Advantages	Disadvantage
Blue	Good speed Reasonable cost	Only 2GB download
Speedynet	Fast Good download Size	Expensive
Zoomweb	Unlimited downloads Cheap	Slow speed

10 a Jim's gym £91 (Strong-Arm would cost £93)
  b Jim's gym   **c** 34

11 a Local Superstore: £64·80
  Online Foties: £63·00
  DIY at home: £40·49
  So, she should choose DIY at home.
  b

	Cost	Advantages	Disadvantages
Local superstore	£64·80	Can be quick	Expensive
Online Foties	£63·00	Convenient	Expensive Have to wait for delivery
DIY at home	£40·49	Cheapest Quick	Less convenient Quality may not be as good

Challenge Pupil's own answer

Exercise 12B

1 CO-OP is the better deal at £140 extra (TS Bank is £150 extra)

2 Bank is the better deal at £1700 extra (Credit Union is £1840 extra).

3 Pupil's own answer, reflecting relative importance of time and cost. Answer should mention the following:
At £20/wk, it will take Paula 8 weeks to save the amount she needs paying £14/wk for 12 weeks will cost £168. Paula can choose between buying the bike now for £168 or waiting 8 weeks and paying £150 (assuming the bike is still available). Another factor is how much she can afford each week (£14/wk compared with £20/wk).

4 Pupil's own answer but should refer to actual cost of each loan, the duration of the loan, flexibility of terms, etc. Offer 3 is probably the best because it is only £1400 more than offer 1 but is a lower monthly repayment, there are no penalties for early payment or for moving home.

Offer 1: £750/month × 20 years = £180 000 (£24 000 more)
Offer 2: £900/month × 18 years = £194 400 (£38 400 more)
Offer 3: £575/month × 26 years = £179 400 plus set-up cost of £2000 = £181 400 (£25 400 more)

Offer	Advantages	Disadvantages
1	Cheapest; medium-sized monthly repayment	Cannot repay early; £3000 penalty for moving
2	No penalties for early payment or for moving	Most expensive (monthly repayment and total)
3	Lowest monthly repayment; no penalties for early payment or for moving	Set-up cost of £2000 Longest term

5 a *Advantages*: You know how long it will take to save the money.
Disadvantages: It will take 12 weeks to save enough money. Price might go up while you are saving (and you have to wait).
  b *Advantages*: You get the money sooner than in a, so you don't have to wait so long to get the game.
Disadvantages: £8 × 10 weeks = £80, so you are paying back £20 more than the cost of the game.
  c *Advantages*: You get the money even sooner than in a or b, so you can get the game immediately.
Disadvantages: £2 × 52 weeks = £104, so you are paying back £44 more than the cost of the game.
  d *Advantages*: You can spend your money on other things. You don't owe money to anyone.
Disadvantages: You don't get the game.

6 a £6274·75   **b** £279·75   **7** £398·50

Challenge

Pupil's own answer

Chapter 13

Exercise 13A

1 Pupil's own answer. Areas of expenditure the family might be able to cut back on, depending on their circumstances, include: clothes, travel, babysitter, leisure and phones. Weekly costs of £50 for babysitting and £80 for leisure seem high; if they could halve these (by going out less and choosing less expensive leisure activities), and if they could reduce phone costs to £30 per week (by choosing cheaper contracts), they would save £75 per week, about £325 per month and £3900 per year.

2 a Income = £50, expenditure = £46
  b 50 weeks
  c, d, e Pupil's own answer

3 a Income = £146·55, expenditure = £104·15
  b £42·40
  c 5 months (or 19 weeks)

Answers

4 See table at bottom of page.

5 Pupil's own answer. Additional possible monthly expenses include: rent or mortgage; contracts for broadband or mobile phone; membership fees (e.g. gym, sports club, film club, etc.); charity donations; loan repayments (e.g. for car); credit card payments.

6 a Mr Laing lost £23·95

   b 1 Make direct debit to savings later so account doesn't go overdrawn (after 4th of the month).

   2 Don't use credit card.

   3 Get an account that doesn't have an account fee.

7 Pupil's own answer.

8 Pupil's own answer.

9 a Pupil's own answer. One possible answer is:

Expense	Choice
Flights	1 Scot Air £2100 for whole group
Lunch	2 Packed lunch £1·50pp
Evening meal	2 Sit down meal £5pp
Ski hire	3 Second-hand ski £16pp/week
Accommodation	1 2 star hotel £300pp
insurance	2 Winter sports £20pp
Bus hire	1 Air-con Special £100/day

These choices would create a holiday costing £474·84 per person. The highlights are luxury bus travel and a full sit-down meal at night. However, the hotel is basic and the skis are second-hand.

   b Pupil's own answer

Challenge Pupil's own answers

Exercise 13B

1 a €56·50   b €146·90   c €352·56

2 a £66·40   b £290·50   c £166·40
  d €112·00   e £627·30   f £1012·37

3 a €510   b €90   c £44·25

4 Money X give $702·00; Change 40 give $696·78; Cash give $701·00. So, Money X is the best value.

5 Money X give £237; Change 40 give £232·80; Cash give £238. So, Cash is the best value.

Note: answer to following questions may vary slightly depending on the sequence in which calculations are done.

6 a £3221·30   b ¥392 644·26

7 a £830   b $108·50

8

	Choice 1	Choice 2	Choice 3
Flights	€2394	€91·20	€34·20
Lunch	€6·84	€1·71	€3·42
Evening meal	€2·28	€5·70	€3·42
Ski hire	€2·28	€34·20	€18·24
Accommodation	€342	€399	€456
Insurance	€17·10	€22·80	€28·50
Bus hire	€114	€570	€11·40

9 a £189   b £180·25

Challenge Pupil's own answer

Chapter 14

Exercise 14A

1 a 5 mph   b 8 km/h   c 30 mph   d 30 m/s

2 60 mph   3 18 km/h   4 480 mph

Exercise 14B

1 a 35 km/h   b 25 m/s   c 24 mph
  d 80 mph   e 4 km/h   f 1 mph (or 24 miles/day)

2 65 mph   3 Duncan is faster at 10 m/s (Calum 8 m/s)

Exercise 13A

Payment type	Description	Advantages	Disadvantage
BACS	Electronic payment directly into account	Automatic payment; credited immediately	If computer systems fail then payment might not be made on time
Direct debit(DD)	Agreement set up by account holder authorising their bank to pay an amount specified by the payee into payee's account	Automatic, so account holder doesn't have to remember to pay	Amount and date is set by payee, so account may go overdrawn if there are insufficient funds
Fixed charge	Fixed charge by service provider	Predictable cost	Some services don't charge fixed charges
CHQ	Cheque: a written document authorising payment of specified amount from holder's account into payee's account	Account holder manages their account	Amount may take a few days to be credited; cheque can 'bounce' (won't be paid) if account has insufficient funds
Standing order (SO)	Agreement set up by account holder authorising their bank to take a fixed amount from their account and pay into another's account	Automatic, so account holder doesn't have to remember to pay; size and date of SO is set by account holder	Account may go overdrawn if there are insufficient funds
Cashpoint	Cash withdrawal at automated teller machine (ATM)	Availability and accessibility	If computer systems fail then funds cannot be accessed; holder may forget PIN; potential security issues
Cashback	Electronic withdrawal from account and given as cash when shopping	Don't have to go to bank or ATM to get cash	If computer systems fail then funds cannot be accessed; holder may forget PIN; potential security issues; can't see balance like at ATM

Answers

4 Less, 58·33 mph   **5** 24·3 cm per day

6 25 metres per hour   **7** 10·9 million miles per minute

Exercise 14C

1. a 10 km   b 10 km/h
2. a 4 miles   b 4 mph
3. a 12 km   b 12 km/h

Exercise 14D

1. a 80 miles   b 175 km   c 48 m   d 9 miles
2. 48 km   **3** 2250 miles
4. Distance column: 20 miles, 40 miles, 200 miles, 10 miles, 5 miles

Exercise 14E

1. Speed = 40 mph
   Time = 30 minutes
   = 0·5 hours
   $D = S \times T$
   = 40 × 0·5
   = 20 miles

2. Speed = 60 km/h
   Time = 15 minutes
   = 0·25 hours
   $D = S \times T$
   = 60 × 0·25
   = 15 km

3. Speed = 20 mph
   Time = 45 minutes
   = 0·75 hours
   $D = S \times T$
   = 20 × 0·75
   = 15 miles

4. Speed = 10 km/h
   Time = 1 h 30 minutes
   = 1·5 hours
   $D = S \times T$
   = 10 × 1·5
   = 15 km

Exercise 14F

1. 6 hours   **2** 6 hours   **3** 3 hours   **4** 50 seconds
5. a 5 hours   b 6 hours   c 13 hours   d 160 seconds
   e 9 hours   f 9 hours

Exercise 14G

1. a 20 mph   b 150 km   c 5 m/s   d 60 miles
   e 2 hours   f 15 seconds   g 210 miles   h 45 mph
   i 40 hours   j 100 mph
2. 4 mph   **3** 400 miles   **4** 3 hours

Exercise 14H

1. a, b, c, d (distance-time graphs)

2. a The car drove at a steady speed of 10 kilometres per hour. Then it stopped for 1 hour before driving at 10 kilometres per hour for the next 2 hours.
   b The car drove at a steady speed of 20 kilometres per hour for the first hour. Then it drove at a steady speed of 10 kilometres per hour for the next 2 hours. It stopped for the next hour.
   c The car drove at a steady speed of 10 kilometres per hour for the first hour. Then it drove at a steady speed of 5 kilometres per hour for the next 2 hours. After that it drove at a steady speed of 20 kilometres per hours for an hour.
   d The car drove at 30 kilometres per hour for the first hour. Then it stopped for an hour before driving at 10 kilometres per hour for the next hour. Then it stopped for an hour.

Exercise 14I

1. a 2 hours 45 minutes   b 3 hours 30 minutes
   c 9 hours   d 4 hours 45 minutes
2. a 12 hours   b 13 hours 30 minutes
   c 7 hours 45 minutes   d 6 hours 15 minutes
3. a 14 hours 30 minutes   b 20 hours 30 minutes
   c 16 hours 45 minutes   d 10 hours 30 minutes
4. a 8 hours 30 minutes   b 48 hours
   c 41 hours 15 minutes   d 51 hours 45 minutes
5. a Rowan 15 hours 15 minutes   Julia 29 hours 30 minutes
   b Rowan 6·2 miles per hour   Julia 3·3 miles per hour

Chapter 15

Exercise 15A

1. a i 12 cm   ii 8 cm²   b i 16 cm   ii 15 cm²
   c i 10 cm   ii 6 cm²   d i 14 cm   ii 6 cm²
   e i 12 cm   ii 9 cm²   f i 22 cm   ii 28 cm²
2. a i 16 cm   ii 12 cm²   b i 16 cm   ii 15 cm²
   c i 80 mm   ii 400 mm²   d i 26 m   ii 40 m²
   e i 54 m   ii 180 m²
3. a 15·6 cm²   b 128 mm²   c 37·8 cm²
4. First bedroom: 16·8 m perimeter;
   Second bedroom: 16·4 perimeter.
   So, the first bedroom has the greater perimeter.
5. a 18 m   b 20 m²
   c i £600   ii £519·80   d £80·20

Challenge

1. Yes. For example, a square with sides of length 4 or a rectangle with sides of length 3 and 6.
2. 15 m × 30 m × 15 m
3. a i m²   ii cm²   iii mm²
   iv km²   v km²   vi mm²
   b i 2400 mm²   ii 6 000 000 m²
   iii 40 cm²   iv 3·456 km²
   c i Russia, Canada, USA   ii Russia

Answers

Exercise 15B

1. a $24\,cm^2$ b $70\,cm^2$ c $12.5\,cm^2$
   d $10\,m^2$ e $6\,m^2$ f $28\,m^2$

2. a $10.5\,cm^2$ b $60\,cm^2$ c $270\,mm^2$ d $250\,mm^2$

3. a $1750\,mm^2$ b $300\,cm^2$ c $120\,mm^2$

4.

Triangle	Base	Height	Area
a	5 cm	4 cm	$10\,cm^2$
b	7 cm	2 cm	$7\,cm^2$
c	9 m	5 m	$22.5\,m^2$
d	12 mm	10 mm	$60\,mm^2$
e	7 m	8 m	$28\,m^2$

Challenge
a, b Pupil's own answer

Exercise 15C

1. a $36\,cm^2$ b $7.5\,m^2$ c $120\,cm^2$
   d $108\,mm^2$ e $150\,cm^2$ f $768\,mm^2$

2. a $80\,cm^2$ b $49\,m^2$ c $80\,cm^2$

3. a $88.2\,cm^2$ b $2350\,mm^2$ (or $23.5\,cm^2$)

4.

Parallelogram	Base	Height	Area
a	8 cm	4 cm	$32\,cm^2$
b	17 cm	12 cm	$204\,cm^2$
c	8 m	5 m	$40\,m^2$
d	15 mm	4 mm	$60\,mm^2$
e	3.5 m	8 m	$28\,m^2$

5. 4.5 cm

Challenge

1. 6 cm

2. a $38.5\,cm^2$ b $96\,cm^2$ c $720\,mm^2$

Exercise 15D

1. a $35\,cm^2$ b $56\,cm^2$ c $8\,m^2$
   d $35\,m^2$ e $160\,mm^2$

2. a $15\,cm^2$ b $66\,cm^2$ c $30\,m^2$
   d 4 cm   e 10 cm   f 10 m

3. A $9\,cm^2$ B $18\,cm^2$ C $6\,cm^2$
   D $18\,cm^2$ E $27\,cm^2$

4. $30\,m^2$

5. $a = 1$ $b = 15$ $h = 1$
   $a = 2$ $b = 14$ $h = 1$
   $a = 3$ $b = 13$ $h = 1$
   $a = 4$ $b = 12$ $h = 1$
   $a = 5$ $b = 11$ $h = 1$
   $a = 6$ $b = 10$ $h = 1$
   $a = 7$ $b = 9$ $h = 1$
   $a = 1$ $b = 7$ $h = 2$
   $a = 2$ $b = 6$ $h = 2$
   $a = 3$ $b = 5$ $h = 2$
   $a = 1$ $b = 3$ $h = 4$

Challenge

1. a $30\,cm^2$ b $135\,cm^2$ c $6.24\,m^2$

2. a

Shape	Number of dots on perimeter	Number of dots inside	Area (cm^2)
i	8	1	$4\,cm^2$
ii	12	3	$8\,cm^2$
iii	8	3	$6\,cm^2$
iv	4	2	$3\,cm^2$
v	9	4	$7.5\,cm^2$
vi	10	4	$8\,cm^2$
vii	11	3	$7.5\,cm^2$
viii	14	4	$10\,cm^2$

b $A = \dfrac{P}{2} + I - 1$ c Pupil's own answer

Task: Design a bedroom

1. a 20 m   b $24.75\,m^2$
   c $24.75\,m^2 \times £32.50/m^2 = £804.38$ but carpet would be bought by the square metre, so $25\,m^2 \times £32.50/m^2 = £812.50$

2. a $45\,m^2$ b 4   c £75.96

3. a 7   b Yes

4. Pupil's own answer   5 Pupil's own answer

Exercise 15E

1. a $320\,cm^3$ b $1680\,cm^3$ c $16\,cm^3$

2. a 9 l   b 1.8 l   c 0.56 l

3. a $24\,cm^3$ b $3.84\,cm^3$ c 3 cm
   d 3 mm   e 2 m

4. a $8\,cm^3$ b $125\,cm^3$ c $1728\,cm^3$

5. $6000\,m^3$ 6 96

7. a $36\,m^3$ b 36 000 l

8. a $9600\,cm^3$

9. a $1080\,cm^3$ b 18

10. 6.2 m

Challenge Pupil's own answers

Exercise 15F

1. a $40\,000\,cm^2$ b $70\,000\,cm^2$
   c $200\,000\,cm^2$ d $35\,000\,cm^2$
   e $8000\,cm^2$

2. a $200\,mm^2$ b $500\,mm^2$ c $850\,mm^2$
   d $3600\,mm^2$ e $40\,mm^2$

3. a $8\,cm^2$ b $25\,cm^2$ c $78.3\,cm^2$
   d $5.4\,cm^2$ e $0.6\,cm^2$

4. a $2\,m^2$ b $8.5\,cm^2$ c $27\,m^2$
   d $1.86\,m^2$ e $0.348\,m^2$

Answers

5 a 3000 mm³   **b** 10 000 mm³   **c** 6800 mm³
  d 300 mm³   **e** 480 mm³

6 a 5 m³   **b** 7·5 m³   **c** 12 m³
  d 0·065 m³   **e** 0·002 m³

7 a 8 litres   **b** 17 litres   **c** 0·5 litres
  d 3000 litres   **e** 7200 litres

8 a 8·5 cl   **b** 120 cl   **c** 84 ml
  d 4·5 l   **e** 2400 ml

9 160

10 a 10 800 m²   **b** 1·08 hectares

11 150 litres   **12** 6 days   **13** 500

Investigation: Units and accuracy

1 a mm² or cm²   **b** litres   **c** hectares
  d m³   **e** cm²   **f** cm³

2, 3 Pupil's own answers

Challenge

1 250

2 An acre is an imperial (non-metric) unit of area, usually used to describe land areas; 1 acre = 4047 square metres.

3 a 1296   **b** 46 656

Chapter 16

Exercise 16A

1 a i 20 cm   **ii** 16 cm²
  b i 40 cm   **ii** 76 cm²
  c i 24 cm   **ii** 19 cm²
  d i 32 cm   **ii** 48 cm²
  e i 40 cm   **ii** 36 cm²
  f i 60 cm   **ii** 88 cm²

2 a i 36 cm   **ii** 58 cm²
  b i 36 m   **ii** 32 m²
  c i 78 cm   **ii** 132 cm²
  d i 16 m   **ii** 9 m²

3 a 336 cm²   **b** 600 cm²   **c** 264 cm²

4 a 192 m²   **b** 32 m²   **c** 18 m²   **d** 142 m²

5 9 m²

6 a should be 10 × 4 + 4 × 5   **b** 60 cm²

7 a 9 × 1 = 9 cm²
  b 8 × 2 = 16 cm²
  c 7 × 3 = 21 cm²
  d 6 × 4 = 24 cm²
  e 5 × 5 = 25 cm²

Exercise 16B

1 a 6 m²   **b** 45 cm²   **c** 12 m²

2 2·4 m²

3 a 2 m²   **b** 4 m²

4 1480 mm²   **5** 172 cm²

Exercise 16C

1 92 cm³

2 a 16 m³   **b** 11 520 m³
  c 135 cm³   **d** 45 cm³

3 30 000 mm³

4 a i 45 cm³   **ii** 47·5 cm³   **iii** 30 cm³
  b Pupil's own answer

Challenge

630 000 cm³

Chapter 17

Pupil's own work, research and presentation.

Chapter 18

Exercise 18A

1 a i 1, 4, 7, 10, 13   **ii** 5, 8, 11, 14, 17
  b i 1, 3, 9, 27, 81   **ii** 5, 15, 45, 135, 405
  c i 1, 6, 11, 16, 21   **ii** 5, 10, 15, 20, 25
  d i 1, 10, 100, 1000, 10 000   **ii** 5, 50, 500, 5000, 50 000
  e i 1, 10, 19, 28, 37   **ii** 5, 14, 23, 32, 41
  f i 1, 5, 25, 125, 625   **ii** 5, 25, 125, 625, 3125
  g i 1, 8, 15, 22, 29   **ii** 5, 12, 19, 26, 33
  h i 1, 2, 4, 8, 16   **ii** 5, 10, 20, 40, 80
  i i 1, 12, 23, 34, 45   **ii** 5, 16, 27, 38, 49
  j i 1, 4, 16, 64, 256   **ii** 5, 20, 80, 320, 1280
  k i 1, 9, 17, 25, 33   **ii** 5, 13, 21, 29, 37
  l i 1, 106, 211, 316, 421   **ii** 5, 110, 215, 320, 425

2 a 8, 10; +2   **b** 12, 15; +3   **c** 1000, 10 000; ×10
  d 8, 16; ×2   **e** 250, 1250; ×5   **f** 21, 28; +7
  g 16, 19; +3   **h** 19, 24; +5   **i** 16, 20; +4
  j 36, 45; +9   **k** 48, 60; +12   **l** 54, 162; ×3

3 a 25, 20; −5   **b** 20, 17; −3   **c** 2, 1; ÷2
  d 1, 0·2; ÷5   **e** 16·5, 15·8; −0·7   **f** 0·01, 0·001; ÷10
  g −5, −8; −3   **h** $\frac{1}{9}, \frac{1}{27}$; ÷3

4 Some possible sequences are:
  a 1, 4, 16, 64; ×4   and   1, 4, 7, 10; +3
  b 3, 9, 27, 81; ×3   and   3, 9, 15, 21; +6
  c 2, 6, 10, 14; +4   and   2, 6, 18, 54; ×3
  d 3, 6, 9, 12; +3   and   3, 6, 12, 24; ×2
  e 4, 8, 12, 16; +4   and   4, 8, 16, 32; ×2
  f 5, 15, 25, 35; +10   and   5, 15, 45, 135; ×3

5 Possible values are:
  a 6, 8; +2   **b** 6, 9; +3   **c** 10, 15; +5
  d 2, 4; ×2   **e** 40, 20; ÷2   **f** 6, 18; ×3

Challenge Pupil's own answer

Answers

Exercise 18B

1. a 12, 102   b 21, 246   c 31, 346
   d 17, 152   e 17, 197   f 34, 394
   g 60, 510   h 46, 451   i 27, 297

2. a 7, 105   b 3, 248   c 7, 203   d 5, 446

3.

Team	1st	2nd	3rd	4th	5th	6th	7th	8th	50th
Sequence A	9	11	13	15	17	19	21	23	107
Sequence B	4	9	14	19	24	29	34	39	249
Sequence C	2	9	16	23	30	37	44	51	345
Sequence D	5	15	25	35	45	55	65	75	495
Sequence E	2	5	8	11	14	17	20	23	149
Sequence F	8	10	12	14	16	18	20	22	106

4. 201   5. 325   6. 694   7. 324

Challenge

1. 104; 599   2. 127

Chapter 18C

1. a i 18, 20   ii $2n$
   b i 55, 60   ii $5n$
   c i 16, 20, 80, 84 ii $4n$
   d i 21, 35, 63, 70 ii $7n$
   e i 40, 50, 500, 510 ii $10n$
   f i 6, 24, 600, 606 ii $6n$

2. a $8n$ b $11n$

3. a i 10, 11   ii $n + 2$
   b i 25, 26   ii $n + 5$
   c i 11, 14, 25, 26 ii $n + 10$
   d i 23, 25, 120, 121 ii $n + 20$

4. a $n + 7$ b $n + 14$

Exercise 18D

1. a i 3, 5, 7   ii 201
   b i 3, 7, 11   ii 399
   c i 2, 7, 12   ii 497
   d i 5, 8, 11   ii 302
   e i 9, 13, 17   ii 405
   f i 11, 21, 31   ii 1001
   g i 6, 13, 20   ii 699
   h i $2\frac{1}{2}, 3, 3\frac{1}{2}$ ii 52
   i i $\frac{1}{4}, \frac{3}{4}, 1\frac{1}{4}$ ii $49\frac{3}{4}$

2. a i $3n + 1$ ii 151
   b i $4n + 2$ ii 202
   c i $12n + 4$ ii 604
   d i $3n$ ii 150

3. a $6n - 3$ b $3n + 7$ c $6n + 1$ d $3n - 2$
   e $7n - 3$ f $2n + 3$ g $4n + 5$ h $8n - 3$
   i $10n + 1$ j $9n - 6$

4. a i $4n$ ii $3n + 1$ iii $7n + 1$
      iv 200   v 151   vi 351
   b i $4n - 4$ ii $4n$ iii $8n - 4$
      iv 196   v 200   vi 396

Challenge

1. a +3   b ×4   c +3, +4, ...   d +3, +5, +7, ...

2. a +1 more; 61, 68   b − 1 more; 69, 62
   c +2, +4, +6; 43, 57   d +4, +6, +8; 56, 72

3. a 1, 3, 9, 27, 81, 243   b 2, 4, 8, 16, 32, 64
   c 1, 0·1, 0·01, 0·001, 0·0001, 0·00001
   d 1, 0·5, 0·25, 0·125, 0·0625, 0·03125
   e 2, 0·8, 0·32, 0·128, 0·0512, 0·02048
   f 1, 0·3, 0·09, 0·027, 0·0081, 0·00243

Chapter 19

Exercise 19A

1. Pupil's own answer using appropriate substitution
2. Pupil's own answer using appropriate substitution
3. Pupil's own answer using appropriate substitution

4. a 75   b 225   c 101   d 441

5. For example: $9(2)^2 = 36$, $(3 \times 2)^2 = 36$.

6. a, b, e

7. a $m \times n = mn$ b $q - p = -p + q$
   c $a \div b = \dfrac{a}{b}$ d $6 + x = x + 6$
   e $3y = 3 \times y$

Challenge

1. $a = 2, b = 2; a = 3, b = 1·5$ 2. $a = 4, b = 2; a = 4.5, b = 1·5$

3. Yes

4. One will be even, the other odd; even × odd = even.

5. One will be a multiple of 3, giving a factor of 3; at least one other will be even giving a factor of 2. So, 3 × 2 = 6 is a factor.

Exercise 19B

1. a $2m$ b $3k$ c $4a$ d $3d$
   e $4q$ f $2t$ g $4n$ h $3g$
   i $3p$ j $4w$ k $5i$ l $4a$

2. a $3p$ b $4m$ c $3 \times k = 3k$
   d $5 \times h = 5h$ e $m+m+m+m+m = 5m$
   f $p+p+p+p+p = 5p$ g $g+g+g = 3g$
   h $n+n+n+n+n+n+n = 7 \times n$ i $y+y+y+y+y = 5 \times y$

3. a $3n$ b $5n$ c $7m$ d $8t$
   e ab f mn g $5p$ h $4q$
   i $\dfrac{m}{3}$ j $\dfrac{5}{n}$ k $7w$ l dk
   m t^2 n $\dfrac{8}{k}$ o $9m$ p g^2

4. a p b t c $q \times r$ d gk
   e a f p g ft h $p \times t$
   i n^2

Answers

5 a $4hp$ b $4st$ c $8mn$ d $25wx$
    e $9bc$ f $24bcd$ g $12afg$ h $60mpq$

6 a $2x$ b $4x$ c $5m$ d $12q$
    e $2m$ f $4p$ g q h n
    i $8m$ j $2k$ k $4p$ l $5t$
    m $\frac{3}{2}p$ n $\frac{8}{5}q$ o $\frac{3}{2}m$ p $\frac{7}{3}t$

7 a $5m$ b $3q$ c $5y$ d $15h$
    e $8r$ f $3a$ g $5t$ h $2bc$

Exercise 19C

1 a $5b$ b $7x$ c $7m$ d $6m$
    e $4d$ f $5g$ g $3k$ h $4t$

2 a $4g$ b $2x$ c $2h$ d $8q$
    e $7h$ f $5x$ g $6y$ h $7d$
    i $3x$ j $2m$ k $3k$ l $2n$

3 a i $3t, 5t, 9t; \ g, 7g, 8g$ ii $17t + 16g$
    b i $m, 4m, 10m; \ 3p, 7p, 9p$ ii $15m + 19p$
    c i $k, 4k; \ 3m, 7m, 8m, -7m; \ 5w, 7w$ ii $5k + 11m + 12w$
    d i $t, 3t, 4t; \ x^2, 3x^2, 5x^2$ ii $8t + 9x^2$
    e i $2y, 3y, 8y; \ y^2, 4y^2, 7y^2, -4y^2$ ii $13y + 8y^2$
    f i $3g, 7g, 10g; \ 3h, 9h; \ 3w, 4w, 7w, -4w$
      ii $20g + 12h + 10w$

4 a $5b + 5$ b $5x + 7$ c $6m + 2$ d $7k + 8$
    e $2x + 7$ f $3k + 4$ g $4p + 3$ h $d + 1$
    i $3m - 3$ j $4t - 4$ k $w - 8$ l $4g - 1$
    m $5t + k$ n $7x + 2y$ o $3g + 7k$ p $3h + 3w$
    q $2t - 2p$ r $n - 2t$ s $p + 2q$ t $2p$

5 a $5g + 7t$ b $6x + 4y$ c $3k + 5m$ d $3x + 4y$
    e $2m + 5p$ f $2n + 7t$ g $4k + 2g$ h 0
    i $2p + q$ j $6g - 4k$ k $-3x$ l $d - 5e$

6 a $6x^2$ b $8k^2$ c $7m^2$ d $4d^2$
    e g^2 f $2a^2$ g $7f^2$ h $2y^2$
    i $4t^2$ j $3h^2$ k 0 l $4m^2$

7 a $10x + 8$ b $5k + 6p$ c $4m + 11t$
    d $4k + 5t$ e $2m + 5p$ f $5d + 6w$
    g $6y + 8x$ h $5p + 7q$ i $2m + 5t$

8 a $11h$ b $5p$ c $6u$ d $-5b$
    e $5j$ f $-12r$ g $6k$ h $8y$
    i $10d$ j $7i$ k $3b$ l $-4b$
    m $9xy$ n $11p^2$ o $-5ab$ p 0
    q $-10fg$ r $-2x^2$

9 a $8h + 5g$ b $2g + 8m$ c $8f + 10d$
    d $11x + 5y$ e $6q + 2r$ f $2s + 4$
    g $3c + 3$ h $14b + 7$ i $14w - 7$
    j $6bf + 5g$ k $7d + 3d^2$ l $4st + 5t$
    m $2t - 3s$ n $2i - 2h$ o $4y - 9w$

10 a $13e + 9f$ b $6u + 7t$ c $4b + 3d$
     d $7a + 7c$ e $6f + 5g$ f $6i$
     g $p + 5q$ h $19j + 4k$ i $2t - 2u$
     j $5s - 4t$ k $q - 2p$ l $-6d - 4e$

11 a $3x^2 + 8x$ b $9ab + 10a$ c $7y^2 - 9y$
     d $11mn - n$ e $3t^2 - 12t$ f $-3q^2 - 2q$

12 a $8x + 7y; \quad 3x + 4y; \quad 4x + y$
     b $5p + 4t; \quad p + 2t; \quad 2p - 4t$
     c $n + 6c; \quad n - 2c; \quad 8c$
     d $2a + 3b; \quad 4b; \quad 2a + 2b$

Exercise 19D

1 a i 7 ii 8 iii 2
    b i 13 ii 9 iii 2
    c i 21 ii 9 iii -15
    d i 12 ii 20 iii -4
    e i 19 ii 16 iii -2
    f i 14 ii 9 iii -31
    g i 14 ii 26 iii 2
    h i 36 ii 40 iii 8
    i i 22 ii 50 iii -6
    j i 133 ii 177 iii 81

2 a 9 b 11 c 21 d 6

3 a 12 b 26 c 41 d 5

4 a 14 b 5 c 24 d 28

5 a -140 b -8 c 23 d -44

6 a i 29 ii -3 iii 93
    b i 168 ii 48 iii 96
    c i 36 ii 4 iii 52
    d i 95 ii 80 iii -10
    e i 42 ii 27 iii -93

Challenge

1 $-1 < n < 1$

2 $3 < n < 5$

3 $n > 1$ and $n < 0$

4 $5x, 10(x - 1), x^2 + 6, 3x + 4, 6x - 2$

Exercise 19E

1 a 56 b 9

2 a 612 units^2 b 180

3 a $900°$ b $1800°$

4 a 200 b 142

5 a 7 b $\frac{8}{5}$

6 a £35   b £55   c £45

7 a 30 cm^2 b 36 cm^2

8 a 72 cm^2 b 45 cm^2

9 a $113°$F   b $104°$F   c $149°$F   d $212°$F

10 a 112 b 180

11 a 45 b 200

12 a i 60 m^3 ii 94 m^2
     b i 27 cm^3 ii 54 cm^2; cube

Answers

Chapter 20A

Exercise 20A

1
a $x = 10$ b $x = 8$ c $x = 15$ d $x = 14$
e $x = 6$ f $x = 15$ g $x = 5$ h $x = 15$
i $x = 45$ j $x = 77$ k $x = 65$ l $x = 72$

2
a $x = 12$ b $x = 8$ c $x = 12$ d $x = 6$
e $m = 13$ f $m = 17$ g $k = 18$ h $p = 3$
i $k = 23$ j $k = 7$ k $m = 16$ l $x = 5$
m $x = 27$ n $n = 11$ o $m = 6$ p $x = 25$

3
a $x = 5$ b $x = 2$ c $x = 3$ d $x = 10$
e $x = 3$ f $x = 3$ g $x = 1$ h $x = 8$
i $x = 15$ j $x = 11$ k $x = 13$ l $x = 8$

4
a $x = 5$ b $m = 3$ c $x = 3$ d $t = 4$
e $x = 8$ f $m = 6$ g $m = 4$ h $x = 4$
i $m = 4$ j $k = 3$ k $x = 2$ l $t = 5$
m $x = 4$ n $y = 4$ o $x = 5$ p $m = 6$

5
a $x = 9$ b $x = 5$ c $x = 6$ d $x = 7$
e $m = 9$ f $m = 9$ g $m = 3$ h $m = 7$
i $k = 4$ j $k = 1$ k $k = 8$ l $k = 5$
m $x = 3$ n $x = 10$ o $x = 5$ p $x = 8$

6
a $x = \frac{1}{2}$ b $x = \frac{1}{5}$ c $x = \frac{5}{2}$ d $x = \frac{13}{4}$
e $x = \frac{12}{5}$ f $x = \frac{17}{2}$ g $x = \frac{11}{4}$ h $x = \frac{11}{10}$

7
a $x = 3$ b $w = 6$ c $g = 3$ d $x = 4$
e $j = 4$ f $x = 5$ g $x = 12$ h $x = 1$
i $m = 2$ j $n = 7$ k $x = 9$ l $x = 9$

Exercise 20B

1
a i $2x + 3$ ii $x = 4$
b i $3x + 1$ ii $x = 5$
c i $5t + 4$ ii $t = 6$
d i $4t - 3$ ii $t = 4$
e i $2y - 1$ ii $y = 7$
f i $7y + 5$ ii $y = 3$

2
a i $x \to \times 3 \to +4 \to 19$ ii $x = 5$
b i $x \to \times 2 \to +4 \to 16$ ii $x = 6$
c i $x \to \times 4 \to -1 \to 23$ ii $x = 6$
d i $x \to \times 5 \to -3 \to 27$ ii $x = 6$
e i $x \to \times 3 \to +1 \to 22$ ii $x = 7$
f i $x \to \times 6 \to -5 \to 7$ ii $x = 2$

3
a $x = 6$ b $x = 7$ c $x = 7$
d $x = 7$ e $x = 14$ f $x = 10$

4
a $x = 7$ b $x = 5$ c $x = 3$ d $y = 12$
e $z = 7$ f $x = 3$ g $b = 2$ h $r = 6$
i $x = 4$ j $p = 11$ k $x = 11$ l $x = 7$

Challenge

a i 33   ii 27   iii 21
b i 6   ii 8   iii 9

Exercise 20C

1
a $x = 5$ b $x = 4$ c $x = 30$ d $x = 7$
e $t = 60$ f $m = 3$ g $k = 18$ h $p = 36$

2
a $x = 4$ b $x = 4$ c $x = 10$ d $x = 4$
e $m = 5$ f $k = 3$ g $n = 2$ h $x = 10$
i $h = 3$ j $t = 7$ k $x = 4$ l $y = 5$
m $x = 1$ n $t = 8$ o $x = 4$ p $m = 2$

3
a $x = 7$ b $m = 7$ c $x = 4$ d $t = 9$
e $x = 11$ f $m = 8$ g $m = 7$ h $x = 5$
i $m = 7$ j $k = 10$ k $x = 4$ l $t = 12$
m $x = 9$ n $m = 5$ o $x = 8$ p $m = 10$

4
a $x = 2$ b $x = 7$ c $x = 10$ d $m = 5$
e $m = 11$ f $m = 7$ g $m = 7$ h $k = 4$
i $y = 10$ j $k = 9$ k $k = 8$ l $x = 6$
m $t = 4$ n $x = 7$ o $x = 7$ p $y = 4$

5
a $x = 2$ b $x = 16$ c $x = 2$
d $x = 25$ e $x = 11$ f $x = 24$
g $x = 4$ h $x = 40$ i $x = 7$

6
a $x = 3$ b $x = 15$ c $x = 4$
d $x = 16$ e $x = 9$ f $x = 12$
g $x = 2$ h $x = 24$ i $x = 4$

7
a $x = 7$ b $x = 6$ c $x = 9$
d $x = 80$ e $x = 10$ f $x = 125$
g $m = 4$ h $b = 8$ i $q = 5$

8
a $3x = 15$; $x = 5$ b $\frac{x}{5} = 5$; $x = 25$
c $\frac{x}{2} = 32$; $x = 64$ d $2x = 22$; $x = 11$

Exercise 20D

1
a $x = 3$ b $x = \frac{5}{2}$ c $x = 7$ d $x = \frac{9}{2}$
e $x = 2$ f $x = 12$ g $x = \frac{3}{2}$ h $x = \frac{6}{5}$
i $x = \frac{11}{2}$ j $x = 5$ k $x = 5$ l $x = 6$

2
a $x = \frac{7}{2}$ b $s = \frac{13}{2}$ c $x = \frac{3}{2}$ d $q = 3.5$
e $p = 7.5$ f $m = 2$ g $n = 7$ h $x = 9$
i $b = 2$ j $t = 9$ k $k = 3$ l $k = 24$
m $n = 14$ n $n = 7$ o $h = 19$

Exercise 20E

1 a $2x + 1 = 33$ b 16

2 a $2x + 5 = x + 12$ b 7

3 a $2x + 10 = 56$ b Jeff 23, Tom 33

4 a $2x + 35 = 89$
b Surjit has 27 CDs, Sanjay has 62 CDs.

5 a $2x + 13 = 129$
b Michelle has 58 DVDs, Gavin has 71 DVDs.

6 a $3x + 5 = 23$; $x = 6$ b $3x + 5 = 38$; $x = 11$

7 a $4x = 52$ b Angus is 13, Paula is 39

8 a $3x = 36$ b Heather scored 12, Moira scored 24

Answers

9 a $5x = 120$ b 96 minutes or 1 h 36 min
10 a $2x + 55 = 207$ b 76 cows, 131 sheep
11 a $2x + 29 = 845$ b 408 girls, 437 boys
12 a $2x + 410 = 528$ b 59

Challenge

1 38, 39   2 4556   3 2405970
4 a 5   b 6   c 7   d 3
   e 5   f 2   g 12   h 6
   i 3   j 9   k 5   l 4
   m 8   n 3   o 3   p 5
   q 8   r 9   s 4

Chapter 21

Exercise 21A

1 a £43   b £27   c £67
2 a £20   b £44   c £92
3 a £55   b £85   c £145
4 a 50 mph   b 40 mph   c 40 mph   d 70 mph
5 a 8·75 ml   b 8 ml   c 6 ml   d 9 ml
6 a £250   b £100   c £210   d £235
7 a 60p   b £2·40   c £3·75
8 a 540°   b 720°
9 a 9   b 43
10 a 26 cm   b 62 cm
11 a 32 m/s   b 58 m/s

Exercise 21B

1 a 20   b 9   c 12
2 a 2   b 5   c 3
3 a 50 m/s²   b 55 m/s²
4 a 3   b 100

Exercise 21C

1 $C = 3m + 5$
2 $S = 5C - 3$
3 $C = 20h + 15$
4 a $y = 3x + 4$ b $y = 4x - 1$ c $y = 2x - 1$
   d $y = 3x + 2$ e $y = 0.5x - 0.2$

Exercise 21D

1 a $C = 2h$ b $d = 300l$
   c $D = J + 40$ d $c = 8p + 50$
   e $j = \dfrac{P}{3} + 5$
2 a $S = a + b + c$ b $P = xy$ c $D = a - b$
   d $D = 7W$ e $A = \dfrac{x + y + z}{3}$
3 a 7   b 21   c $7w$

4 a i 14   ii 18   iii $13 + t$
   b i 12   ii 10   iii $13 - m$
5 a 30 miles   b 60 miles   c $30t$ miles
6 a 1000   b 5000   c $1000x$
7 $60m$
8 a 10   b 3   c $\dfrac{b}{2}$
9 a 8   b 14   c $2T$
10 a bk miles
11 a 210 pence   b $35k$ pence   c kq pence
12 a $6b$ b $7b + 2y$

Challenge

2 All even   3 b $A = \dfrac{D}{2} - 1$ 5 $A = \dfrac{D}{2}$
6 b $A = \dfrac{D}{2} + 1$ 7 b $A = \dfrac{D}{2} + 2$

Chapter 22

Exercise 22A

1 b $\angle P = 73°$, $\angle R = 50°$ c $PR = 87$ mm
11 b The sum of the two shorter sides must be greater than the longest side, but 3 cm + 4 cm = 7 cm < 8 cm

Exercise 22B

1 b $\angle B = 132°$
   c $AB = 68$ mm; $BC = 85$ mm
3 b $\angle P = 89°$, $\angle Q = 121°$ c $PQ = 39$ mm

Exercise 22C

Pupil's own answers

Chapter 23

Exercise 23A

1 a $\angle JKL$ b $\angle HIG$ c $\angle EDF$
   d $\angle OMN$ e $\angle MNO$ f $\angle WVX$
   g $\angle ABC$ h $\angle HGI$ i $\angle PQR$
2 $\angle PQR$, $\angle QRP$, $\angle RPQ$
3 $\angle ABC$ (or $\angle CBA$), $\angle EBC$ (or $\angle CBE$), $\angle BCD$ (or $\angle DCB$), $\angle CDE$ (or $\angle EDC$), $\angle DEB$ (or $\angle BED$), $\angle DEF$ (or $\angle FED$)

Exercise 23B

1 a 28°   b 47°   c 31°   d 45°
2 a 60°   b 125°   c 90°   d 147°
3 a 130°   b 140°   c 102°   d 221°
4 a 57°   b 122°   c 21°
   d 46°   e 134°
5 a 51°   b 129°   c 116°   d 113°
6 a 60°   b 50°

Answers

Exercise 23C

1 a, c, e, g   2 Pupil's own answers   3 a, b, f, h

4 a YZ   b RS, PQ   c AC and BD
   d EF, GH; HE, GF

5 a 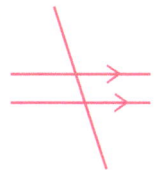   b

   c 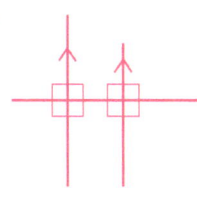   d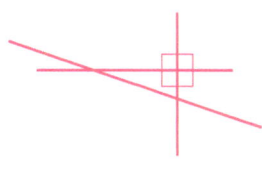

Challenge Pupil's own answers

Exercise 23D

1 a, c, e, f

2 a, b, e

3 a 50° alternate   b 62° corresponding
   c 108° alternate   d 52° corresponding
   e 94° alternate   f 151° corresponding

4 a 70°   b $b = 82°$, $c = 89°$
   c $d = 90°$, $e = 53°$
   d $f = 65°$, $g = 115°$, $h = 65°$, $i = 115°$
   e $j = 100°$, $k = 40°$, $l = 140°$, $m = 100°$
   f $n = 29°$, $o = 89°$, $p = 118°$, $q = 118°$

Challenge

1 $x = 48°$

2 $x = 70°$

3 $x = 115°$

Exercise 23F

1 a 55°   b 42°   c 38°
   d 42°   e 95°   f 30°

2 a 65°   b 35°   c 62°, $d = 56°$
   d $e = 45°$, $f = 90°$ e $g = 48°$, $h = 48°$

3 a 71°   b 24°   c 35°
   d $d = 66°$, $e = 48°$ e $f = 47·5°$, $g = 47·5°$
   f $h = 60°$, $i = 60°$, $j = 60°$

4 a $a = 25°$, $b = 155°$ b $c = 116°$, $d = 64°$
   c $e = 26°$, $f = 154°$ d $g = 27°$, $h = 153°$
   e $i = 73°$, $j = 107°$ f $k = 39°$, $l = 141°$
   g $m = 64°$, $n = 116°$ h $o = 58·5°$, $p = 121·5°$

5 a $a = 45°$ b $b = 30°$ c $c = 72°$ d $d = 21°$

Challenge

1 a 316°   b 315°   c 147°   d 24°
   e 46°   f 77°

2 42° and 96°, or 69° and 69°

Exercise 23F

1 a 109°   b 108°   c 68°   d 94°
   e 56°   f 145°

2 a 71°   b 135°   c 105°   d 61°

3 a 125°   b 73°   c 122°
   d 74°   e 154°

4 a $a = 58°$ (angles in a quadrilateral)
      $b = 122°$ (angles on a straight line)
   b $c = 72°$ (angles on a straight line)
      $d = 98°$ (angles in a quadrilateral)

5 a 92°   b 63°

6 a   b
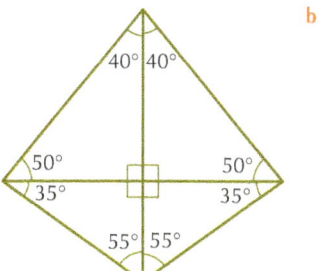

Challenge

1 a Two pairs of equal sides; two pairs of equal angles;
      opposite angles equal; opposite sides parallel.
   b Use the fact that opposite angles are equal and all of the
      interior angles total 360°.
   c ∠ABC = 63°, ∠BCD = 117°

2 a   b They are equal: $p = q$

   c $p = 110°$, $q = 110°$

3 a 110°   b None are equal; two pairs add up to 180°

4 The sum of angles in a triangle is 180°, so the sum of angles
   in two triangles is 360°.

331

Answers

Exercise 23G

1 b i 720° **ii** 900° **iii** 1080°
  c

Name of polygon	Number of sides	Number of triangles inside polygon	Sum of interior angles
triangle	3	1	180°
quadrilateral	4	2	360°
pentagon	5	3	540°
hexagon	6	4	720°
heptagon	7	5	900°
octagon	8	6	1080°
n-sided polygon	n	$n-2$	$180(n-2)°$

2 a i 5 sides **ii** 540° **iii** 105°
  b i 6 sides **ii** 720° **iii** 130°
  c i 5 sides **ii** 540° **iii** 281°

3 120°

4 a 120° **b** 135°

5 a 1800° **b** 150° **c** 30°

6 a 135° **b** 140° **c** 60°

Challenge

1 a 80° **b** 114° **c** 52°

Chapter 24

Exercise 24A

1 a 20 m **b** 50 m **c** 35 m
  d 78 m **e** 63 m

2 a 25 m **b** 15 m **c** 29 m

3 a 1 cm to 2 m **b** 5·8 m **c** 46·4 m²

4 a 16 cm **b** 6 cm **c** 2 cm
  d 3 m **e** 2·5 m **f** 1·2 m

5 a i 5·8 m × 3·9 m **ii** 3·9 m × 1·9 m
   iii 5·8 × 3·7 m **iv** 5 m × 3·7 m
  b 88 m²

6 Pupil's own answer

Challenge

1 Pupil's own answers **2** Pupil's own answers

3 a 1:100 **b** 1:400 **c** 1:25
  d 1:100 000 **e** 1:50 000

Exercise 24B

1 a 1:200 **b** 1:500 **c** 1:20
  d 1:50 000 **e** 1:250 000

2 a 45 km **b** 25 km **c** 12 km

3 2 km **4** 12·5 km **5** 5·5 cm

6 a 260 m **b** 650 m

7 11 km **8 a** 10·5 km **b** 12·75 km (2·25 km further)

Challenge

1 a 1:63 360 **b** 1:36
  c 1:10 560 **d** 1:1800

2 Pupil's own answers **3** Pupil's own answers

Exercise 24C

1 a 180° **b** 270° **c** 045° **d** 225°

2 a 064° **b** 018° **c** 097° **d** 300°

3 a 045° **b** 020° **c** 258° **d** 321°

4 a 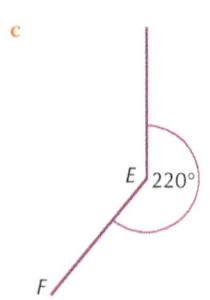 **b**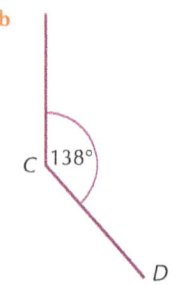

  c 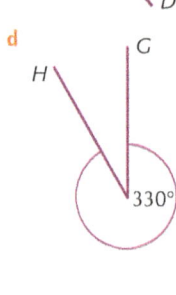 **d**

5 a i 254° **ii** 074°
  b i 222° **ii** 042°

6 a 10·9 km
  b i 080° **ii** 150° **iii** 283°

7 Pupil's own answer

8 Culloden

Challenge

1 a

  b i 164 nautical miles
    ii 027°

Answers

2 a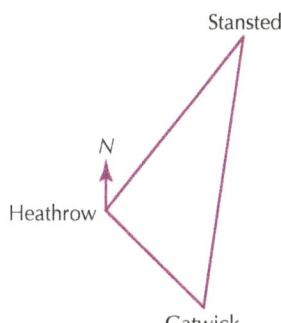

b 136°

Note: drawing not to scale

3 Pupil's scale drawing

Chapter 25

Exercise 25A

1 a 4   **b** $\frac{1}{2}$ **c** 1·5   **d** 0·8

2 a Rectangle B

  b Rectangle C; $\frac{20}{5} = 4$ but $\frac{32}{9} = 3\frac{5}{9}$

3 0·005 or $\frac{1}{200}$

4 0·1 or $\frac{1}{10}$

Challenge

$x = 2$

Exercise 25B

1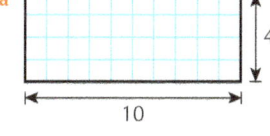

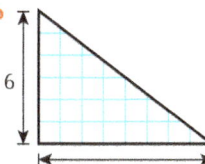

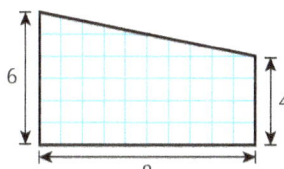

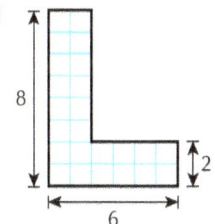

2 a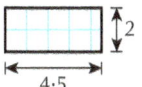

b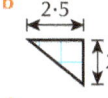

c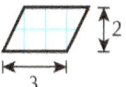

d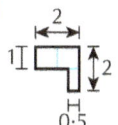

3 a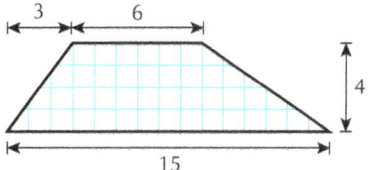

b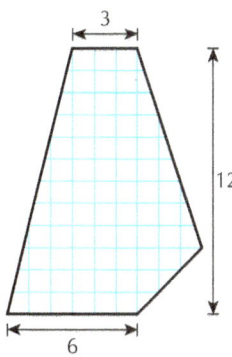

4 a 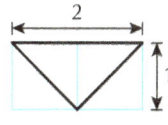   **b**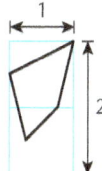

5 a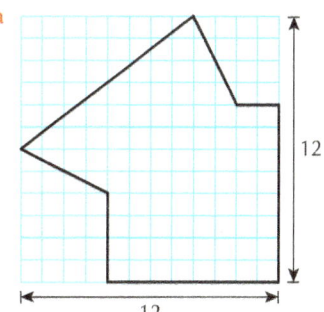

b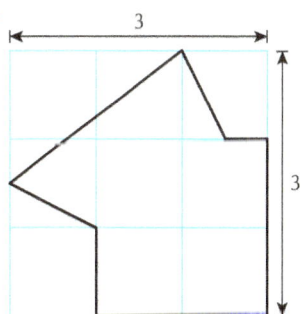

Challenge

a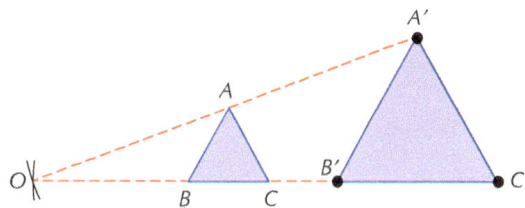

333

Answers

b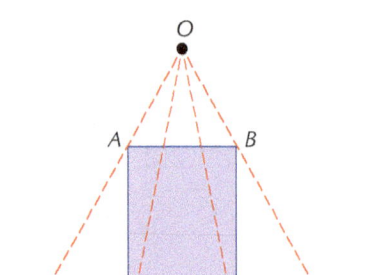

$OA' = 3OA$
$OB' = 3OB$
$OC' = 3OC$
$OD' = 3OD$

3 a $A\,(1, 4)$; $B\,(4, 5)$; $C\,(5, 2)$
  b $(2, 1)$

4 a $L\,(1, 3)$; $M\,(2, 1)$; $N\,(5, 3)$
  b $(6, 1)$ or $(4, 5)$

5 b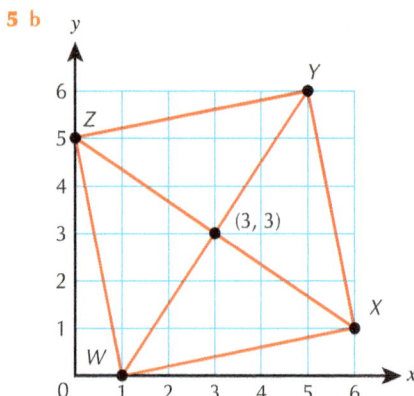

  c $t\,(0, 5)$
  d $(3, 3)$

6 a i $(4, 2)$ **ii** $(4, 6)$ or $(4, 7)$
     iii $(4, 0)$ or $(4, 1)$; $(4, 3)$; $(4, 4)$
  b isosceles triangle

7 a $(1, 1)$, $(4, 4)$
  b $(4, 7)$, $(7, 4)$

c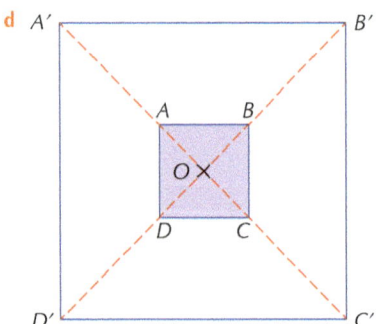

$OA' = 2OA$
$OB' = 2OB$
$OC' = 2OC$
$OD' = 2OD$

Exercise 26B

1 a 6 cm^2 **b** 10 cm^2 **c** 6 cm^2 **d** 12 cm^2

2 a 20 cm^2 **b** 15 cm^2 **c** 24 cm^2

3 a 24 cm^2 **b** 8 cm^2 **c** $17{\cdot}5\text{ cm}^2$

4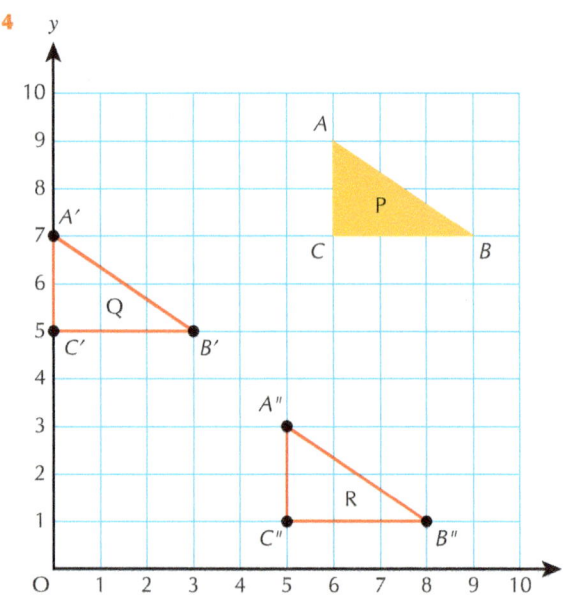

d

$OA' = 3OA$
$OB' = 3OB$
$OC' = 3OC$
$OD' = 3OD$

Exercise 25C

Pupil's own answers

Chapter 26

Exercise 26A

1 $E\,(1, 2)$; $F\,(2, 5)$; $G\,(5, 6)$;
   $H\,(0, 0)$; $I\,(0, 5)$; $J\,(4, 2)$

2 a rectangle   **b** kite
  c isosceles triangle   **d** square
  e pentagon   **f** rhombus

a $A\,(6, 9)$; $B\,(9, 7)$; $C\,(6, 7)$
b Triangle Q above
c $(0, 7)$, $(3, 5)$, $(0, 5)$
d Triangle R above
e $(5, 3)$, $(8, 1)$, $(5, 1)$
f 1 unit right, 6 units up

Answers

5 a A (2, 4);   B (1, 1);   C (4, 1)
  b P (2, 6);   Q (1, 9);   R (4, 9)
  c S (8, 6);   T (9, 9);   V (6, 9)
  d S (8, 4);   T (9, 1);   V (6, 1)
  e reflection in B

6 a Vertices (4, 2), (8, 2), (8, 6)

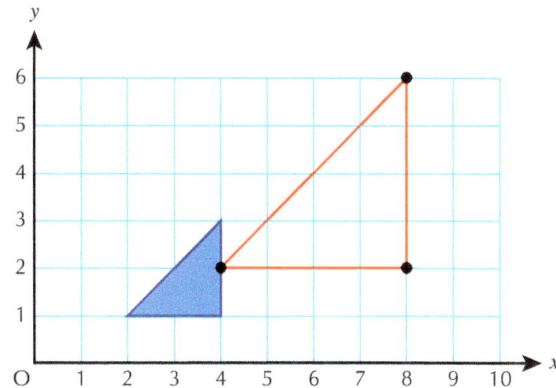

  b Vertices (0, 4), (4, 2), (8, 4) and (4, 6)

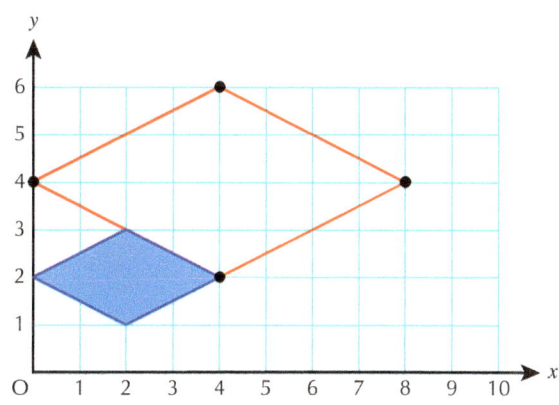

  c Vertices (3, 3), (8, 2), (12, 9), (9, 9), (9, 6), (6, 6), (6, 9), (3, 9)

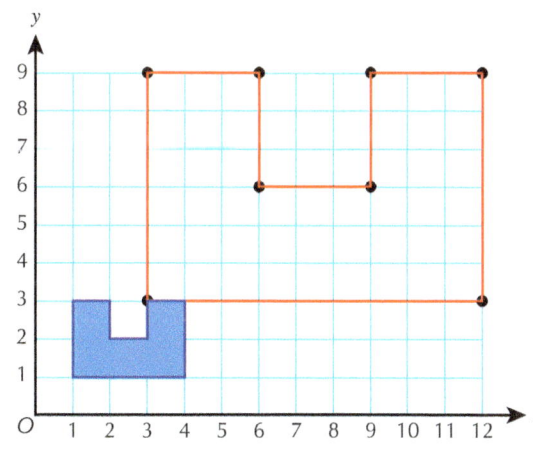

7

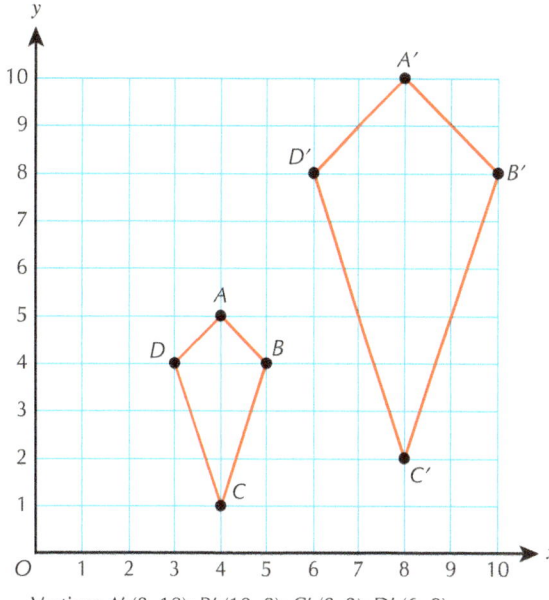

Vertices A' (8, 10); B' (10, 8); C' (8, 2); D' (6, 8)

8 a

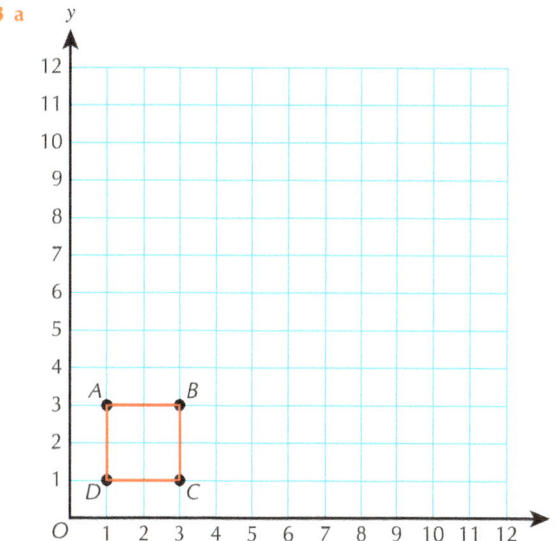

  b 4 cm²
  c Vertices (2, 6), (6, 6), (6, 2), (2, 2)
     Area 16 cm²
  d Vertices (5, 9), (9, 9), (9, 3), (3, 3)
     Area 36 cm²
  e Vertices (4, 12), (12, 12), (12, 4), (4, 4)
     Area 64 cm²
  f $A = f^2 \times 4$ where A = Area and
         f = Scale factor

Challenge Pupil's own answer

335

Answers

Chapter 27

Exercise 27A

1 a 1
  b 3
  c 4
  d 2
  e 0
  f 1

2 a 1   b 2   c 8   d 1
  e 5   f 4   g 6   h 3
3 a 1   b 0   c 2   d 0
  e 1   f 0
4 a 1   b 1   c 0   d 1
  e 1   f 1

Challenge Pupil's own answer

Exercise 27B

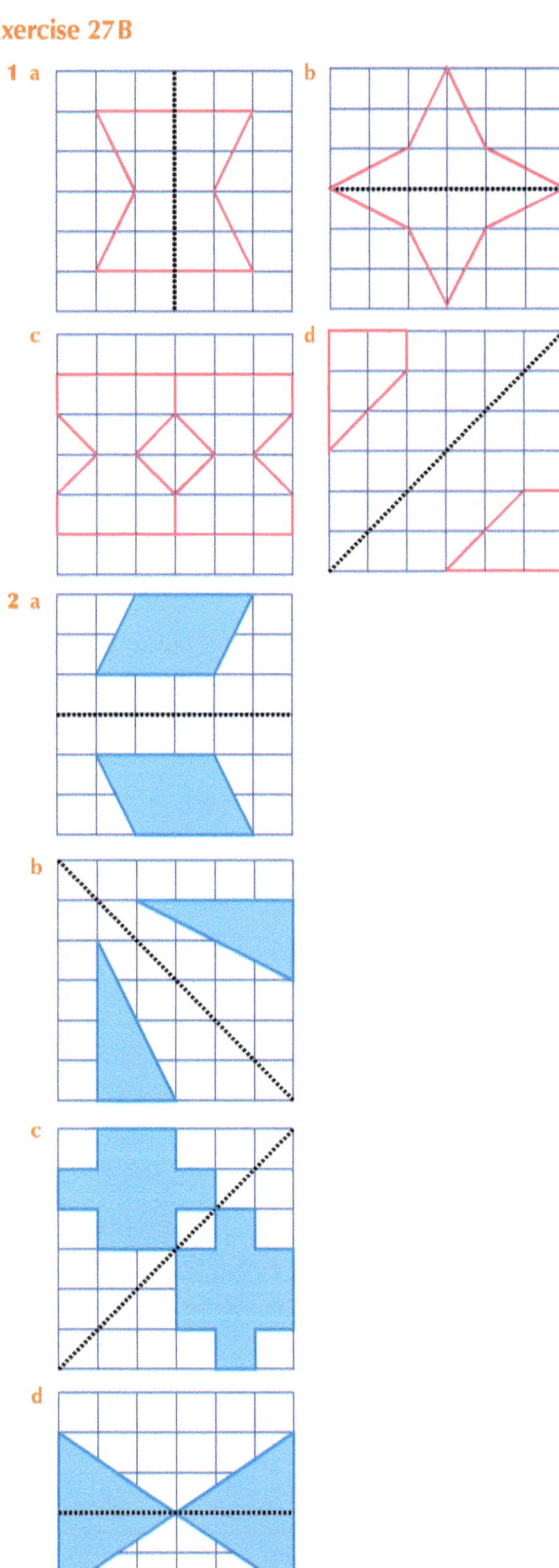

Answers

3 a
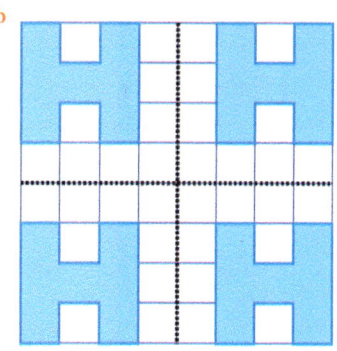

b

4 Pupil's own answer

Challenge

1 a
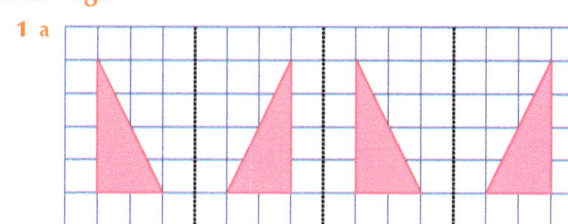

2 b

Exercise 27C

1 a
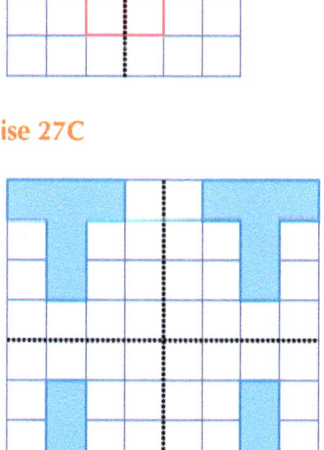

b
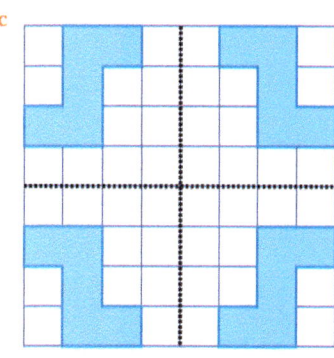

c
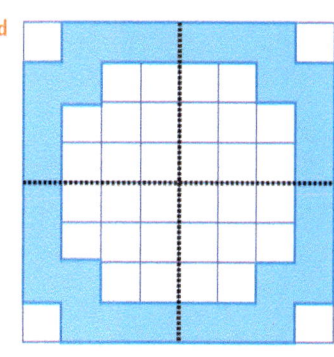

d

2 a
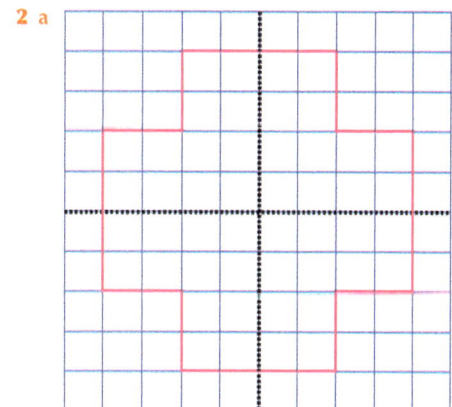

337

Answers

b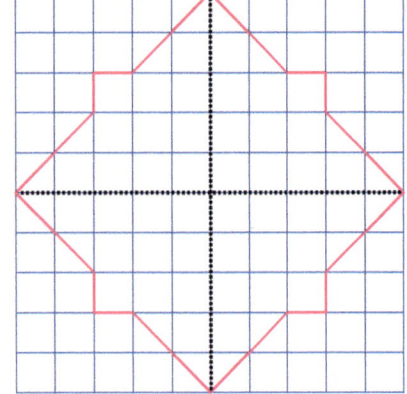

3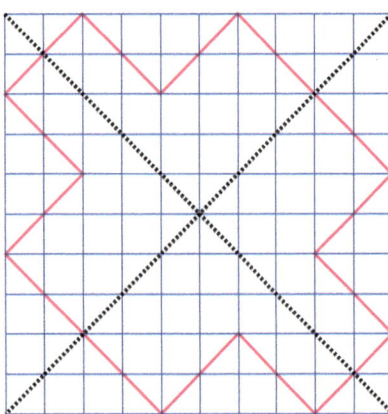

4 a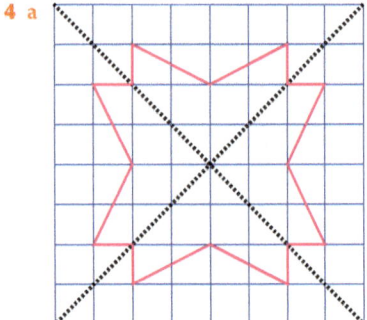

b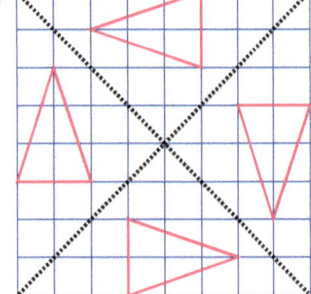

c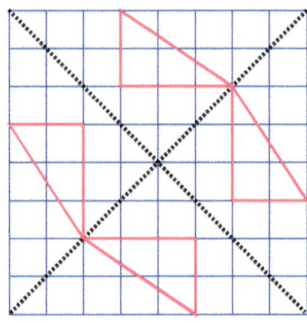

d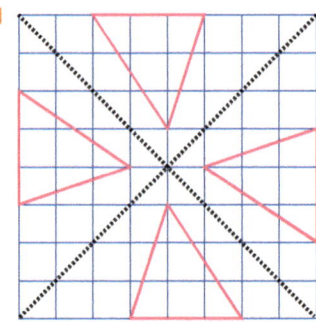

e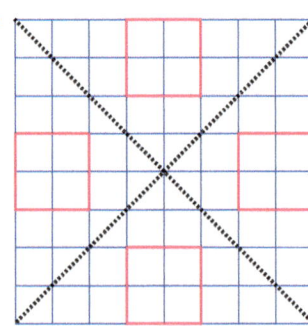

f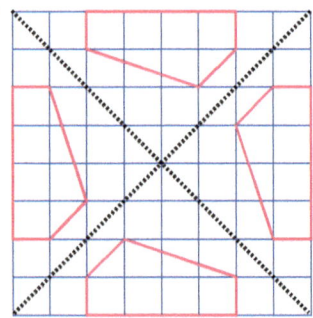

5 Pupil's own answer

Challenge Pupil's own answer

Answers

Chapter 28

Exercise 28A

1. **a** 331 km **b** 169 km
   c Newcastle **d** Newcastle and Aberdeen
   e 252 km

2. **a** Tuesday, Thursday and Saturday
   b Davinder and Teresa
   c Davinder
   d Tuesday and Thursday
   e Friday and Saturday
   f Davinder

3. **a** 3 **b** Maths, History, Literature
   c Gill **d** Gill or Eve

4. **a** 32 **b** 11

5. **a** 6 **b** 8 **c** 4 **d** Toyota

6. **a** Reha **b** Jake
   c Easier to see who won most
   d Tells who won what

7. **a** Fewer pupils have school lunches as they get older.
   b S1 and S2; 87 − 64 = 23; 104 − 87 = 17
   c 115

8. **a** Age 10: boys 4% more; age 15: girls 2% more.
   b At age 10, a lower percentage of girls have mobile phones than boys. By age 15, a higher percentage of girls than boys have phones.
   c Increases every year for boys and girls. Big jump from 11 to 12 for both.

9. 20 boys over 160 cm, compared to 16 girls, so claim is correct.

10. **a** There are more younger members under 40. Generally, the older the age group, the fewer the number of runners.
    b

Age	Male	Female
20–29	48%	40%
30–39	20%	30%
40–49	15%	20%
50+	17%	10%

   c

Age	Male
20–29	96
30–39	40
40–49	30
50+	34

11. **a** 22 **b** 7 **c** February

12. Protein blue; carbohydrate green; fat orange; fibre red; sodium yellow.

Challenge

a

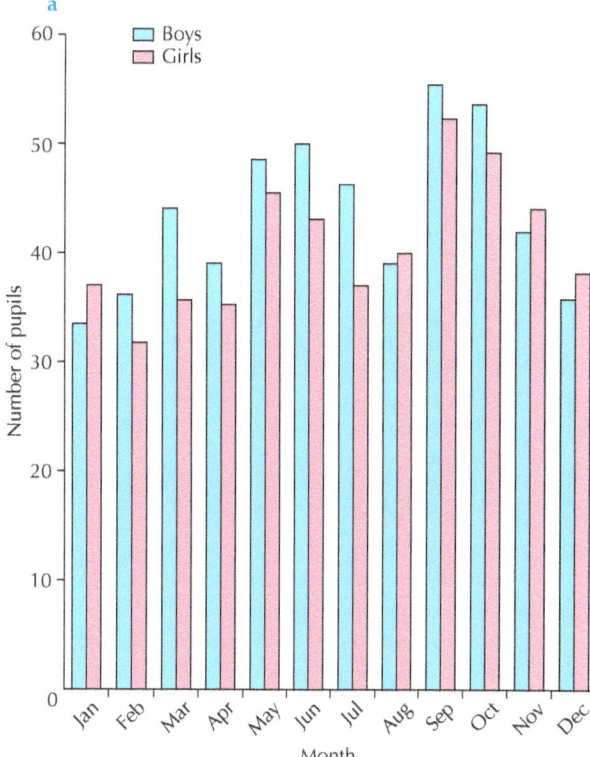

b The boys are slightly higher in June, July and August than in December, January and February. However, the differences are not very obvious.

Exercise 28B

1. **a** 17 **b** 5 **c** 40
   d No, he is not correct. To calculate the total distance thrown, take the middle value of each interval and multiply by the number of pupils who threw in that interval, and then add for all the intervals:
   $2 \times 0.5 + 5 \times 1.5 + 12 \times 2.5 + 16 \times 3.5 + 8 \times 4.5 + 3 \times 5.5 = 150.5$m

2. **a** 7 **b** 100 **c** 60% of crime is not theft.
   d For example:
   Theft was the most frequent crime; 7% of crime was drug related; violent crime made up 10% of crime.

3. **a** Farm A **b** Farm B
   c Out of 4 farms surveyed, 3 had a decreasing population.

4. **a** Ms Archibold is correct. The pie charts only show the relative proportion of the economy made up by manufacturing, and not the actual amounts.
   b Add a key to show the total size of the economy

5. Jools's graph is better at showing the annual increases.

6. Sales in 2010 were £3000. Sales in 2011 were £4000. This is a 33% increase. This is good, but probably not 'dramatic'. Without knowing what the 2009 sales were, you cannot tell the longer trend.

7. George needs to decide how to judge what 'most successful' means – is it most games won, most trophies won, etc. He should use official websites, and not use club or fan websites, which might be biased.

Challenge Pupil's own answer

Answers

Chapter 29

Exercise 29A

1 Pupil's own answers

Challenge Pupil's own answers

Exercise 29B

1 Pupil's own answers

Exercise 29C

1 a Carry out an experiment: Get a group of pupils to read the same number of words from two different newspapers. Time them. Find average time for each newspaper.
  b Carry out an experiment: Show someone various cards. Ask them to recall as many as possible. Note how many are correct.
  c Questionnaire: Ask a variety of people what they think about smoking. Give several options.

2 a Too time consuming
  b Too precise; use larger age intervals.
  c Biased; does not reflect population.
  d Only one variable should be changed. For example;
  - **test:** give boys or girls different test at the same time
  - **gender:** give boys and girls same test at the same time
  - **time:** give boys or girls same test at different times

3 Pupil's own answers

Challenge Pupil's own answers

Chapter 30

Exercise 30A

1 a

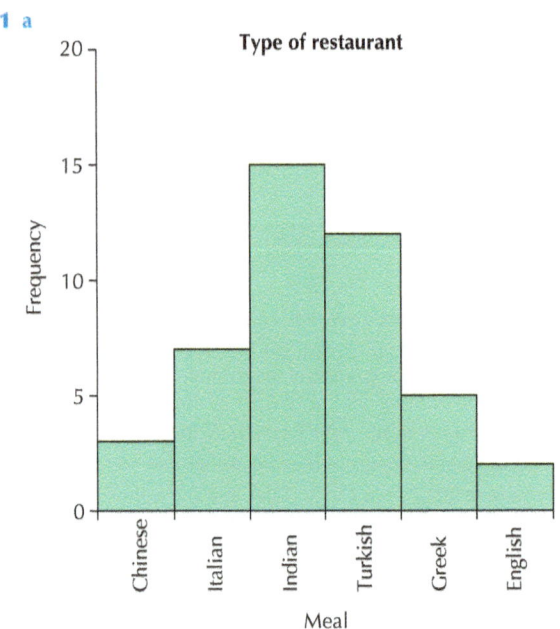

b

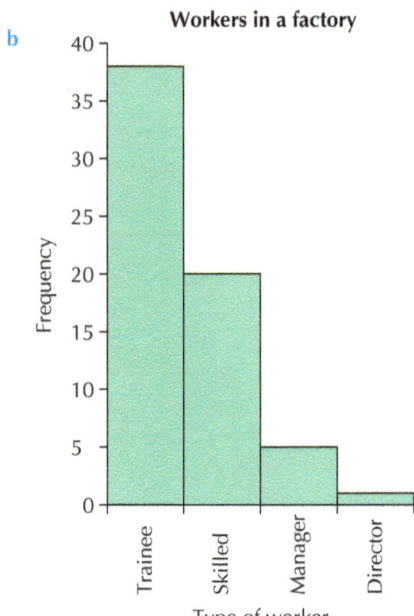

c

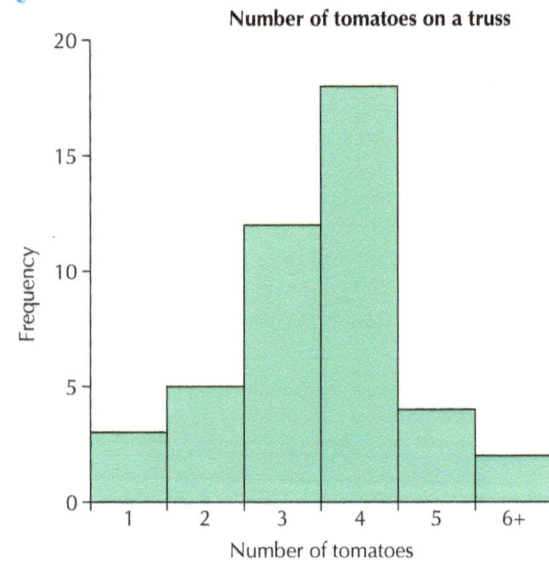

2 a

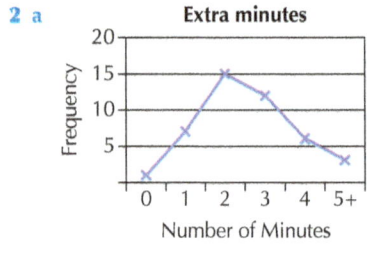

Answers

b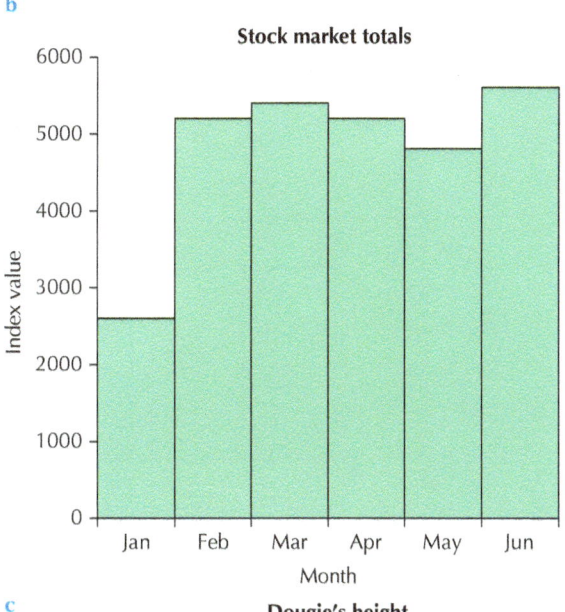

4 a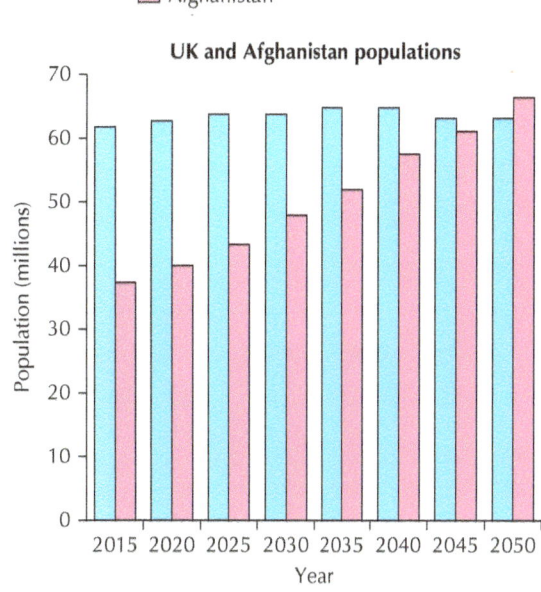

b 2048
c 2037, 65 million
d Pupil's own answer

5 a Pupil's own answer

c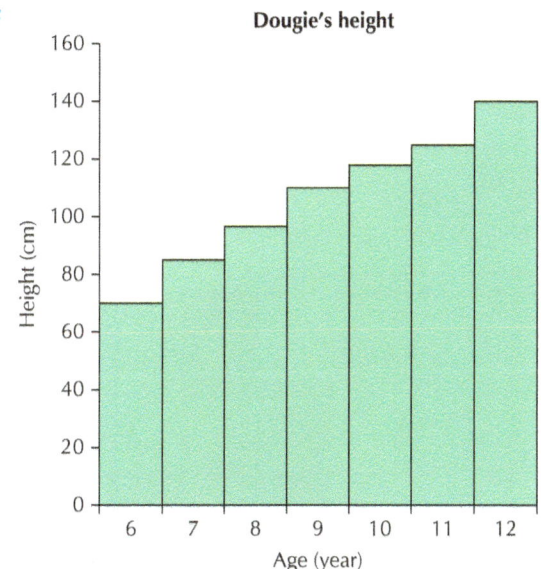

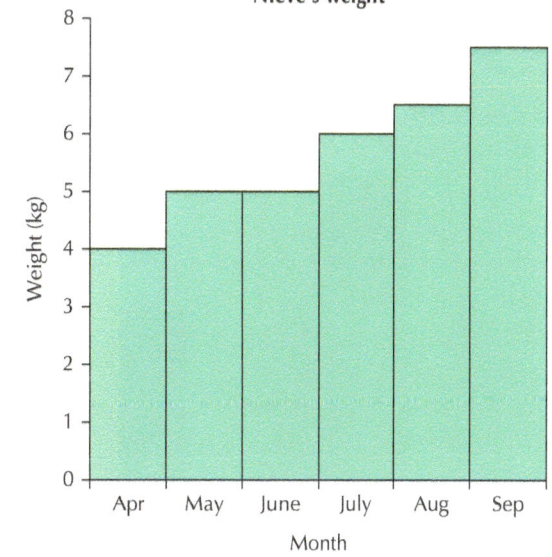

3 a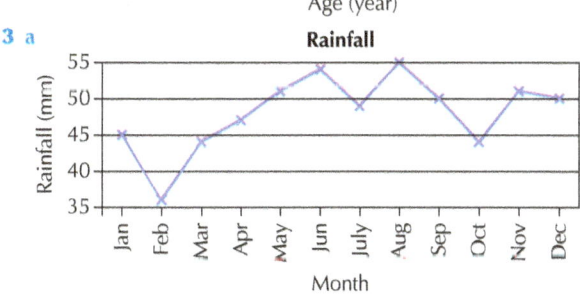

b August
c February
d 3
e 6 mm
f January, February

Challenge

1 a Pupil's own answer
  b Pupil's own answer for data display. Trend shows higher maximum temperatures in summer and lower maximum temperatures in winter.

341

Answers

Exercise 30B

1 a

Pencils	0–5	6–10	11–15	16–20
Frequency	7	7	6	11

b

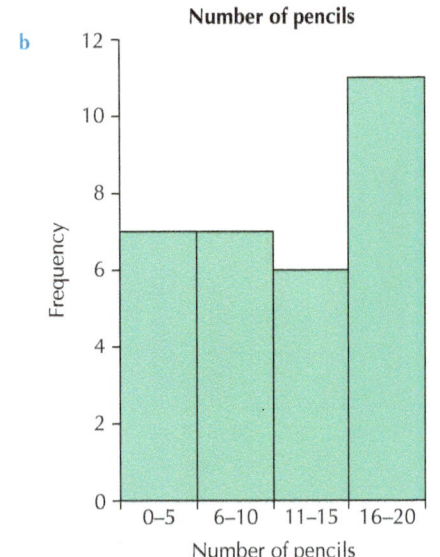

2 a

Time on Computer (h/wk)	0–4	5–9	10–14	15–19	20–24	25–29
Frequency	8	9	1	5	6	3

b

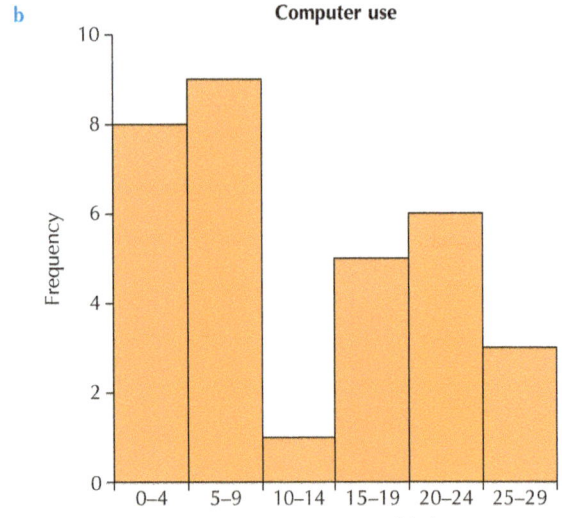

3 a

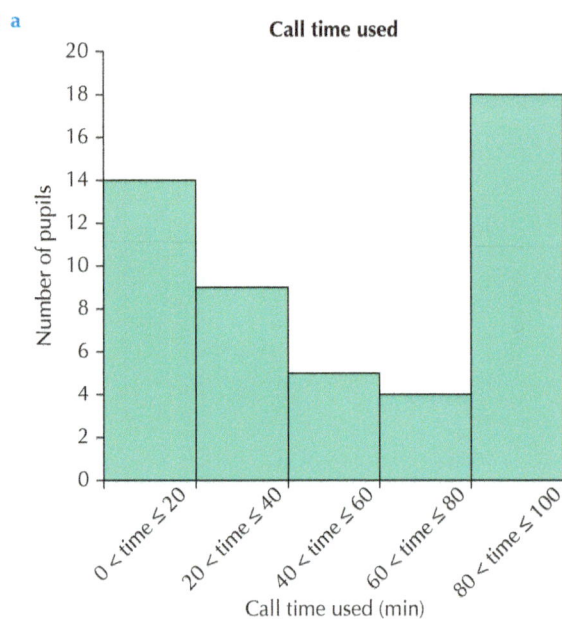

b 22
c Pupil's own answers

4 a 60 < time ≤ 75
b No team scored between 75 minutes and 85 minutes.
c

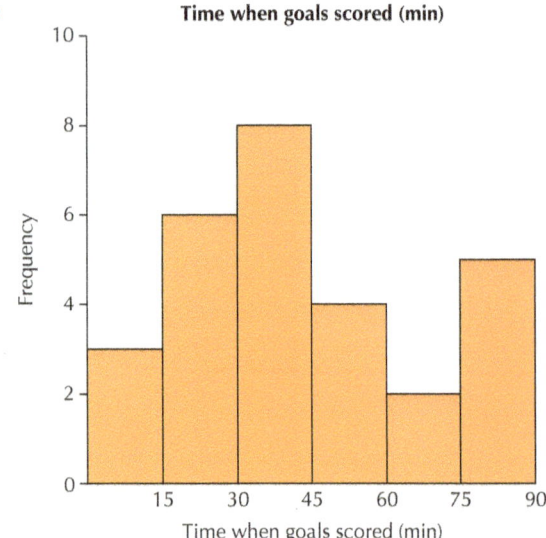

Challenge Pupil's own answers

Exercise 30C

1 a i £1·80 **ii** £4·20
  b i 5 kg **ii** 4 kg

2 a 0·9 km **b** 3 min 18 seconds

3 a i 4·8 km **ii** 7·2 km
  b i 1·25 miles **ii** 2·5 miles
   iii 3·75 miles

Answers

4 a

Euros (€)	1	5	10	15	20
Pounds (£)	0·8	4	8	12	16

b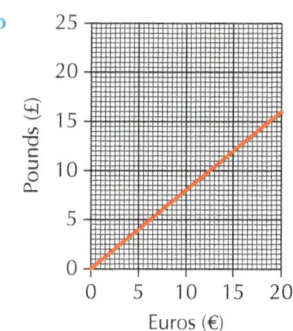

c i £5·60 ii £12·80 iii £14
d i €11·25 ii €15 iii €13·50

5 a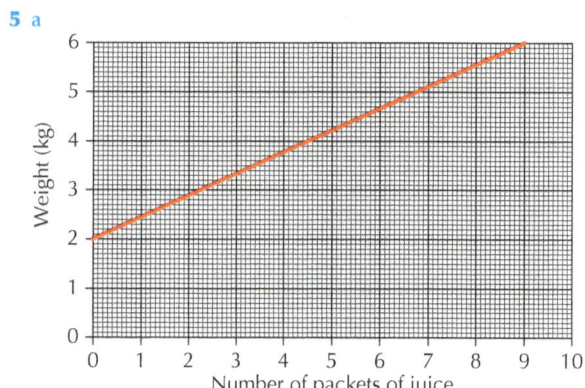

b 7

6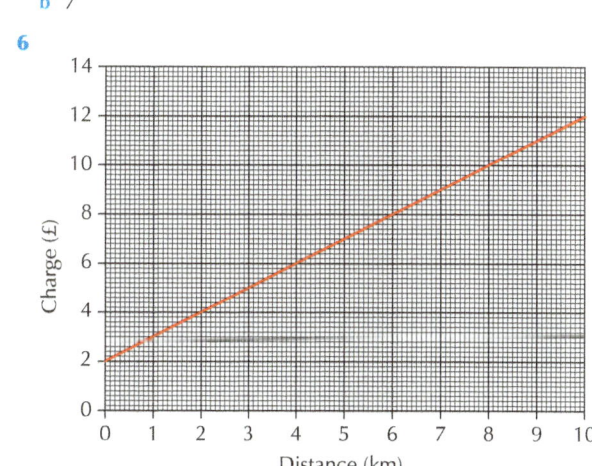

7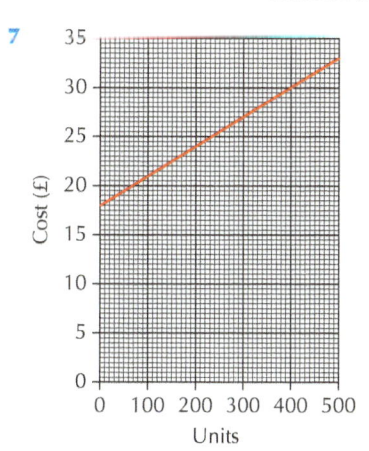

Exercise 30D

1 a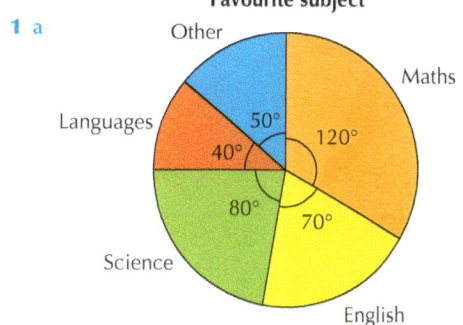

b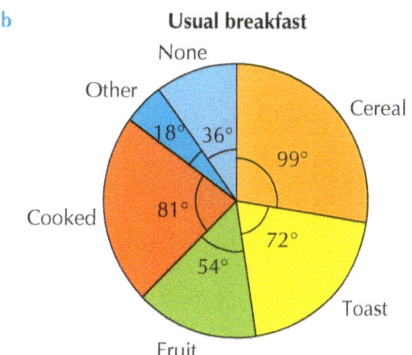

c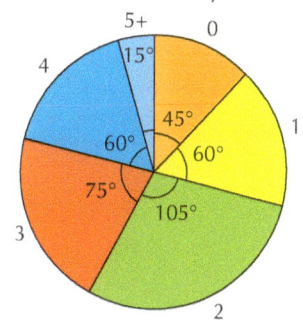

d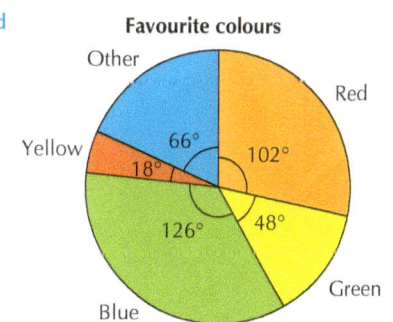

2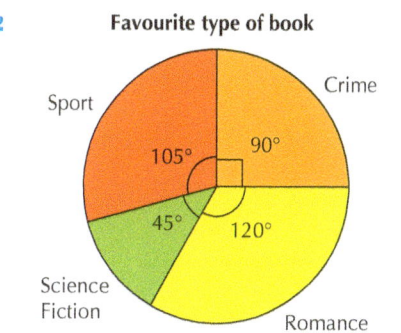

343

Answers

3 a £120

b

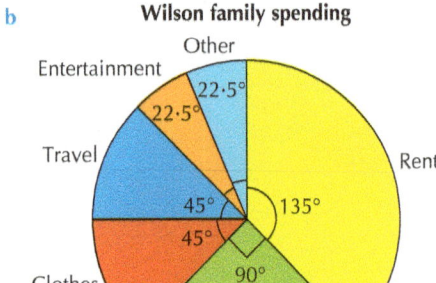

Wilson family spending

4

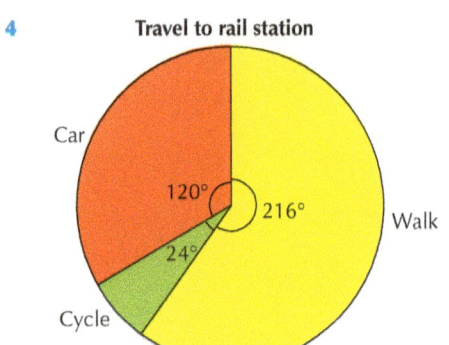

Travel to rail station

5 Pupil's own answers

Challenge

1 a
Protein 7·5%
Carbohydrate 71%
Fat 6%
Fibre 6%
Sodium 0·3%

b Porridge oats

2 Pupil's own answers

3 Pupil's own answers

Exercise 30E

1 a S1

Time to start
7 pm   ⊮ II
7:30 pm   III

Time to finish
9 pm   IIII
9:30 pm   II
10 pm   II
10:30 pm   I
11 pm   I

How much to charge
25 p   I
50 p   III
75 p   I
£1   IIII
£1·50   I

Food
Crisps   ⊮ ⊮
Beef burgers   III
Chips   II
Ice pops   ⊮
Chocolate   I
Hot dogs   I
Pizza   II

S2

Time to start
7 pm   II
7:30 pm   IIII
8 pm   IIII
8:30 pm

Time to finish
9 pm   I
9:30 pm   III
10 pm   IIII
10:30 pm   II

How much to charge
75p   I
£1   II
£1·25   I
£1·50   III
£2   III

Food
Crisps   ⊮ I
Beef burgers   I
Chips   ⊮ II
Ice pops   II
Chocolate   III
Hot dogs   ⊮
Pizza

S3

Time to start
7:00 pm   I
7:30 pm   II
8:00 pm   ⊮ I
8:30 pm   I

Answers

Time to finish
9:30 pm   II
10:00 pm I
10:30 pm III
11:00 pm III
11:30 pm I

How much to charge
£1·50   II
£2   IIII
£2·50   II
£3   II

Food
Crisps   III
Beef burgers
Chips   I
Ice pops
Chocolate   III
Hot dogs   II
Pizza   NN III

b

S1
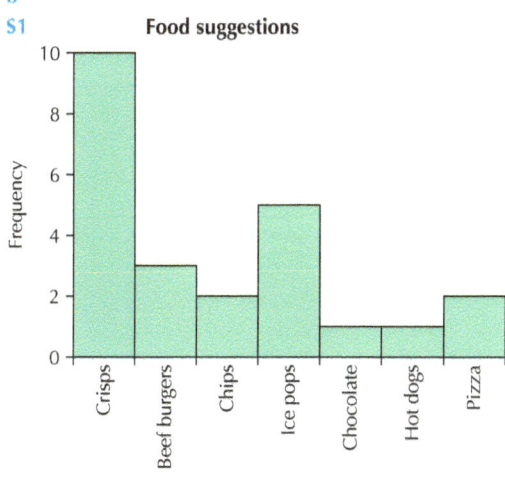

S2
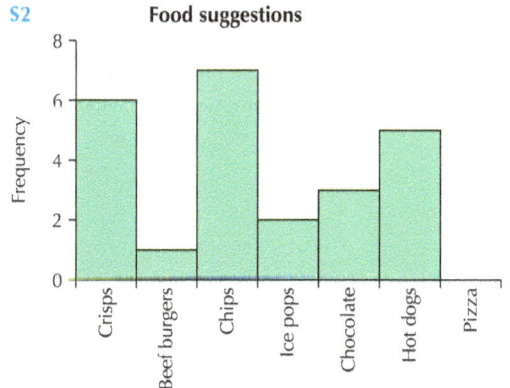

S3
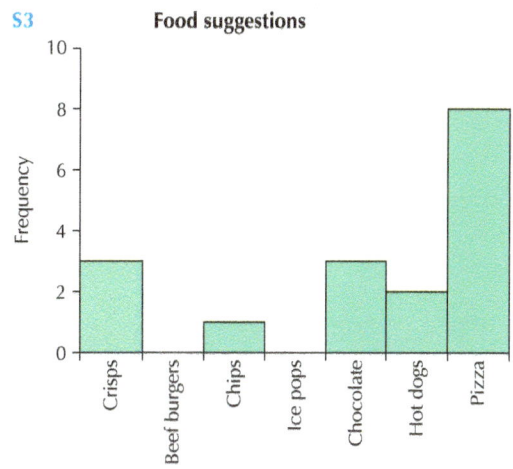

c Pizza is very popular with S3 but not with S1 or S2. Crisps are very popular with S1, quite popular with S2, but not very popular with S3.

Chips are most popular with S2, but not so popular with S1 or S3.

2 a S1 87; S2 93; S3 85; S4 90; S5 85

b S1
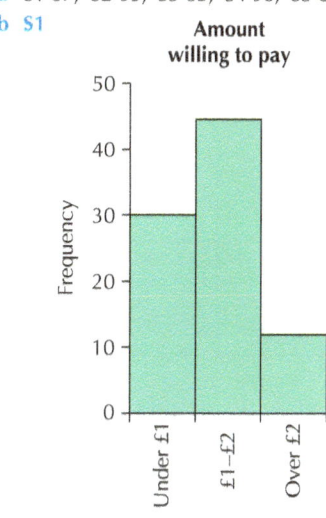

S2
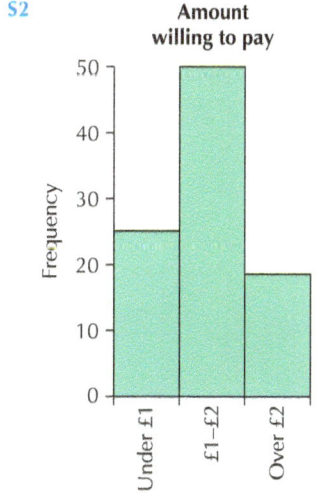

345

Answers

S3

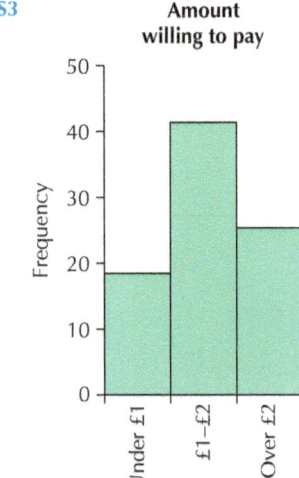

S4

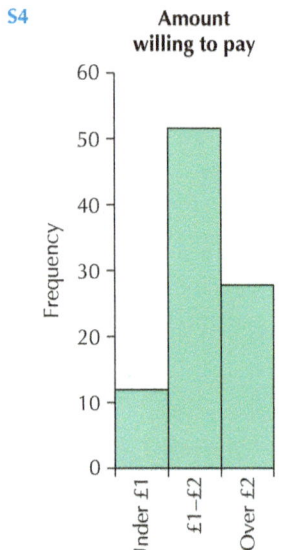

S5

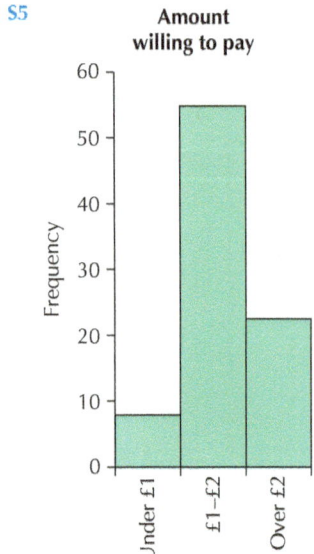

c The older pupils, those in S4 and S5, are willing to pay more for dinner.

d

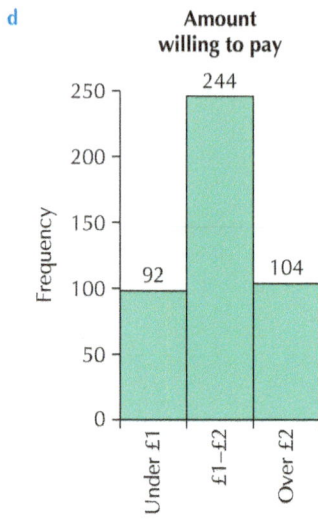

3 Pupil's own answers
4 Pupil's own answers
5 Pupil's own answers
6 Pupil's own answers

Chapter 31

Exercise 31A

1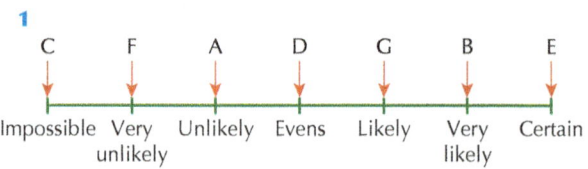

2 a $\frac{1}{10}$ b $\frac{1}{2}$ c $\frac{3}{10}$
  d $\frac{1}{5}$ e 1 f $\frac{2}{5}$

3 a $\frac{1}{13}$ b $\frac{1}{4}$ c $\frac{3}{13}$
  d $\frac{1}{52}$ e $\frac{2}{13}$ f $\frac{1}{13}$

4 a $\frac{1}{2}$ b $\frac{3}{10}$ c $\frac{1}{5}$
  d 0 e $\frac{4}{5}$

5 a i $\frac{3}{8}$ ii $\frac{1}{4}$ iii $\frac{1}{4}$ iv $\frac{1}{8}$
  b
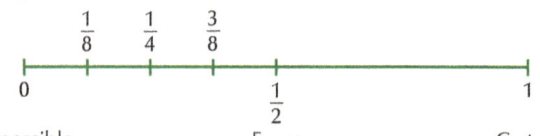

7 a $\frac{1}{6}$ b $\frac{1}{2}$ c $\frac{1}{3}$
  d $\frac{1}{2}$ e $\frac{1}{6}$ f $\frac{1}{3}$

Answers

8 $\frac{1}{5}$

9 $\frac{1}{5}$

10 a i B. For bag A $P(\text{red}) = \frac{1}{2}$; for bag B $P(\text{red}) = \frac{4}{5}$

    ii A. For bag A $P(\text{blue}) = \frac{1}{4}$; bag B $P(\text{blue}) = \frac{1}{5}$

    iii A. For bag A $P(\text{green}) = \frac{1}{4}$; bag B $P(\text{green}) = 0$

  b

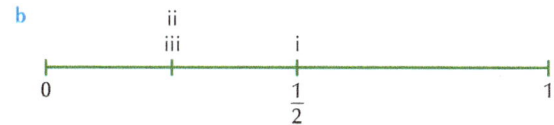

11 a $\frac{1}{11}$ **b** $\frac{2}{11}$ **c** $\frac{2}{11}$ **d** $\frac{4}{11}$

12 a $\frac{5}{12}$ **b** $\frac{1}{3}$ **c** 0   **d** $\frac{1}{6}$

13 a 24   **b** 10   **c** 10

  d i $\frac{5}{12}$ **ii** $\frac{5}{12}$

14 a 31   **b** 6   **c** 18

  d i $\frac{6}{31}$ **ii** $\frac{18}{31}$

15 a $\frac{9}{40}$ **b** $\frac{31}{80}$ **c** $\frac{1}{10}$

  d $\frac{67}{80}$ **e** $\frac{11}{16}$

Challenge Pupil's own answer

Exercise 31B

1 a Red

  b i $\frac{1}{10}$ **ii** $\frac{1}{5}$ **iii** $\frac{7}{10}$

2 a 17

  b i $\frac{17}{20}$ **ii** $\frac{3}{20}$

  iii

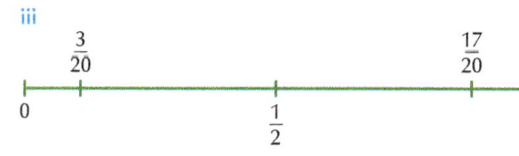

3 a $\frac{1}{10}$ **b** $\frac{9}{10}$ **c** $\frac{1}{2}$ **d** $\frac{1}{2}$

  e $\frac{3}{10}$ **f** $\frac{7}{10}$ **g** $\frac{1}{5}$ **h** $\frac{4}{5}$

4 a $\frac{1}{4}$ **b** $\frac{3}{5}$ **c** $\frac{3}{4}$ **d** 0

  e $\frac{3}{4}$ **f** $\frac{2}{5}$ **g** $\frac{1}{4}$ **h** 1

5 a 50

  b i $\frac{17}{50}$ **ii** $\frac{3}{10}$ **iii** $\frac{1}{5}$ **iv** $\frac{4}{25}$

  c

6 24; $\frac{1}{4}$ of 32 is 8, so 8 black and $32 - 8 = 24$ white.

7 a $\frac{1}{4}$ **b** $\frac{1}{5}$ **c** $\frac{2}{5}$

  d $\frac{1}{10}$ **e** $\frac{1}{20}$

8 a $\frac{4}{5}$ **b** $\frac{3}{8}$

9 a 0·55   **b** 0·85   **c** 0·15

10 a $\frac{61}{100}$ **b** $\frac{39}{100}$ **c** Winning a prize

  d 300

Exercise 31C

1 a 10   **b** 7   **c** 2   **d** 8

  e 10   **f** 3   **g** 3   **h** 3

  i 4   **j** 2

2 a $\frac{1}{2}$ **b** $\frac{7}{20}$ **c** $\frac{1}{10}$ **d** $\frac{2}{5}$

  e $\frac{1}{2}$ **f** $\frac{3}{20}$ **g** $\frac{3}{20}$ **h** $\frac{3}{20}$

  i $\frac{1}{5}$ **j** $\frac{1}{10}$

3 a $\frac{1}{7}$ **b** $\frac{1}{4}$ **c** $\frac{1}{10}$

Challenge Pupil's own answers

Exercise 31D

1 a The probability of rolling a six is 1/6 and is the same for everyone.

  b The weather is not identical or consecutive days or it would always be the same.

  c It is more likely not to snow than to snow. So the statement is incorrect.

  d The probability of picking a chocolate

$$\text{Sweet} = \frac{\text{number of chocolate sweets}}{\text{total number of sweets}}$$

2 a Incorrect

  b Correct

  c Correct

  d Incorrect

3 a No

  b Grid 1

  c For grid 2 the probability of winning is $\frac{1}{3}$, for grid 3 it is $\frac{1}{2}$.

4 Fiya plays twice a week, every week of the year, so the probability of her winning is $104 \times 0.0095 = 0.988$. She can expect to win about once a year.

5 It is quite safe to go to a firework display as the probability of being hit is very small.

6 Dougie should buy the 'Boiler' Kettle; it is cheaper to buy 3 'Boiler' kettles than 1 'Hot stuff' kettle.

7 Pupil's own answer.   **8** 8 days

9 20   **10** 15 days

11 10 goals   **12** 7 prizes

13 1254

Number, money and measure

	Second	Third	Fourth
Estimation and rounding	I can use my knowledge of rounding to routinely estimate the answer to a problem then, after calculating, decide if my solution is reasonable, sharing my solution with others. **MNU 2-01a**	I can round a number using an appropriate degree of accuracy, having taken into account the context of the problem. **MNU 3-01a**	Having investigated the practical impact of inaccuracy and error, I can use my knowledge of tolerance when choosing the required degree of accuracy to make real-life calculations. **MNU 4-01a**
Number and number processes including addition, subtraction, multiplication, division and negative numbers	I have extended the range of whole numbers I can work with and having explored how decimal fractions are constructed, can explain the link between a digit, its place and its value. **MNU 2-02a**		
	Having determined which calculations are needed, I can solve problems involving whole numbers using a range of methods, sharing my approaches and solutions with others. **MNU 2-03a**	I can use a variety of methods to solve number problems in familiar contexts, clearly communicating my processes and solutions. **MNU 3-03a**	Having recognised similarities between new problems and problems I have solved before, I can carry out the necessary calculations to solve problems set in unfamiliar contexts. **MNU 4-03a**
	I have explored the contexts in which problems involving decimal fractions occur and can solve related problems using a variety of methods. **MNU 2-03b** Having explored the need for rules for the order of operations in number calculations, I can apply them correctly when solving simple problems. **MTH 2-03c**	I can continue to recall number facts quickly and use them accurately when making calculations. **MNU 3-03b**	I have investigated how introducing brackets to an expression can change the emphasis and can demonstrate my understanding by using the correct order of operations when carrying out calculations. **MTH 4-03b**
	I can show my understanding of how the number line extends to include numbers less than zero and have investigated how these numbers occur and are used. **MNU 2-04a**	I can use my understanding of numbers less than zero to solve simple problems in context. **MNU 3-04a**	

Number, money and measure (continued)

	Second	Third	Fourth
Multiples, factors and primes	Having explored the patterns and relationships in multiplication and division, I can investigate and identify the multiples and factors of numbers. MTH 2-05a	I have investigated strategies for identifying common multiples and common factors, explaining my ideas to others, and can apply my understanding to solve related problems. MTH 3-05a I can apply my understanding of factors to investigate and identify when a number is prime. MTH 3-05b	
Powers and roots		Having explored the notation and vocabulary associated with whole number powers and the advantages of writing numbers in this form, I can evaluate powers of whole numbers mentally or using technology. MTH 3-06a	I have developed my understanding of the relationship between powers and roots and can carry out calculations mentally or using technology to evaluate whole number powers and roots, of any appropriate number. MTH 4-06a Within real-life contexts, I can use scientific notation to express large or small numbers in a more efficient way and can understand and work with numbers written in this form. MTH 4-06b

Number, money and measure (continued)

	Second	Third	Fourth
Fractions, decimal fractions and percentages including ratio and proportion	I have investigated the everyday contexts in which simple fractions, percentages or decimal fractions are used and can carry out the necessary calculations to solve related problems. **MNU 2-07a** I can show the equivalent forms of simple fractions, decimal fractions and percentages and can choose my preferred form when solving a problem, explaining my choice of method. **MNU 2-07b** I have investigated how a set of equivalent fractions can be created, understanding the meaning of simplest form, and can apply my knowledge to compare and order the most commonly used fractions. **MTH 2-07c**	I can solve problems by carrying out calculations with a wide range of fractions, decimal fractions and percentages, using my answers to make comparisons and informed choices for real-life situations. **MNU 3-07a** By applying my knowledge of equivalent fractions and common multiples, I can add and subtract commonly used fractions. **MTH 3-07b** Having used practical, pictorial and written methods to develop my understanding, I can convert between whole or mixed numbers and fractions. **MTH 3-07c**	I can choose the most appropriate form of fractions, decimal fractions and percentages to use when making calculations mentally, in written form or using technology; then use my solutions to make comparisons, decisions and choices. **MNU 4-07a** I can solve problems involving fractions and mixed numbers in context, using addition, subtraction or multiplication. **MTH 4-07b**
		I can show how quantities that are related can be increased or decreased proportionally and apply this to solve problems in everyday contexts. **MNU 3-08a**	Using proportion, I can calculate the change in one quantity caused by a change in a related quantity and solve real-life problems. **MNU 4-08a**

Number, money and measure (continued)

	Second	Third	Fourth
Money	I can manage money, compare costs from different retailers, and determine what I can afford to buy. **MNU 2-09a**	When considering how to spend my money, I can source, compare and contrast different contracts and services, discuss their advantages and disadvantages, and explain which offer best value to me. **MNU 3-09a**	I can discuss and illustrate the facts I need to consider when determining what I can afford, in order to manage credit and debt and lead a responsible lifestyle. **MNU 4-09a**
	I understand the costs, benefits and risks of using bank cards to purchase goods or obtain cash and realise that budgeting is important. **MNU 2-09b**	I can budget effectively, making use of technology and other methods, to manage money and plan for future expenses. **MNU 3-09b**	I can source information on earnings and deductions and use it when making calculations to determine net income. **MNU 4-09b**
	I can use the terms profit and loss in buying and selling activities and can make simple calculations for this. **MNU 2-09c**		I can research, compare and contrast a range of personal finance products and, after making calculations, explain my preferred choices. **MNU 4-09c**
Time	I can use and interpret electronic and paper-based timetables and schedules to plan events and activities, and make time calculations as part of my planning. **MNU 2-10a**	Using simple time periods, I can work out how long a journey will take, the speed travelled at or distance covered, using my knowledge of the link between time, speed and distance. **MNU 3-10a**	I can research, compare and contrast aspects of time and time management as they impact on me. **MNU 4-10a**
	I can carry out practical tasks and investigations involving timed events and can explain which unit of time would be most appropriate to use. **MNU 2-10b**		I can use the link between time, speed and distance to carry out related calculations. **MNU 4-10b**
	Using simple time periods, I can give a good estimate of how long a journey should take, based on my knowledge of the link between time, speed and distance. **MNU 2-10c**		

Number, money and measure (continued)

	Second	Third	Fourth
Measurement	I can use my knowledge of the sizes of familiar objects or places to assist me when making an estimate of measure. **MNU 2-11a** I can use the common units of measure, convert between related units of the metric system and carry out calculations when solving problems. **MNU 2-11b** I can explain how different methods can be used to find the perimeter and area of a simple 2D shape or volume of a simple 3D object. **MNU 2-11c**	I can solve practical problems by applying my knowledge of measure, choosing the appropriate units and degree of accuracy for the task and using a formula to calculate area or volume when required. **MNU 3-11a** Having investigated different routes to a solution, I can find the area of compound 2D shapes and the volume of compound 3D objects, applying my knowledge to solve practical problems. **MTH 3-11b**	I can apply my knowledge and understanding of measure to everyday problems and tasks and appreciate the practical importance of accuracy when making calculations. **MNU 4-11a** Through investigating real-life problems involving the surface area of simple 3D shapes, I can explore ways to make the most efficient use of materials and carry out the necessary calculations to solve related problems. **MTH 4-11b** I have explored with others the practicalities of the use of 3D objects in everyday life and can solve problems involving the volume of a prism, using a formula to make related calculations when required. **MTH 4-11c**
Mathematics – its impact on the world, past, present and future	I have worked with others to explore, and present our findings on, how mathematics impacts on the world and the important part it has played in advances and inventions. **MTH 2-12a**	I have worked with others to research a famous mathematician and the work they are known for, or investigated a mathematical topic, and have prepared and delivered a short presentation. **MTH 3-12a**	I have discussed the importance of mathematics in the real world, investigated the mathematical skills required for different career paths and delivered, with others, a presentation on how mathematics can be applied in the workplace. **MTH 4-12a**

Number, money and measure (continued)

	Second	Third	Fourth
Patterns and relationships	Having explored more complex number sequences, including well-known named number patterns, I can explain the rule used to generate the sequence, and apply it to extend the pattern. MTH 2-13a	Having explored number sequences, I can establish the set of numbers generated by a given rule and determine a rule for a given sequence, expressing it using appropriate notation. MTH 3-13a	Having explored how real-life situations can be modelled by number patterns, I can establish a number sequence to represent a physical or pictorial pattern, determine a general formula to describe the sequence, then use it to make evaluations and solve related problems. MTH 4-13a I have discussed ways to describe the slope of a line, can interpret the definition of gradient and can use it to make relevant calculations, interpreting my answer for the context of the problem. MTH 4-13b Having investigated the pattern of the coordinate points lying on a horizontal or vertical line, I can describe the pattern using a simple equation. MTH 4-13c I can use a given formula to generate points lying on a straight line, plot them to create a graphical representation then use this to answer related questions. MTH 4-13d

Number, money and measure (continued)

	Second	Third	Fourth
Expressions and equations		I can collect like algebraic terms, simplify expressions and evaluate using substitution. **MTH 3-14a**	Having explored the distributive law in practical contexts, I can simplify, multiply and evaluate simple algebraic terms involving a bracket. **MTH 4-14a** I can find the factors of algebraic terms, use my understanding to identify common factors and apply this to factorise expressions. **MTH 4-14b**
	I can apply my knowledge of number facts to solve problems where an unknown value is represented by a symbol or letter. **MTH 2-15a**	Having discussed ways to express problems or statements using mathematical language, I can construct, and use appropriate methods to solve, a range of simple equations. **MTH 3-15a** I can create and evaluate a simple formula representing information contained in a diagram, problem or statement. **MTH 3-15b**	Having discussed the benefits of using mathematics to model real-life situations, I can construct and solve inequalities and an extended range of equations. **MTH 4-15a**

Shape, position and movement

	Second	Third	Fourth
Properties of 2D shapes and 3D objects	Having explored a range of 3D objects and 2D shapes, I can use mathematical language to describe their properties, and through investigation can discuss where and why particular shapes are used in the environment. **MTH 2-16a** Through practical activities, I can show my understanding of the relationship between 3D objects and their nets. **MTH 2-16b** I can draw 2D shapes and make representations of 3D objects using an appropriate range of methods and efficient use of resources. **MTH 2-16c**	Having investigated a range of methods, I can accurately draw 2D shapes using appropriate mathematical instruments and methods. **MTH 3-16a**	I have explored the relationships that exist between the sides, or sides and angles, in right-angled triangles and can select and use an appropriate strategy to solve related problems, interpreting my answer for the context. **MTH 4-16a** Having investigated the relationships between the radius, diameter, circumference and area of a circle, I can apply my knowledge to solve related problems. **MTH 4-16b**

Shape, position and movement (continued)

	Second	Third	Fourth
Angle, symmetry and transformation	I have investigated angles in the environment, and can discuss, describe and classify angles using appropriate mathematical vocabulary. **MTH 2-17a** I can accurately measure and draw angles using appropriate equipment, applying my skills to problems in context. **MTH 2-17b** Through practical activities which include the use of technology, I have developed my understanding of the link between compass points and angles and can describe, follow and record directions, routes and journeys using appropriate vocabulary. **MTH 2-17c** Having investigated where, why and how scale is used and expressed, I can apply my understanding to interpret simple models, maps and plans. **MTH 2-17d**	I can name angles and find their sizes using my knowledge of the properties of a range of 2D shapes and the angle properties associated with intersecting and parallel lines. **MTH 3-17a** Having investigated navigation in the world, I can apply my understandings of bearings and scale to interpret maps and plans and create accurate plans, and scale drawings of routes and journeys. **MTH 3-17b** I can apply my understanding of scale when enlarging or reducing pictures and shapes, using different methods, including technology. **MTH 3-17c**	Having investigated the relationship between a radius and a tangent and explored the size of the angle in a semi-circle, I can use the facts I have established to solve related problems. **MTH 4-17a** I can apply my understanding of the properties of similar figures to solve problems involving length and area. **MTH 4-17b**

Shape, position and movement (continued)

Second	Third	Fourth
	I can use my knowledge of the coordinate system to plot and describe the location of a point on a grid. **MTH 2-18a / MTH 3-18a**	I can plot and describe the position of a point on a 4-quadrant coordinate grid. **MTH 4-18a** I can apply my understanding of the 4-quadrant coordinate system to move, and describe the transformation of, a point or shape on a grid. **MTH 4-18b**
	I can illustrate the lines of symmetry for a range of 2D shapes and apply my understanding to create and complete symmetrical pictures and patterns. **MTH 2-19a / MTH 3-19a**	Having investigated patterns in the environment, I can use appropriate mathematical vocabulary to discuss the rotational properties of shapes, pictures and patterns and can apply my understanding when completing or creating designs. **MTH 4-19a**

Information handling

	Second	Third	Fourth
Data and analysis	Having discussed the variety of ways and range of media used to present data, I can interpret and draw conclusions from the information displayed, recognising that the presentation may be misleading. **MNU 2-20a** I have carried out investigations and surveys, devising and using a variety of methods to gather information and have worked with others to collate, organise and communicate the results in an appropriate way. **MNU 2-20b**	I can work collaboratively, making appropriate use of technology, to source information presented in a range of ways, interpret what it conveys and discuss whether I believe the information to be robust, vague or misleading. **MNU 3-20a** When analysing information or collecting data of my own, I can use my understanding of how bias may arise and how sample size can affect precision, to ensure that the data allows for fair conclusions to be drawn. **MTH 3-20b**	I can evaluate and interpret raw and graphical data using a variety of methods, comment on relationships I observe within the data and communicate my findings to others. **MNU 4-20a** In order to compare numerical information in real-life contexts, I can find the mean, median, mode and range of sets of numbers, decide which type of average is most appropriate to use and discuss how using an alternative type of average could be misleading. **MTH 4-20b**
	I can display data in a clear way using a suitable scale, by choosing appropriately from an extended range of tables, charts, diagrams and graphs, making effective use of technology. **MTH 2-21a / MTH 3-21a**		I can select appropriately from a wide range of tables, charts, diagrams and graphs when displaying discrete, continuous or grouped data, clearly communicating the significant features of the data. **MTH 4-21a**
Ideas of chance and uncertainty	I can conduct simple experiments involving chance and communicate my predictions and findings using the vocabulary of probability. **MNU 2-22a**	I can find the probability of a simple event happening and explain why the consequences of the event, as well as its probability, should be considered when making choices. **MNU 3-22a**	By applying my understanding of probability, I can determine how many times I expect an event to occur, and use this information to make predictions, risk assessment, informed choices and decisions. **MNU 4-22a**